69.50 80K

SPACECRAFT CONTAMINATION: SOURCES AND PREVENTION

Edited by
J. A. Roux
The University of Mississippi
University, Mississippi

T. D. McCay
NASA Marshall Space Flight Center
Marshall Space Flight Center, Alabama

Volume 91
PROGRESS IN
ASTRONAUTICS AND AERONAUTICS

Martin Summerfield, Series Editor-in-Chief
Princeton Combustion Research Laboratories, Inc.
Monmouth Junction, New Jersey

Technical papers selected from the AIAA 21st Aerospace Sciences Meeting, January 1983, and the AIAA 18th Thermophysics Conference, June 1983, and subsequently revised for this volume.

Published by the American Institute of Aeronautics and Astronautics, Inc.
1633 Broadway, New York, N.Y. 10019

American Institute of Aeronautics and Astronautics, Inc.
New York, New York

Library of Congress Cataloging in Publication Data
Main entry under title:

Spacecraft contamination.

(Progress in astronautics and aeronautics; v. 91)
Technical papers selected from the AIAA 21st Aerospace Sciences Meeting, January 1983, and the AIAA 18th Thermophysics Conference, June 1983, and subsequently revised for this volume.
Includes index.
1. Space vehicles – Contamination – Congresses. I. Roux, J. A. II. McCay, T. D., III. American Institute of Aeronautics and Astronautics. IV. Series.
TL507.P75 vol. 91 (TL945) 629.1 s (629.47'74) 84-12401
ISBN 0-915928-85-X

Copyright © 1984 by the American Institute of Aeronautics and Astronautics, Inc. All rights reserved. Printed in the United States of America. No part of this publication may be reproduced, distributed, or transmitted, in any form or by any means, or stored in any data base or retrieval system, without the prior written permission of the publisher.

Progress in Astronautics and Aeronautics
Series Editor-in-Chief
Martin Summerfield
Princeton Combustion Research Laboratories, Inc.

Series Associate Editors

Burton I. Edelson
*National Aeronautics
and Space Administration*

Allen E. Fuhs
Naval Postgraduate School

J. Leith Potter
Vanderbilt University

Norma J. Brennan
Director, Editorial Department
AIAA

Camille S. Koorey
Series Managing Editor
AIAA

Table of Contents

Preface xi

Chapter I. Contamination Overview 1

Improved Methods for Characterizing Material-Induced Contamination 3
A.P.M. Glassford, R.A. Osiecki, and C.K. Liu, *Lockheed Palo Alto Research Laboratories, Palo Alto, California* and M. Hitchcock, *Air Force Materials Laboratory, Wright-Patterson Air Force Base, Ohio*

Potential for Cross Contamination for Payloads in the STS Bay 29
R.G. Moss, *Ford Aerospace and Communications Corporation, Palo Alto, California*

Chapter II. Sources and Prevention of Contamination 37

Debris from Spallation of Foam Insulation of Cryogenic Fuel Tanks in Space Launch Sytems 39
E.P. del Casal, *Energy Incorporated, Idaho Falls, Idaho*

Particle Dispersion around a Spacecraft 54
A.L. Lee, *Lockheed Missiles & Space Company, Inc., Sunnyvale, California*

Impact of the STS Ground/Launch Particle Contamination Environment on an Optical Sensor 73
L.E. Bareiss and F.J. Jarossy, *Martin Marietta Denver Aerospace, Denver, Colorado*

Analysis of Contamination Degradation of Thermal Control Surfaces on Operational Satellites 96
J.E. Ahern, R.L. Belcher, and R.D. Ruff, *Aerojet ElectroSystems Company, Azusa, California*

Abatement of Gaseous and Particulate Contamination in a Space Instrument 108
J.J. Scialdone, *NASA Goddard Space Flight Center, Greenbelt, Maryland*

Chapter III. Properties and Effects of Contamination 137

Infrared Optical Properties of Thin CO, NO, CH_4, HCl, N_2O, O_2, N_2, and Ar Cryofilms 139
    B.E. Wood, *Calspan Field Services, Inc., Arnold Air Force Station, Tennessee*, and J.A. Roux, *University of Mississippi, Oxford, Mississippi*

Infrared Optical Properties of Solid Mixtures of Molecular Species at 20 K 162
    K.F. Palmer, *Westminster College, Fulton, Missouri*, J.A. Roux *University of Mississippi, Oxford, Mississippi*, and B.E. Wood, *Calspan Field Services, Inc., Arnold Air Force Station, Tennessee*

Measurements of Infrared Optical Properties of Al_2O_3 Rocket Particles 180
    W.L. Konopka, R.A. Reed, and V.S. Calia, *Grumman Aerospace Corporation, Bethpage, New York*

Improvements in Rocket Engine Nozzle and High Altitude Plume Computations 197
    S.D. Smith, *Lockheed Huntsville Research & Engineering Center, Huntsville, Alabama*

α_s/ϵ_H Measurements of Thermal Control Coatings over Four Years at Geosynchronous Altitude 215
    D.F. Hall and A.A. Fote, *The Aerospace Corporation, El Segundo, California*

Calorimetric Measurements of Thermal Control Surfaces on Operational Satellites 235
    J.E. Ahern and K. Karperos, *Aerojet ElectroSystems Company, Azusa, California*

Experimental Investigation of Bipropellant Exhaust Plume Flowfield, Heating, and Contamination and Comparison with the CONTAM Computer Model Predictions 261
    H. Trinks, *Technical University, Hamburg-Harburg, Hamburg, Federal Republic of Germany* and R.J. Hoffman, *Science Applications, Inc., Los Angeles, California*

Particle Sampling of Solid Rocket Motor Exhausts in High-Altitude Test Cells 293
    P.T. Girata Jr. and W.K. McGregor, *Sverdrup Technology, Inc./AEDC Group, Arnold Air Force Station, Tennessee*

Postfire Sampling of Solid Rocket Motors for Contamination Sources in High-Altitude Test Cells 312
    P.T. Girata Jr. and W.K. McGregor, *Sverdrup Techology, Inc./AEDC Group, Arnold Air Force Station, Tennessee*

Author Index for Volume 91 328
List of Series Volumes 329

Table of Contents for Companion Volume 92

Chapter I.   CARS..1

   CARS Diagnostics of High Pressure and Temperature Gases 3
      J.H. Stufflebeam, R.J. Hall, and J.F. Verdieck
      United Technologies Research Center, East Hartford, Connecticut

   **CARS Thermometry and N_2 Number-Density Measurments in a
   Turbulent Diffusion Flame**..24
      L.P. Goss, D.D. Trump, G.L. Switzer, and B.G. MacDonald,
      Systems Research Laboratories, Inc., Dayton, Ohio

   **Comparison of CARS Combustion Temperatures with
   Standard Techniques**...45
      R.R. Antcliff, *Systems Research Laboratories, Inc., Dayton, Ohio* and
      O. Jarrett Jr., *NASA Langley Research Center, Hampton, Virginia*

   Electronically Resonant CARS Detection of OH 58
      J.F. Verdieck, R.J. Hall, and A.C. Eckbreth,
      United Technologies Research Center, East Hartford, Connecticut

   **Simultaneous CARS and Luminosity Measurements in a Bluff-Body
   Combustion**...82
      G.L. Switzer, D.D. Trump, and L.P. Goss, *Systems Research
      Laboratories, Inc., Dayton, Ohio,* and W.M. Roquemore,
      R.P. Bradley, J.S. Stutrud, and C.M. Reeves, *Air Force Wright
      Aeronautical Laboratories, Wright-Patterson Air Force Base, Ohio*

Chapter II.   Laser-Induced Fluorescence...105

   **Nonintrusive Pressure Measurements with Laser-Induced Iodine
   Fluorescence** 107
      J.C. McDaniel, *University of Virginia, Charlottesville, Virginia*

   Laser-Induced Schlieren Effect in Sodium-Nitrogen Mixtures 132
      J.W.L. Lewis and J.D. Selman, *The University of Tennessee
      Space Institute, Tullahoma, Tennessee*

   **Use of Laser-Induced Fluorescence for Fundamental Gas-Phase
   Kinetic Measurements**...147
      A. Fontijn, *Rensselaer Polytechnic Institute, Troy, New York*

Chapter III.   Particle Diagnostics 175

   Nonintrusive Laser-Based Particle Diagnostics - Invited Review..........................177
      E.D. Hirleman, *Arizona State University, Tempe, Arizona*

   Interpretation of Optical Measurements of Soot in Flames 208
      R.A. Dobbins, R.J. Santoro, and H.G. Semerjian, *National Bureau
      of Standards, Washington, D.C.*

   **In Situ Measurement of the Complex Refractive Index of Combustion
   Generated Particulates**..238
      E.A. Powell and B.T. Zinn, *Georgia Institute of Technology,
      Atlanta, Georgia*

Chapter IV. Combustion Diagnostics Applications 253

**Temperature and Concentration Measurements in an Internal
Combustion Engine Using Laser Raman Spectroscopy** 255
    A. zur Loye and D.A. Santavicca, *Princeton University,
Princeton, New Jersey*

Rayleigh Thermometry with Low-Power Laser Sources 270
    D. Benhachmi, N. Younes, H. Yakout, P.E. Emmerman, and
R. Goulard, *The George Washington University, Washington, D.C.*

**Laser Tomography for Simultaneous Concentration and
Temperature Measurement in Reacting Flows** 300
    S.R. Ray and H.G. Semerjian, *National Bureau of Standards,
Washington, D.C.*

Flow Measurement in a Model Combustion Chamber 325
    P. Magre, J. Labbé, and G. Collin, *ONERA, Châtillion, France*

Preface

As a scientific discipline theromphysics encompasses a broad range of engineering sciences. The rapid development of man's exploration of both the "inner" and "outer" space of the universe has been accompanied by a requirement for in-depth understanding and analysis of the thermophysical aspects of these domains. Thermophysics includes materials and optics from both the macroscopic and microscopic points of view. Although previous thermophysics volumes in this series were usually composed of papers representing the broad range of thermophysics topics of current interest, for this volume it was decided to select a very specific topic of current and on-going interest – namely spacecraft contamination. The problem of spacecraft contamination has been a significant one for some time and is now receiving accelerated attention with the deployment of advanced satellite systems and the successful operation of the Space Shuttle.

This volume is dedicated to a variety of topics dealing with the central theme of spacecraft contamination. It has been realized that the useful life of many of today's spacecraft is determined by the degradation of the thermal control surfaces and sensor systems caused by a variety of contamination sources. This is especially true for satellites requiring cryogenically cooled sensor systems. Spacecraft contamination has been identified as the limiting factor in many otherwise long term missions. In some cases, it is the thermal balance of the spacecraft that the contamination upsets, thus affecting the onboard electronics. In other cases, it is the sensor systems that suffer from serious optical degradation caused by contamination. This volume contains papers providing state-of-the-art analyses of contamination sources, contamination effects (optical and thermal), contamination measurement methods (simulated environments and orbital data), and contamination prevention techniques.

The types of contamination can be either molecular or particulate. The sources of contamination can be from outgassing of materials, plume effluents from attitude control thrusters, ground preparation, and even cross contamination of one payload by another in the payload bay. The advent of the Space Shuttle has produced unique opportunities for space exploration but it has also introduced a unique set of contamination problems associated with a very large and complex spacecraft system. This requires as great an understanding as possible of the mechanisms and effects of contamination. Only then will it be possible to determine when a given scenario may or may not be harmful for a particular mission. Finally, the means necessary to avoid contamination must be established in order to achieve the goals of a given space mission.

This volume contains the results of current research in the field of spacecraft contamination, drawn primarily from the thermophysics sessions of the AIAA 21st Aerospace Sciences Meeting in Reno, Nevada in January 1983 and the AIAA 18th Thermophysics Conference in Montreal, Canada in June 1983. The papers selected were reviewed, revised, updated, and organized into three chapters covering contamination, the sources and prevention of contamination, and the properties and effects of contamination.

Chapter I is an overview of the various sources of spacecraft contamination. The two papers show the many and diverse mechanisms associated with the degradation of thermal and optical performance of spacecraft surfaces and optics. *Glassford, Osiecki, Liu,* and *Hitchcock* discuss contamination related to outgassing and condensation as well as current activities initiated by the Air Force Materials Laboratory to develop standardized approaches for measuring various contaminants and the use of this information for selecting design materials. In the second paper *Moss* presents some of the unique contamination problems associated with the Space Shuttle payload bay.

Chapter II deals with some of the causes of spacecraft contamination and concludes with a paper aimed at the ultimate goal of contamination prevention. In the first paper *del Casal* presents the potential for spallation of foam insulation on cryogenic fuel tanks for space launch systems. Spallation results from the orbital low pressures outside the insulation and the near-atmosphere pressures inside the insulation. *Lee* discusses the dispersion of particle contamination about a spacecraft. This is especially meaningful for the Space Shuttle, which is a typical launch vehicle, i.e., a relatively dirty spacecraft; the spacecraft may intercept these particles during later orbits. In a very timely paper, *Bareiss* and *Jarossy* assess the particle contamination of the optical sensors during ground and launch operations of the Shuttle. The degree of sensor contamination could be severe. *Ahern, Belcher,* and *Ruff* show how the location of spacecraft venting systems can cause substantial changes in spacecraft thermal control absorptances. This chapter concludes with a very creative paper that reaches to the heart of the contamination issue, i.e., contamination prevention. *Scialdone* presents a detailed analysis on preventing both gaseous and particle contamination of a telescope system to be flown on the Space Shuttle.

In Chapter III, the properties and effects of various contaminants are presented. These properties have been determined from measurements made primarily in ground test facilities. Such measurements are important because they underlie the anlysis of how much a thermal control surface, an optical system, or a sensor will be affected by both molecular and particulate contamination. In the first paper, *Wood* and *Roux* present the infrared transmittance and the infrared optical properties of a variety of individual species common in the exhaust products from typical attitude control thrusters. Next, the effects and properties of mixtures of exhaust species are discussed by *Palmer, Roux,* and *Wood*. It is shown that the behavior of mixtures of gases, such as exist in plume exhaust, is significantly different from that of pure species. This is especially true for compounds containing hydrogen, where hydrogen bonding has strong effects upon the optical infrared transmittance. *Konopka, Reed,* and *Calia* present the experimentally determined infrared optical properties of Al_2O_3. These properties are important because solid-propellant rocket thrusters are being considered for some attitude control systems.

The work of *Smith* details how the plume boundary-layer flow expands about the nozzle and can actually impinge on objects located upstream of nozzle exhibit plane. The in-orbit measurements of *Hall* and *Fote* show the degradation of "clean" thermal control surfaces over four years at geosynchronous attitude. The work of *Ahern* and *Karperos* also shows the in-orbit measurement of the solar absorptance. These samples are considered to be clean; any degradation is indicated by departures from the baseline values before contamination by foreign substances. An excellent work by *Trinks* and *Hoffman* provides both experimental and theoretical analyses (using the CONTAM computer program) of the contamination associated with the bipropellant attitude control thrusters proposed for the GALILEO spacecraft. The details of the engine valving sequence, combustion, pulse length, and gasdynamics are studied to show their influence on the engine cleanliness with respect to spacecraft contamination.

The last two papers in the volume are by *Girata* and *McGregor*. The first shows the results of particle sampling probes in the near-field region of a solid-propellant rocket motor fired in a high-altitude ground test facility; their results show the chemical composition and size distribution of potential particle contaminants. Their second paper examines the contamination associated with the postfire efflux of a solid-propellant rocket motor.

As coeditors we acknowledge the assistance of Mrs. Norma Brennan, Editorial Department Director, Mrs. Camille Koorey, Progress Series Managing Editor, and Dr. Martin Summerfield, Editor-in-Chief of the *AIAA Progress in Astronautics and Aeronautics* series. We also thank Dr. Fred Nelson of the University of Missouri – Rolla for organizing the thermophysics sessions of the AIAA 21st Aerospace Sciences Meeting. Finally, we express our appreciation to the entire AIAA Thermophysics Committee for the team effort and outstanding work that was done to solicit excellent papers for both the AIAA 21st Aerospace Sciences Meeting and the AIAA 18th Thermophysics Conference.

J.A. Roux
T.D. McCay
April 1984

ing # Chapter I. Contamination Overview

Improved Methods for Characterizing Material-Induced Contamination

A. P. M. Glassford,* R. A. Osiecki,† and C.-K. Liu*
Lockheed Palo Alto Research Laboratories, Palo Alto, California
and
Lt. M. Hitchcock‡
Air Force Materials Laboratory, Wright-Patterson Air Force Base, Ohio

Abstract

It is current practice to assess material outgassing contamination potential using standardized total mass loss (TML) and collected volatile condensible material (CVCM) data, together with the criteria that TML and CVCM should be less than 1.0% and 0.1%, respectively. This procedure has proven to be satisfactory for screening materials, but TML and CVCM data do not permit the migration of outgassed contaminants in a system to be modeled quantitatively and related to actual system performance requirements. At this time no standardized methods have been developed for measuring the type of data required by the contamination migration models. The Air Force Materials Laboratory instituted a program to survey currently available procedures and technology, identify appropriate new test methods, and develop these methods to the point at which they can be standardized. This paper reviews the initial activities of this Air Force program. The physical phenomena involved in outgassing and condensation, the types of analytical expressions used to represent these phenomena, and the types of data required by the migration models are discussed. Current apparatuses

---

Presented as Paper 83-1496 at the AIAA 18th Thermophysics Conference, Montreal, Canada, June 1-3, 1983. Copyright © 1983 by Lockheed Missiles and Space Company, Inc. Released to AIAA to publish in all forms.
  *Staff Scientist, Materials Sciences Laboratory.
  +Research Scientist, Materials Sciences Laboratory.
  ‡Project Engineer, Nonmetallic Materials Division.

capable of measuring these data are reviewed. Test methods capable of being developed to provide more detailed information on the controlling phenomena are presented.

Introduction

Contamination of space system surfaces by condensation of outgassing products from materials of construction has long been recognized as a potential mechanism for degrading lifetime or performance. As a result, procedures have been developed by the industry to control contamination at the design stage. The usual first step in these procedures is to select only materials with a low percentage of outgassing products, and whose outgassing products are only minimally condensible at normal spacecraft operating temperatures. In the early days of the aerospace industry, many organizations conducted tests to obtain the kind of material property data needed to make this selection. One of the most extensive of these test programs was conducted by Stanford Research Institute (SRI) for the Jet Propulsion Laboratory.[1] The procedure used by SRI was formalized into a standardized method as ASTM E595 in 1977. In the ASTM E595 procedure, a material specimen is exposed to a vacuum at 125°C in an effusion cell for 24 h. The percentage total mass loss (TML) is found from the weight of the specimen in the laboratory atmosphere before and after the test. The outgassing species condensible at 25°C are collected by a 25°C plate held in the flux field leaving the effusion cell orifice. This collector is weighed before and after the test, and the percentage of outgassed material condensible at 25°C, which is called the collected volatile condensible material (CVCM), is determined. A very large amount of this type of data has been generated, and has been tabulated by NASA Goddard Space Flight Center.[2] It has become customary to accept a material for use in space systems only if its TML and CVCM are below 1.0% and 0.1%, respectively.

The use of ASTM E595 data together with the 1.0% and 0.1% criteria clearly only screens or ranks materials. This procedure takes no account of the amount of material used in the spacecraft, the operating temperature of the material, the amount of outgassed material that impinges on a critical surface, the temperature of the critical surface, or how much condensed outgassed material can be tolerated by the critical surface before its properties become excessively degraded.

In recent years, space systems have become more contamination sensitive. With the longer system lifetimes that are now sought, thermal control properties must stay within design limits for a longer period. Current and future high-resolution cryogenic and other optical sensor systems are especially vulnerable to formation of condensed deposits on optical surfaces.

Another type of problem stems from the fact that many materials that satisfy the 1.0% and 0.1% criteria are high-cost, special-order products produced in small lots specially for the aerospace industry. When a large amount of these materials is needed, the cost may be excessive, and low-cost alternative materials not passing the screening test may well be adequate. Also, suppliers have removed some of these special-order materials from the market because of low profits or environmental concerns. These materials may have to be replaced by materials which do not pass the screening test.

Because of the limitations of screening procedures in the face of increasingly severe system requirements, the aerospace industry has begun to address contamination assessment more systematically. Computer models have been developed to represent the flow of outgassed species from source material to specific critical surfaces in the vacuum environment. Using these models the amount of contaminant migration that will occur in a specific application can be predicted. Material acceptability is then determined by comparing allowable contamination levels computed from system performance requirements with contamination levels predicted by the flow model. Modeling of contaminant migration is a relatively costly procedure. Hence, it is normal practice when assessing a new design to accept materials in the first instance using the screening procedure. Modeling is used when detail system design reveals areas of special concern.

Most aerospace organizations have developed a contamination flux modeling capability, and most have developed test apparatuses and methods to generate the data needed to support the models. Although most models are basically similar in approach, the types of apparatus and the test methods used by the various organizations differ significantly. As a result, many of the data generated by the industry have been obtained under nonstandard conditions and cannot be used with confidence by any organization other than that of the originator. To make a start towards rectifying this situation, the Air Force Materials Laboratory has instituted a program to develop standardized methods for measuring the type of material properties needed to support contaminant flux modeling. The program, in two phases, will cover measurement of the outgassing rate, the condensation rate, and the morphology of the condensed deposit. In Phase I, an extensive assessment of methods for measuring outgassing and condensation properties was made by reviewing the literature and surveying the industry. A report on this work is currently being prepared. On the basis of this assessment, test methods that appear to be suitable for development into standardized procedures have been selected. In Phase II, the selected methods will be developed to a point at which preliminary standard test procedures can be written.

The migration of outgassing products involves a wide variety of physical phenomena, each with its own set of governing relationships and relevant material properties. Ideally, the contamination model would account for all of these phenomena in a rigorously correct manner. However, this may not be technically possible, because some of the controlling mechanisms are not yet well enough understood; nor practical, since measurement of the detailed property data required to characterize all of these phenomena could be quite costly and time consuming. A basic problem to be addressed in developing models is therefore selection of a practical compromise between realism and cost. The selected form of the model determines the type of data that is needed, and hence the requirements for the standard test method. This issue has not yet been adequately addressed by the industry. The industry review conducted in this program reveals that the model approaches currently in use range from very simplified to very complex. Fortunately it is possible to proceed toward development of standard test methods without completely resolving this issue at this time. For the present purposes, a standard test method is defined as consisting of a standard apparatus plus a standard procedure for using the apparatus to obtain the standard data. This paper shows that standard apparatuses can be built that will accommodate a range of test procedures, ranging from simple engineering tests to basic material-property measurements. The preferred standard test procedures can then be determined later by industry concensus.

This paper reviews the main technical inputs and alternatives involved in selecting new methods for measuring outgassing, condensation, and deposit morphology phenomena. The discussion considers, in turn, the nature of the basic phenomena involved; the various types of models used to represent these phenomena and hence the type of property data needed for the models; and the types of measurement apparatuses that can be used to make these measurements. Based upon these considerations, new apparatus concepts are proposed.

Outgassing Data Requirements and Measurement

Sources and Mechanisms

Outgassing is the process of release of molecules from a material into the gaseous phase under the influence of reduced pressure or elevated temperature or both. Most outgassed species are relatively high-volatility, loosely bound molecules, which were initially either distributed through the bulk material or adsorbed onto the outer surface. For example, polymeric materials, which are the main outgassing sources in space systems, may contain solvents, catalysts, incompletely polymerized fractions, plasticizers, etc., distributed throughout their bulk. They may also contain water, either as a result of the manufacturing process or sorbed from the atmosphere. As their temperature is raised, polymers may undergo thermally induced transitions that result in the creation of additional bulk-distributed low-volatility species by breakdown of the basic material. Exposure to the space environment and to oxygen in the upper atmosphere can cause near-surface material to degrade into high-volatility species. Also, all materials will initially have several layers of atmospheric gases adsorbed on their outer surfaces.

For surface adsorbed molecules, outgassing occurs by simple desorption. Bulk sorbed material must first diffuse through to the surface, where it is then desorbed. When additional high-volatility species are formed by thermally induced transitions, then the outgassing process will consist of a chemical process, followed by bulk diffusion, followed by surface desorption. When the outer layers of material are degraded by the space environment, outgassing will occur by a chemical degradation process followed by diffusion over a short path, followed by desorption. Outgassing from capillary adsorbed sources occurs by a succession of physical desorptions and re-adsorptions on capillary walls.

All of the mechanisms described are some combination of physical desorption, bulk diffusion, and formation of higher volatility species from the bulk material by chemical changes induced by some combination of environmental effects. It can be shown that, for most materials, the time required for a surface molecule to desorb is very short compared to the time required to diffuse even a short distance in the bulk material. Thermally-induced chemical changes usually occur very rapidly, so much so that in thermal methods of material analysis it is important to use small samples so that the rate of the chemical change being measured is not masked by diffusion effects. This means that, in all but the thinnest materials, diffusion rates rather than desorption or chemical rates of change will probably control the rate of outgassing in most cases. However, the actual controlling mechanism cannot be determined without performing an outgassing test.

Modeling Data Requirements

The data needed by the model are the outgassing rate and/or total mass loss for each component in the space system as a function of time, with the system at operating temperature. These data are needed separately for each individual outgassing species, because the different species will have different surface condensibilities and will have different effects on surface properties. Assuming that systems are made from several different materials, three basic methods might be used to assemble the needed data:

1) Direct measurement of the outgassing rate of the actual system at the actual operating temperature.
2) Summation of the outgassing rates of each of the materials measured separately at the temperature and in the geometry of the system application.
3) Summation of the outgassing rates of each material calculated from basic transport property data (such as diffusion coefficients) substituted into solutions of the basic transport equations (e.g., Fick's Law) for each geometry and temperature that each material has in the system application.

The most generally applicable and hence most desirable approach would be method 3. The theoretical bases for physical adsorption, bulk diffusion, and chemical rate processes are well established and have been treated at length in the literature. It is, in principle, a straightforward matter to derive analytical expressions for the rate of outgassing due to each of these processes. These expressions define the basic transport properties which control outgassing. If a data base of these basic properties were available, it would be a simple matter to predict outgassing rate for any practical situation. The case of diffusion-controlled outgassing provides an example of this approach.

Bulk diffusion is controlled by Fick's law,

$$\frac{\partial C}{\partial t} = -D\nabla^2 C \qquad (1)$$

where C is the local concentration and D is the diffusion coefficient. D is given by

$$D = D_o \exp(-E_d/RT) \qquad (2)$$

where E_d is the activation energy for diffusion. These equations can be solved for given geometric and temporal boundary conditions. As an example, for isothermal outgassing into vacuum from both sides of an infinite slab of thickness L cm, with initial uniform concentration of outgassing species of C_o g/cm^3, the concentration distribution is given by the following one-dimensional solution:

$$C(x, t) = \frac{4C_o}{\pi} \sum_{m=0}^{\infty} \frac{1}{2m + 1}$$

$$\times \sin\left[\frac{(2m + 1)\pi x}{L}\right] \exp\left[\frac{-D(2m + 1)^2 \pi^2 t}{L^2}\right] \qquad (3)$$

The outgassing rate Q g/cm^2 is given by

$$\dot{Q} = -D \left(\frac{\partial C}{\partial x}\right)_{x=0,L}$$

$$= \frac{4DC_o}{L} \sum_{m=0}^{\infty} \exp\left[\frac{-D(2m + 1)^2 \pi^2 t}{L^2}\right] \qquad (4)$$

If C_o, E_d, and D_o are known, $\dot{Q}$ can be predicted as a function of time for any one-dimensional flow situation (such as outgassing from a glue line) at any constant value of temperature using Eqs. (2) and (4). Equations similar to Eq. (4) can be generated for any other geometry and temperature history so that once C_o, E_d, and D_o are known, Q can be predicted for all possible situations. These data will, of course, be needed separately for each outgassing species.

Expressions similar to Eq. (4) can be derived from published analytical models for thermally induced chemical changes, physical desorption, and capillary desorption. For these cases, as for diffusion, the expressions will define material properties such as adsorption energies, rate constants, etc., which, once known, can be used in the appropriate equations to predict outgassing rate by each mechanism for any temperature or geometry.

The problem with the approach of method 3 is that it is difficult and expensive to generate the required property data base. It can be shown that transport properties can be deduced from outgassing data in appropriately controlled experiments for materials which outgas a few easily distinguishable species.[3] However, some materials outgas many species of similar behavior, and it may not be possible to identify in advance which mechanisms are controlling and hence how the transport properties are to be deduced from the measured outgassing data.

Model data obtained by method 1 relate to a particular combination of materials, geometries, and temperature histories and cannot usefully be included in a generally applicable data base. Since this method requires that the system be already designed and built, at least in representative form, the data obtained are more useful for modeling the in-service performance of an existing system than for supporting material selection during the design process.

Method 2 is appropriate to the design stage, since it permits the behavior of different materials to be compared. However, in order for the appropriate data base to be truly general, it must contain outgassing data for each material in all possible combinations of geometry and temperature history. The number of these possible combinations can be greatly diminished, without serious loss of generality, by considering only isothermal temperature histories.

One method for obtaining basic property data is to deduce them from isothermal outgassing rate data. An example of this is given below. Since this is possible, a practical strategy for developing a data base can be proposed. First, isothermal outgassing data will be generated, in support of a method 2 approach. These measurements will be performed in such a manner as to permit an attempt to be made to extract basic property data from the isothermal outgassing data. This strategy provides a fall-back position of generating at least an isothermal data base for specific geometries for those materials too complex to analyze further.

An example of how basic property data can be deduced from isothermal outgassing data for specified geometries can be given for the case of isothermal outgassing by one-dimensional diffusion. For evacuation times $> (0.1\ L^2/D)$, all terms in Eq. (4) for $m > 0$ become negligible. Eq. (4) can then be rewritten as follows:

$$\dot{Q} = \frac{4DC_o}{L} \exp\left(\frac{-D\pi^2 t}{L^2}\right), \quad t > 0.1\ L^2/D \quad (5)$$

If the isothermal outgassing rate of a specimen in which outgassing occurs by one-dimensional diffusive flow is plotted against time on log-linear paper, then at evacuation times greater than $0.1\ L^2/D$ a straight line should be obtained with a slope of $-D\pi^2/L^2$ and a vertical intercept of $4DC_o/L$. From these data, D and C_o can be deduced. By repeating the test over a range of temperatures, E_d can be found from the temperature dependence of D [Eq. (2)]. This method was used in Ref. 3 to find C_o, D, and E_d for two groups of species outgassed from RTV 560.

In principle, basic property data can be obtained more quickly from transient measurements than from isothermal measurements, if the controlling outgassing mechanism is known. Taking the diffusion controlled outgassing situation as an example, it has been shown that solutions to Fick's Law can be obtained for variable temperature if the experimental temperature variation follows certain analytically convenient forms.[4] The transport properties can then be deduced from the shape of the curve of outgassing rate versus temperature. This test is difficult to perform because of the problem of transferring heat to the sample to achieve the desired heating rate while minimizing temperature variations within it. When more than one species are present, it is more difficult to separate the behavior of each species than it is in isothermal outgassing measurements. Another more serious problem in performing transient heating tests on polymers is that raising the temperature will sooner or later cause thermally induced chemical changes in the polymer. Since the objective of transient heating tests is to accelerate the test, it is fundamental to the approach that the material be heated substantially above its normal operating temperature in order to signficantly shorten the testing time. In this case, species which would not be found at normal operating temperatures might be created, which would lead to erroneous conclusions.

Methods for Measuring Outgassing Rate

Major Techniques. There are three basic types of methods for measuring outgassing rate or total mass loss as a function of time.

1) Direct measurement of mass loss: The sample is placed on the pan of a vacuum microbalance, and its mass m_s is measured as a function of time. The outgassing rate is equal to dm_s/dt.

2) Collection of condensed outgassing species: The sample is placed at some location in the vacuum system, or in a specially constructed holder. The outgassing flux leaving the sample is condensed on a cooled collector surface whose weight can be measured, such as a quartz crystal microbalance (QCM). The outgassing rate is found from the amount condensed on the cooled collector surface, the sticking coefficient of the outgassing flux at the collector temperature, and the view factor from the material sample to the collector. To measure total outgassing rate, the collector should be cold enough for the sticking coefficient of all incident species to be unity. For practical purposes this means the collector temperature should be less than about 100 K.

3) Sample chamber pressure measurement: There are at least four variations of this method, all of which basically involve placing the sample in a vacuum chamber and measuring the pressure history.

a) Rate of rise method: The vacuum chamber holding the sample is periodically closed off, and the instantaneous rate of pressure rise dP/dt is measured. The sample outgassing rate is determined from dP/dt and the known volume of the vacuum chamber.

b) Rate of fall method: The vacuum chamber holding the sample is evacuated with and without the sample present, while the rate of pressure fall dP/dt is measured. The outgassing rate is calculated from the difference in chamber dP/dt measured with and without the sample present.

c) Throughput method: The vacuum chamber holding the sample is evacuated through tubulation of known conductance. The sample outgassing rate is found by measuring the pressure difference across the known conductance.

d) Collection method: The vacuum chamber holding the sample is periodically disconnected from its pumping system and connected immediately to a relatively large collection chamber of known volume, previously evacuated to a very low pressure. The sample outgassing rate is found from the rate of pressure rise in the known collection volume.

Comparison of Main Techniques. The vacuum industry has successfully used pressure measurement techniques to measure outgassing rates for nearly three decades. A substantial body of information obtained by this method, primarily at 25°C, is available.

The ability to measure pressure as a function of time is basic to vacuum chamber operation. Outgassing data can be obtained quickly and economically, using standard vacuum technology pressure measurement techniques, and can be applied directly and unambiguously to system design. However, the approach has several disadvantages which make it unsuitable for generating the type of outgassing data needed to support space system modeling: it is difficult, if not impossible, to perform tests reliably at temperatures other than 25°C; the range of measurable data and the accuracy and resolution of the data are restricted by limitations in pressure gauge and vacuum chamber design; and system characteristics such as readsorption of outgassed products by the chamber walls, uncertainty in flow conductance, and background outgassing by the chamber walls affect measurement accuracy. For these reasons, pressure measurement techniques are not considered to be suitable for the present purposes.

The two main candidate techniques are therefore measurement of mass loss and measurement of collected mass. The accuracy, resolution, and sensitivity of these techniques and their suitability for use as part of a standardized outgassing test method are compared below.

Measurement Accuracy. Although vacuum microbalances can be built that can measure masses as low as 10^{-8} to 10^{-9} g, this degree of accuracy is very difficult to achieve because of various spurious effects that appear in the range of 10^{-8} g and below. Maximum practical sensitivities are more likely to be in the range of $10^{-6} - 10^{-7}$ g, and even these values can be achieved only with considerable experimental care. Microbalance sensitivity is related to scale pan capacity. Manufacturers' data suggest that the maximum ratio of capacity to sensitivity achievable in practice is of the order of 10^7, corresponding to changes of about $10^{-7} - 10^{-6}$ g in sample sizes of about 1 - 10 g.

In practice, a QCM used in a collection apparatus can be expected to have a frequency stability of better than 1 Hz per day. In an apparatus using a cooled 10-MHz QCM to collect flux from a sample held in an effusion cell roughly 8 cm away, a QCM frequency change of 1 Hz corresponds to about 10^{-6} g lost by the sample. Typical sample sizes used in such an apparatus are 1 - 10 g. Hence, this method also has a capacity-to-sensitivity ratio of roughly 10^7. While this comparison is far from rigorous, it would appear that in practical situations both the mass loss method using a vacuum microbalance and the mass collection method using a QCM as a collector have generally similar measurement accuracies and sensitivities.

In principle, the mass collection technique could be executed by suspending a cooled collector plate from a vacuum microbalance, but this approach is not considered to be attractive. Apart from the problem of adequately cooling the collector plate without interfering with weight measurement, the sensitivity is significantly less than the QCM method. The QCM can detect mass changes of the order of 2×10^{-9} g/cm^2. To achieve this with the suspended plate technique with a balance sensitivity of 10^{-6} g, the plate area would have to be of the order of 500 cm^2.

Application to a Standardized Test Method. Measurement of the outgassing rate by monitoring the weight of a material sample suspended from a microbalance has two main advantages. First, the mass loss is measured directly and the data are therefore unambiguous. Second, vacuum microbalances are available commercially in standard configurations, both as separate units and as part of thermogravimetric analysis apparatus. Hence, an organization could acquire the capability to perform standarized isothermal or transient outgassing measurements in a short time with a high degree of confidence. On the other hand, high-precision microbalances are relatively large and cumbersome, must be installed in a vibration-free environment, and require a certain amount of operator experience in order to obtain the high levels of performance of which they are capable. Because the sample materials must be either placed on a pan or suspended directly from the balance, there are constraints on the sample size and configuration. It is also difficult to heat or cool the sample and to measure its temperature.

The collection method using a cooled QCM as a collector has three major advantages. First, there are no restrictions on how the sample is configured or held in the vacuum chamber, and it can be readily instrumented and heated or cooled. Second, the number and proportions of outgassed species condensed on the collector QCM can be determined by heating the QCM in a controlled manner and observing the different evaporation characteristics of the species. Third, the QCM sensor is relatively sturdy, and no special skill or care in application is required to obtain the very high sensitivity of which it is capable.

A disadvantage of this collection method is that determination of the sample mass loss is indirect. To relate the mass collected on the QCM to the mass leaving the sample, it is necessary to know the QCM sensitivity calibration, the distribution of outgassing flux leaving the sample or sample holder, and the sticking coefficient of the outgassed species at the QCM temperature. These factors can be readily established and can be verified by comparing mass changes measured by both direct mass loss and mass collection techniques. It has been satisfactorily demonstrated that the sensitivity of QCMs to evenly distributed deposits is closer to the theoretical value than calibration methods can measure.[5] The sticking coefficients of all practically significant outgassing products can be assumed to be essentially unity if the QCM is cooled to liquid-nitrogen temperature. The view factor between sample and QCM depends on the apparatus design and must be determined by calibration. For a sample placed in an effusion cell, the flux distribution from the orifice will be nearly cosine, but somewhat modified by the shape of the cell orifice. This modified distribution can be measured experimentally and should be constant. For a sample placed freely in the vacuum chamber, the view factor must be calculated separately for each sample.

Methods for Distinguishing Outgassed Species

Most materials outgas more than one species, each of which has a different outgassing rate, condensibility (which is related to vapor pressure), and effect on the properties of the contaminated surface. These data must be obtained separately for each species if the effects of contamination migration are to be modeled realistically. It is not necessary to know the chemical identity of each outgassed species for this purpose. It is necessary only to be able to label the species in some convenient manner so that measured outgassing rate, condensation kinetics, and induced surface effects data can be correctly assigned.

There are many possible techniques for distinguishing between chemical species. From the multitude of possible techniques, Wendlandt[6] has identified several as being specially applicable to evolved gas analysis (EGA) during the heating of materials. These are gas chromatography, mass spectrometry, infrared absorption spectroscopy, use of cold traps or chemical absorbents, thin-layer chromatography, chemical detectors, titrimetry, and condensation nuclei. Of these techniques, only mass spectrometry and infrared absorption spectroscopy can perform real-time species analysis in vacuum. All others require removal of a sample of evolved gas from the test location to an analysis location. Infrared absorption spectroscopy in vacuum is better suited to analysis of condensed films than of low-pressure gas molecules, because it requires enough of the analyzed material to be present to produce measurable amounts of absorption. On the other hand, mass spectrometry is well suited to continuous analysis of low concentrations of outgassing species in a vacuum.

Two additional methods of separating species have evolved from aerospace outgassing rate measurement programs. A natural by-product of the QCM-based mass collection technique for measuring outgassing rate is that a condensed deposit of outgassed species is obtained on the QCM. If the QCM is then heated in a controlled manner, the different species will evaporate separately in unique and different temperature regimes. This technique gives the number of different species present, their relative amounts, and their different reevaporation rates as a function of temperature. As with other analytical procedures, there is some difficulty in resolving the data when there are species present with similar evaporation properties. However, the technique is relatively new, and many of these difficulties seem resolvable with experience and development. Even in its present state of development, it has been used successfully as a stand-alone method for species separation and labeling.

In some cases the presence of more than one outgassing species can be inferred from the time dependence of the outgassing rate. This method is usually applied to outgassing rates that have exponential time dependencies. The presence of a different outgassing species is inferred from portions of the outgassing rate data which have different slopes when plotted in log-linear fashion. This approach is limited in usefulness to cases of a small number of outgassed species. It cannot be applied successfully to resolve the presence of a small amount of one component in a large amount of another. Also, the information used to identify the species is not a basic property of that species, but is rather an effect which could result from a range of combinations of species, material geometries, and temperatures. It is thus not strictly usable as a stand-alone method, but should be used to supplement another more definitive technique.

An Improved Outgassing Measurement Apparatus

In an earlier section it was argued that the preferred type of data base would consist of basic transport properties for each outgassed species. Since this type of data will be difficult and costly to obtain, the type of data sought in the first instance should be the isothermal outgassing rates, from which property data may or may not be deducible. Transient methods could possibly be used eventually to obtain property data more economically. It is therefore desirable that a new measurement apparatus should be capable of generating isothermal outgassing data. The apparatus must be able to separate and label species so that the outgassing rates of each species should be determinable from the total rate. The degree of resolution should be high so as to maximize the chances of being able to deduce basic property data. It should also be capable of making both isothermal and transient outgassing measurements. On the basis of the considerations of the two previous sections, it is believed that the QCM-based mass-collection technique combined with a mass spectrometer best fulfills these needs. The basic accuracy and application advantages of the QCM collection method have already been demonstrated by using it to obtain basic transport properties when more than one species is being outgassed.[3] Addition of mass spectrometry will strengthen its capability for resolving the more complex multispecies situations. Development of this technique should concentrate on two areas - rigorous examination of the accuracy with which outgassing rates can be measured and development of data analysis techniques to resolve the behavior of individual species and basic property data from the outgassing rate data. The accuracy of the QCM-based mass collection approach has been partially examined and verified in previous work.[7] However, in order for the technique to be accepted as a standard approach, it must now be reexamined in a more complete and rigorous manner.

The relationship between the QCM collector output and the mass loss by the sample depends upon a number of factors. These include the QCM sensitivity, the sticking coefficient of the outgassing flux at the temperature of the collector, and the flux distribution from the sample to the QCM. If an effusion cell is used to hold the sample and its orifice is not large enough, the pressure in the cell may become high enough to affect the outgassing rate. These factors should all be thoroughly investigated in a development program.

A direct overall verification of the accuracy of the collection method can be performed by measuring the outgassing rate simultaneously by both mass loss and mass collection techniques. This can be achieved by using a collection type apparatus in which the sample is suspended from a vacuum microbalance.

It is proposed to develop a capability to resolve the outgassing rates of each separate species by using a combination of QCM and mass spectrometer data. The species separation data according to volatility, determined from periodic QCM warmup, will be combined with mass spectrometer data taken during QCM warmup and throughout the test. Data from the mass spectrometer scans taken during the test will be stored in a computer. These scans will in general include more than one species and will not necessarily be simple to interpret without additional data. At the end of the test the QCM will be warmed up while the mass spectrometer is scanned. The mass spectrometer fingerprint of each separate species will be determined and stored as it is evaporated separately from the QCM. The separate species fingerprint data will then be compared with the mixed species data obtained for various times in the outgassing test, from which the relative proportions of each species in the outgassing flux can be deduced as a function of time. This, in turn, will permit the outgassing rate of each individual species to be obtained as a function of time. Mass spectrometer data will not be adequate by themselves to quantify the relative amounts of each species directly because of differences in ionization constants and possibly incomplete mass spectrometer fingerprints. Relative amounts of each species can, however, be deduced from the QCM warmup test. One important practical point to be resolved is to determine how sophisticated a mass spectrometer must be to fulfill the requirements of this test. In general, mass spectrometers range from low-cost residual gas analyzer versions with ranges of about 100 - 200 amu, up to very expensive units with amu capabilities in the thousands, interfaced with computerized data banks to facilitate species identification. Polymers will outgas high-molecular-weight species, so a capability to resolve high amu values would be desirable. A possible compromise approach is to use a lower cost mass spectrometer as the unit dedicated to the test apparatus, but to include a separate analysis in a high-resolution gas chromatograph/mass spectrometer (GC/MS) as a part of the test procedure, since central GC/MS capability is now available at most laboratories. The GC/MS data and access to its library would be valuable in interpreting the mass spectrometer data obtained during outgassing and QCM warmup.

Condensation Kinetics and Deposit Morphology Data Requirements and Measurements

Basic Processes

Many physical processes are involved in the growth of a condensed deposit, and the following description is a much simplified picture. When an impinging flux strikes an initially bare surface, the impinging molecule may be reflected or adsorbed. If it is adsorbed, it will diffuse across the surface until it either desorbs or strikes another adsorbed molecule or cluster of molecules. If the cluster of molecules is not large enough for stability, then the molecule will eventually leave the cluster and once more diffuse across the surface. If the cluster of molecules is larger than the critical value for nucleation, the incident molecule will remain in the cluster and bulk condensation will occur. A molecule may leave the nucleus by bulk evaporation. Molecules incident on the cluster will condense directly into the bulk phase.

Many workers have studied the phenomena of adsorption, surface diffusion, nucleation, and bulk condensation. Many theories have been proposed for adsorption. Bulk behavior can be represented satisfactorily by the Langmuir equation. Representation of surface diffusion and nucleation is a more difficult task, and improved theories are still being proposed. Zinsmeister[8] proposed that a one-atom cluster be assumed stable, which implies that surface diffusion and nucleation effects are ignored. This is probably reasonable for condensation on a relatively cold surface. Rhodin and Walton[9] proposed a model in which surface diffusion kinetics were considered and a two-molecule nucleus was assumed to be stable. Hirth and Pound[10] described a model which assumes that isolated clusters of bulk condensed phase having positive free energy of formation were formed over a dilute adsorbed layer. A two-phase model given by Hudson and Sandejas[11] includes a primary, highly bound adlayer as a substrate, with the formation of bulk nuclei within a secondary, mobile phase. Many other models have been proposed though none has been entirely verified by experiment.

The shape of the condensed deposit depends strongly upon the mobility of the condensed molecules on the surface. On rough surfaces, at lower surface temperatures, or with larger molecules, the surface mobility will be relatively low and the deposit will have a smooth film-like form. On smoother surfaces, at higher temperatures, or with smaller molecules, the condensed molecules can diffuse over large distances on the surface and can collect at a discrete number of nucleation sites, producing an island or droplet morphology. An island or droplet morphology will usually produce more scattering and less loss of reflectance or transmittance than a smooth film deposit of the same average mass per unit area.

The condensation process and the morphology of the deposit can be significantly affected by environmental effects. It is well known that ultraviolet radiation can cause chemical changes in surface deposits, altering their volatility and mobility and hence their net condensation rate and morphology. Other environments, such as oxygen in low Earth orbit, charging at high altitudes, electrons, and protons, may also affect the condensation process. In the aerospace application, these environments must be considered to be major variables. Further, their effects should be considered together as well as separately, because there may be synergism.

Model Data Requirements

As in the case of outgassing data, the condensation data needed to predict the rate and form of condensed deposit accumulation on a given surface when exposed to a given impinging flux can be assembled in more than one way. Again, three main methods of obtaining the data can be identified.

1) Condensation rate and deposit morphology can be measured for the mixed species outgassing flux from the specific system of materials of interest at the specific operational temperature, impinging on a specific surface at the same rate as would be experienced in the actual application.

2) Condensation rate and morphology can be extracted from a general data base for single species outgassed from specific materials impinging on specific surfaces, for a range of values of constant impinging flux rate and temperature, surface type and temperature, and amount already condensed.

3) Condensation rate and morphology can be calculated for the situation of interest from a data base of basic properties, such as adsorption energies, latent heat, surface diffusion constants, accommodation coefficients, etc., measured for all possible outgassed species and condensing surfaces.

As in the case of outgassing rate prediction, method 3 is the preferred approach because of the greater generality, but the required data base will be very difficult to generate. The phenomena involved in condensation are more numerous, complex, and currently less well understood than those involved in outgassing. As a result, it is even harder to characterize all of the controlling mechanisms, their interactions, and the basic material properties that relate to them. Hence the task of generating a basic material property data base will be even more difficult than for outgassing.

Data obtained by method 1 are applicable only to one practical situation and are hence not suitable for inclusion in a generally applicable data base. The data required to support method 2 are more generally applicable and can be generated relatively inexpensively. However, for generality the data have to be generated separately for each combination of incident flux rate, temperature, and species, and surface type, temperature, and finish, etc. Continuing the analogy with the case of outgassing data, the basic property data required for method 3 can in principle be extracted from the method 2 data base. It thus seems reasonable to seek, in the first instance, a method 2 data base and to attempt to generate basic property data from it.

Because of the complex nature of the condensation process, it is not a simple matter to assemble a physically correct model expressing net condensation rate as a function of all controlling variables. Indeed, such a model, even if attainable, might be impractically complex for aerospace engineering design purposes. The industry has therefore developed a number of approximate models. The property data that these models require is thus at least partially empirical and varies from model to model. It will therefore not be possible to define standard data, and methods to obtain them, until a basic model has been standardized. Some of the more popular models are reviewed below. Fortunately the data required by all of them fall into the method 2 category, and all can be obtained by some form of isothermal condensation rate measurement.

The most simple model is the assumption that all incident flux $\dot{m}_i$ condenses on the surface, and that desorption and/or bulk reevaporation are negligible. The net condensation rate $\dot{m}_c$ is then given by

$$\dot{m}_c = \dot{m}_i \tag{6}$$

In general, only a fraction of the impinging flux will condense. This situation is represented by introducing an experimentally measured sticking coefficient, S, such that

$$\dot{m}_c = S \times \dot{m}_i \tag{7}$$

Equation (7) is strictly valid only if the reevaporation rate is zero. It has sometimes been applied when the reevaporation rate is not zero, in which case S includes the effect of reevaporation and should not be called "sticking coefficient."

The reevaporation or desorption rate may not be negligible by comparison with $\dot{m}_i$. In this case, if the reevaporation or desorption rate is denoted by $\dot{m}_e$, $\dot{m}_c$ is given by

$$\dot{m}_c = S \times \dot{m}_i - \dot{m}_e \tag{8}$$

Equation (8) assumes that the surface conditions and the condensed deposit are uniformly distributed over the surface. In fact, the mobility of the condensed species on the surface is usually sufficiently high to produce some island or droplet type structure, at least in the early stages of deposition. In this case, the impinging flux might strike an island of bulk condensate, or the adsorbed layer between the islands. The value of S and $\dot{m}_e$ will in general be different for these two regions. The fractional area covered by bulk deposit can be denoted f_b, while the sticking coefficient of the impinging flux on adsorbed and bulk condensate are S_a and S_b, respectively. The bulk reevaporation rate is $\dot{m}_b$, and the desorption rate is $\dot{m}_d$. The net condensation rate is then given by

$$\dot{m}_c = \dot{m}_i [S_a(1 - f_b) + S_b f_b] - [\dot{m}_b f_b + \dot{m}_d(1 - f_b)] \tag{9}$$

In Eq. (9), S_b and $\dot{m}_b$ are properties of the outgassing species alone. S_b may be assumed to be close to unity.

The term $\dot{m}_b$ is related to the vapor pressure of the species by the Langmuir equation, and thus can be predicted from available data for known species. The terms S_a, f_b, and $\dot{m}_d$ are highly dependent on the interaction between the condensing species and the surface, and on the temperature, and must be determined experimentally for each situation. The term f_b depends upon the mobility of the condensate on the surface.

Since optical scattering by contaminant deposits is one of the contamination effects of most concern, a model should account for noncontinuous deposits in order to be acceptable. Of the four engineering models described above, only Eq. (9) is able to do this, yet Eq. (9) is itself very crude and must be made even more complex if it is to be realistic. The effect of surface diffusion is not accounted for explicitly, nor is the presence of a nucleation barrier. The model accounts for only one species, whereas it is expected that in most space system applications more than one species of incident flux will be encountered.

Methods for Measuring Condensation Kinetics and Deposit Morphology

An apparatus for measuring the rate of condensation on a surface is basically quite simple, and has only two major systems of components. One system consists of a source of molecular flux of known species, a method for controlling the flux magnitude, temperature, and direction, and instrumentation for measuring its magnitude. The other system consists of a collector surface, with methods for measuring the condition of the clean surface, the surface temperature, the incident flux, the rate of condensation, and the amount of material condensed. There are several alternative types of components that can be incorporated into these systems.

Molecular Source. Although various types of molecular source have been used for general condensation studies, in the present context the molecular source will necessarily be species outgassed from a material sample. The sample can either be placed freely in the test chamber or in an effusion cell. The use of an effusion cell is preferred because it makes temperature control easier, and the flux can be concentrated and directed in a selected direction. The flux from the effusion cell can be collimated or allowed to follow a natural near-cosine distribution into the chamber. There are several ways to measure the flux magnitude. If a cooled plate suspended in the flux field from a vacuum microbalance is cold enough for all incident flux to condense, the flux mass per unit area at the measurement location is equal to the change of mass recorded by the microbalance, divided by the area of the collector plate. A more convenient and accurate method is to place a cooled quartz crystal microbalance (QCM) in the flux field. The QCM indicates mass flux per unit area directly. Use of a mass spectrometer or ion gauge to measure molecular density is feasible but less accurate because of the uncertainties involved in relating gauge current to mass flux per unit area.

Collector Surface Measurements. Measurement of the collector surface temperature and the angle of incidence are relatively routine. Characterization of a surface is a difficult task, but many suitable sophisticated surface inspection methods are now available. The major experimental tasks are thus measurement of the impinging flux, the net rate of condensation, the amount condensed, and the morphology of the deposit. The impingement rate can be measured by placing a cooled collector or ion gauge in the flux field, as described in the section on Molecular Source.

The condensation rate and the total amount condensed on the collector can be measured directly by suspending the collector surface from a vacuum microbalance. This method is unambiguous but cumbersome, and it is difficult to control and measure the collector surface temperature without interfering with the mass measurement. A more convenient method is to simulate the collector surface with a QCM. This is done by placing the QCM in a position in the flux field where the incident flux on the QCM is the same as that on the collector surface. The QCM temperature is controlled to be the same as that of the collector to be simulated. It is then assumed that the rate of condensation and the total mass condensed on the collector are the same as that on the QCM. Although this assumption will not be entirely correct because the QCM will have a different surface, the error may be small enough to justify using this highly convenient method for engineering measurements. A third method of measuring condensation rate is to measure the flux leaving the collector surface. The net condensation rate is then the difference between the impinging flux and the leaving flux. The leaving flux can be measured using a cooled collector or an ion gauge. The flux leaving the surface is the sum of the reflected and reevaporated fluxes. A major disadvantage to this approach is that it does not indicate that total mass condensed. This can only be found by integrating the condensation rate data.

Deposit Morphology Measurement. At the present time there is no developed method suitable for measuring deposit morphology in vacuum. Pictures of contaminated surfaces have been taken ex situ using photomicrography and the scanning electron microscope. These pictures have provided a great deal of information about the deposit morphology and strongly suggest that optical methods would be well suited to this task. At this time, such methods remain to be developed.

An Improved Apparatus for Condensation Kinetics Measurement

It will not be possible to finally select standard test procedures until the type of model that the industry will use to represent condensation, and hence the type of data needed, has been selected. Fortunately, it is possible to select an apparatus concept that appears to be adaptable to a wide range of test procedures, so the problem of model selection need not be addressed at this time.

The first requirement of the apparatus is that it should be capable of measuring the rate of condensation of an impinging flux on an isothermal surface as a function of those parameters represented by the model used. Since the model has not yet been defined, this means that the apparatus should be capable of varying all controlling parameters, so that the data requirements of the most sophisticated model types can be satisfied. Hence the apparatus should be capable of measuring condensation rate as a function of impingement rate, species, and temperature; surface type, condition, and temperature; amount already deposited; etc. A second requirement is that the data should be obtained with sufficient resolution that it would be possible to deduce basic properties such as adsorption energies, etc., from them. A third requirement is that the apparatus be capable of providing environmental exposure to the test surface during deposition. Since there are a number of environments of interest, this requirement may excessively increase the cost and complexity of a test apparatus. However, the capability for ultraviolet exposure is considered to be a minimum requirement. Finally, a method for measuring the surface morphology is required.

Based upon these requirements, an apparatus concept has been selected for further development. In this concept, an effusion cell holding a material outgassing sample is used as the source. The total flux impinging on the test surface is measured using a cooled QCM. The distance between the effusion cell and the test surface can be changed during a test so that the impinging flux can be varied. The rate of condensation and total amount condensed on the test surface is measured by monitoring a QCM held at the same temperature as the test surface. A device for selecting specific outgassed species for study from the total outgassing flux, yet to be developed, is located between the source and the test surface. A shutter is provided between the source and the test surface so that the impinging flux can be interrupted and the reevaporation rate measured. Capability is provided for exposing the test surface to ultraviolet or other types of environment. An optical device for imaging the contaminated surface, yet to be developed, is provided.

The proposed apparatus is very similar to existing systems. The developments that are to be added are a device to select one species from a multispecies flux and a device to measure surface morphology. The technique of using a QCM to simulate a test surface is widely used in the industry, but the validity of this simulation has not been vigorously examined. In developing the test method, this will be done by exposing QCMs with different surface coatings and various other test surfaces at the same temperature to the same impinging flux and comparing the accumulation rates and amounts.

A means for separating individual species from a multispecies flux must be developed. Different samples of the same material will, generally speaking, contain the same initial amounts of each species to be outgassed. However, because each species will have different transport properties in the parent material, their outgassing rates will vary differently with temperature and time. The relative proportions of each species in the total outgassing flux will therefore vary with time. The form of the time dependence will depend on the specific combination of temperature and geometry. The composition of the flux striking a surface will therefore not be constant. It is thus desirable to obtain condensation kinetics data for each outgassed species both separately and in combination with other outgassed species. This means that the experimental method should incorporate some technique for separating out the different outgassed species.

Two general techniques may be conceived for separating out a single species for study. The first uses the fact that for most material there is generally some combination of geometry, temperature, and time since the beginning of outgassing for which the outgassing flux consists predominantly of one given species. For a given material geometry and temperature, the outgassing species will generally be composed of more volatile species at early times and progressively less volatile species at later times as the more volatile species becomes depleted. During the outgassing process, there will be periods in which outgassing flux consists of predominantly a single species, separated by periods in which the flux consists of a mixture of species. By intelligent selection of material geometries and temperatures, plus possibly using some elevated temperature preconditioning to remove the more volatile components, it should be possible to achieve a situation in which essentially only one species is being outgassed from the source material at a given time.

In the second technique, a specific species is selected from a mixed outgassing flux by using one or more intermediate surfaces held at temperatures that condense out some species but not the selected one. There are several ways to implement this approach. In one method, the total outgassing flux is first condensed on a cooled collector. The collector is then warmed to a temperature at which evaporation is due solely or primarily to the species of interest, the more volatile components having been completely evaporated, while the evaporation rate of the less volatile species is still negligibly low. In the second technique, the total outgassing flux is directed first at an intermediate surface adjusted to a temperature at which those species less volatile than the species of interest will condense. The reflected flux will contain the species of interest, plus those species of higher volatility. The flux is then directed at another surface, which is held at a temperature at which only the species of interest will condense. This second surface will then acquire a deposit of predominantly the species of interest. It can then be warmed in a controlled manner to provide a single species source for condensation kinetics measurements.

Development of a method for determining surface deposit morphology will be the most challenging task. The most promising techniques for development use some type of optical approach. Optical methods have been used successfully to image surface deposits in the laboratory environment, and would seem to be adaptable to being used in a vacuum chamber. Laser scanning devices used in conjunction with conventional microscopy or with scattered light detection are now becoming available. It is hoped that a continuing search of this currently very active technical field will lead to a satisfactory solution to this problem.

References

[1] Muraca, R. F. and Whittick, J. S., Polymers for Spacecraft Applications. Final Report, June 1964 to August 1967, NASA Contract No. NAS7-100, Stanford Research Institute, Menlo Park, Calif., Sept. 1967.

[2] Campbell, W. A. Jr., Marriott, R. S., and Park, J. J., Outgassing Data for Spacecraft Materials, NASA Reference Publication 1061, Aug. 1980.

[3] Liu, C.-K. and Glassford, A. P. M., "Kinetics Data for Diffusion of Outgassing Species from RTV560 Silicone Rubber," J. Vac. Sci. Technol., Vol. 15 (5), Sept./Oct. 1978, 1762-68.

[4] Farrell, G. and Carter, G., "Diffusive Processes in a Solid During Tempering," Vacuum, Vol. 17 (1), Jan. 1967, 15-19.

[5] Lu, C. S., "Theory and Practice of the Quartz Crystal Microbalance," Applications of Piezoelectronic Quartz Crystal Microbalances, Methods and Phenomena, edited by C.S. Lu and A. W. Czanderna, Vol. 8, Elsevier, New York, 1983, Chap. 2.

[6] Wendlandt, W. W., Thermal Methods of Analysis, 2nd ed., Wiley, New York.

[7] Glassford, A. P. M. and Liu, C.-K., "Outgassing Rate of Multilayer Insulation Materials at Ambient Temperature," J. Vac. Sci. Technol., Vol. 17 (3), May/June 1980, 693-704.

[8] Zinsmeister, G., "A Contribution to Frenkels Theory of Conddensation," Vacuum, Vol. 16 (10), Oct. 1966, 529-535.

[9] Rhodin, T. and Walton, D., Metal Surfaces, American Society for Metals, Metals Park, Ohio, 1963.

[10] Hirth, J. P. and Pound, G. M., Condensation and Evaporation, The Macmillan Co., New York, 1963.

[11] Hudson, J. B. and Sandejas, J. S., J. Vac. Sci. and Technol., Vol. 4 (5), 1968, 230 - 238.

Potential for Cross-Contamination for Payloads in the STS Bay

Robert G. Moss*
Ford Aerospace & Communications Corporation, Palo Alto, California

Abstract

Since the Space Shuttle (STS) has become operational for commercial and military payloads, some issues which had been identified but not addressed intensively are receiving more attention. One of these is the potential for payload contamination by the materials in and around the STS bay, including other payloads. Contamination from the materials of construction of the STS Orbiter itself, the waste products from the crew cabin, and plume contamination from attitude control and maneuvering jets has been recognized as a potential problem for years. An induced environment contamination monitor (IECM) was flown on STS-2, STS-3 and STS-4 to provide data on the quantity and nature of contamination in and around the STS payload bay. Although contamination was observed, it was not as severe as most worst-case predictions. Contamination from unexpected sources occurred. The amounts noted and some effects of contamination could be significant and have adverse effects if appropriate adjustments are not made. In addition if appropriate contamination from the STS and products of the cabin and propulsion system, payloads are exposed to contamination from other payloads in the STS bay. This introduces a variable of uncertain dimensions which requires careful review and may necessitate special protective measures for contamination sensitive payloads. Potential sources of contamination from the STS bay are discussed and possible precautions and preventative measures are discussed.

Presented as Paper 83-1562 at the AIAA 18th Thermophysics Conference, Montreal, Canada, June 1-3, 1983. Copyright © American Institute of Aeronautics and Astronautics, Inc., 1983. All rights reserved.

*Senior Engineering Specialist, Western Development Laboratory.

Until the advent of the STS for launching payloads into space, there was little need to consider possible contamination from other payloads launched at the same time, or from the launch vehicle itself. When multiple payloads were launched at once they almost always were from a single organization, and normally were multiple versions of the same payload. Therefore cross contamination potential from other payloads was essentially the same as self contamination. The STS cargo bay may contain payloads from several different organizations. The launch vehicle itself provided few possible sources for contamination. The STS has several materials of construction with significant contaminant potential. Propellants and reaction products were left behind during the ascent. The Orbiter creates a cloud of propellant products around itself as it maneuvers in space. The time between launch and payload separation normally was a few hours at most. However, for STS missions the payload could remain with the Orbiter for several days. Contamination associated with the launch vehicle has not been a concern until plans for regular use of the STS became more advanced.

There are a number of potential sources of contamination from the materials of construction of the STS Orbiter itself, hydraulic fluids, the waste products from the crew cabin, and plume contamination from the attitude control and maneuvering jets have been recognized as a potential problem for years. (Refs. 1-3) NASA is not imposing strict requirements for contamination control or use of low outgassing materials on payload organizations using the STS. While NASA personnel do encourage use of materials and designs with low contamination potential, the primary responsibility for maintaining payload cleanliness is the using organizations.

An induced environment contamination monitor (ICEM) was developed, manufactured, and flown on STS-2, STS-3, STS-4 to provide flight data on the quantity and nature of contamination in and around the STS payload bay (Refs. 4 and 5). Although contamination was observed, it was not as severe as most worst-case predictions. There appeared to be less contamination from outgassing of paints or other materials of construction of the Orbiter than had been expected. However, initial flights did reveal a significant amount of larger solid debris in the cargo bay, mainly associated with STS contamination and launch preparation. Fibers, films, even washers were observed in the bay during STS-1 (Refs. 5 and 6).

Particulate contamination from the payload integration areas is a more significant source. The vertical payload facility and payload change room are relatively clean, with 75 μ particulate counts of 5513, 9956/ft^2 and 7309/ft^2 respectively (Ref. 7). The particulate count varies with the payload, type of activity, and number of people in the area. The Orbiter Payload Facility requires considerable modification before it can be considered a clean area. Particulate counts were 200, 000/ft^2 and up during early STS missions. Even after installation of HEPA filters and improved cleanliness procedures are implemented the OPF will be less of a clean area than the VPR or PCR. Organizations whose payload will be in the OPF should be aware of the particulate contamination potential and take appropiate precautionary measures.

The Orbiter itself is a potential source of contamination. There are several features of construction and operation of the Shuttle which are not present in conventional launch vehicles. These features are possible or actual sources of contamination. The assembly and integration areas for STS are different. Equipment and facilities capable of servicing a manned launch vehicle are not the same as those used for rocket launches. Particulate and organic contaminants such as hydraulic fluid, thermal protective coatings and insulation are present. At the least some sort of protective cover or enclosure is advisable for any payload in the OPF or VPF when there is no work being done on the payload itself.

The Orbiter bay doors are not sealed as tightly as nose covers and fairings on rocket launches. It is possible that particles or contaminants could be introduced into the cargo bay after payload integration and before launch. Recently the TRDSS payload was contaminated by a severe storm while in the cargo bay at the launch pad. While storms and launch delays cannot be predicted, the possibility of contamination from the environment before launch should be considered.

There are several contamination sources associated with the cargo bay. Hydraulic lines run the entire length of the bay. Leakage of hydraulic fluid during flight is possible. Fortunately NASA has demonstrated that teflon coated beta glass cloth is an effective barrier to hydraulic fluid (Ref. 8). Use of this barrier cloth is optional at the request of the using organization.

The A276 white thermal control paint on the inside of the cargo bay can be a source of contamination. Despite the potential for contamination from materials of construction, actual flight experience during STS-2 and STS-3 did not indicate significant deposition of outgassing products as measured by the IECM (Refs. 5 and 6).

RTV 560 is used to bond the thermal tiles of the Orbiter. This high outgassing silicone apparently has caused some contamination of the windows but has not been detected in the payload bay as yet.

There are 2 major in-orbit events which create significant amounts of gaseous and particulate contaminants: water and waste dumping and firing of the RCE system. Both events create significant quanitites of materials which remain near the Orbiter for some time before they disperse into space. If the Orbiter is carrying a greater than normal charge, as when it is entering or leaving eclipse, the reaction products or waste water may be electrostatically attracted back towards the Orbiter. In addition, both direct and return flux from the water dump or RCE may cause significant contamination. Various models for gaseous contamination in the STS environments have been proposed (Refs. 9 and 10). Actual flight experience from early STS missions did show significant increases in material detected in the IECM coincident with water dumping or RCE firing. Increased mass levels persisted for up to an hour after the event (Ref. 5).

The environment around the Orbiter contains atomic oxygen, which has been found to react with some spacecraft materials and change optical properties (Ref. 11). This effect is a type of contamination in that it adversely effects thermal control and optical surfaces.

In addition to potential contamination from the STS and products of the cabin and propulsion system, payloads are exposed to contamination from other payloads in the STS bay. When the payload for a particular STS flight is manifested, the potential for cross contamination is a minor consideration if it is considered at all. Weight, physical size, and ability to load and eject payloads without physical interference are primary concerns. Users have very little opportunity to object to being launched with a particular payload because it may cause contamination. Some payloads may not be designed or built with contamination as a consideration. Other adjacent payloads may be low cost science experiments sponsored by colleges or high schools. It is probable that such organizations would be unconcerned about the potential for contamination from their equipment.

Despite all these potential contamination sources it should be possible to prevent significant payload contamination by taking appropiate precuations. Basic spacecraft design can be modified to minimize contamination related to the STS. Special precautions or protective devices may be used. Operational procedures may be modified to minimize the risk of contamination. A variety of techniques have been proposed by others (Refs. 9 and 12). The longest list of possible precautions was given by Scialdone (Ref. 9); some of them are most useful for sensitive instruments located in the cargo bay and are not of general interest. The more general precautions are noted below, with some of our modifications and suggestions appended.

(1) Protect sensitive surfaces with dust covers and doors to inhibit ingestion of gases and particles and to limit the exposure to periods when the environment is acceptable.

(2) Provide energy sources that can be used to maintain a surface at elevated temperatures and to sublime existing contaminant deposits.

(3) Use cryogases to purge the system. This may prevent some contaminant molecules from reaching the surface.

(4) Sufficiently vent insulation materials, lubricated motors, and other sources away from the critical surfaces.

(5) Expose systems and materials to vacuum bake, and purging before using them in space, preferably a short time before.

(6) Select materials with low outgassing, as determined from actual tests on their rates of outgassing versus time, or, if this is not possible, Volatile Cendensable Material (VCM) selection tests. Note that the VCM test does not provide outgassing rates or ensure that the material will be acceptable in actual use.

(7) Use flights with lower bay temperatures because outgassing is strongly dependent on temperature.

(8) Use flights with lower power requirements to reduce the amount of water generated by the power cells, and, hence, the number of flash evaporator operations, or use flights that carry ample water storage capabilities.

(9) When acceptable, use high-altitude flights because the ambient scattering return fluxes are an order of magnitude lower at 400 km than at 200 km, thus there is less chance of contamination.

(10) Avoid critical surface exposure during Reaction Control System/Vector Control System and evaporator operations or other venting. Use protective covers.

(11) Delay uncovering or exposing payloads as long as possible. For example, exposing a surface 40 hours after flight initiation can reduce contaminant deposits by about 90%.

(12) Use dedicated flights that include instruments and payloads concerned with contamination.

(13) Use previously flown Shuttles, preferably those that have been recently flown. They have been exposed to vacuum and degassed, especially of highly volatile materials. Avoid recently retiled Orbiters.

(14) Use the results of the induced environment contamination monitor (IECM). This package of instruments, designed for mapping and measurement survey during several early flights, may indicate a best location for the payload.
(15) Investigate the effect of other payloads on your payload. Perform detailed analysis on the environment at the payload location. A knowledge of the sources outgassing, their temperatures, their view factors, contaminant transport mechanisms, operational modes, etc., is necessary.
(16) Be informed in advance of flight plans or obtain real time information on RCS/VCS/evaporator and other venting operations so that protective actions can be taken.
(17) Position or protect the payload so that it is not exposed to atomic oxygen which may be present near the top of the cargo bay.
(18) Provide disposable protective covers or canistors for the entire payload.
(19) Keep the cargo bay doors closed whenever possible, to protect against contamination from water dumps, RCE firings and atomic oxygen activated by radiation.
(20) Determine the contamination potential from other payloads in the bay, if possible. If not possible, expect the worst and act accordingly.
(21) Orient the payload so that the more sensitive surfaces are protected from possible contamination sources.

Obviously all of these precautions are not likely to be possible, or even needed, for a particular mission. Careful selection and implementation of the best and most appropiate actions should greatly reduce cross-contamination potential for STS-launched payloads. An added benefit is that many activities which protect your payload from other payloads in the cargo bay and from the Shuttle environment will also protect the outer payloads from contamination by your payload.

References

1. Hueser, J. E. and Brock, F. J., "Shuttle Flow Field Analysis Using Direct Simulation Monte Carlo Technique", AFML-TR-78-190, NASA-CP-2039, Proceedings of the USAF/NASA International Spacecraft Contamination Conference, March 1978, pp. 250-273.

2. Stechman, R. C., "Space Shuttle Plume Contamination", pp. 401-411.

3. MeKeown, D. and Claysmith, C. R., "Quartz Crystal Microbalance Systems for Shuttle Contamination", pp. 605-628.

4. Miller, E. R., "Shuttle Induced Environment Contamination Monitor", pp. 534-566.

5. Leger, L. J. and Jacobs, S. and Ehlers, H. K., "Space Shuttle Preliminary Contamination Assessment from STS-1 and STS-2", NASA Conf. Pub. 2229, 12th Space Simulation Conference, May 1982, pp. 281-301.

6. Leger, L., NASA/JSC personal communication, April 1982.

7. Whitehead, V., "Relationship Between Air and Cleanliness Classes in Kennedy Space Center Cargo Processing Facilities", 7th Guidance and Navigation Community Contamination Control Seminar, October, 1982.

8. Richmond, R. G. and Kelso, R. M., "Effectiveness of the Shuttle Orbiter Payload Bay Liner as a Barrier to Molecular Contamination from Hydraulic Fluids", AFML-TR-78-190, NASA-CP-2039, March, 1978, pp. 846-862.

9. Scialdone, J. J., "Assessment of Shuttle Payload Gaseous Environment Contamination and Its Control", ESA-SP-145, Spacecraft Materials in Space Environment", October, 1979.

10. Bareiss, L. E. and Hetrick, M. H., "Verification Approach for the Shuttle/Payload Contamination Evaluation Computer Program", CP788, AIAA/IES/ASTM 19th Space Simulation Conference, October, 1978, pp. 37-47.

11. Leger, L. J., "Oxygen Atom Reaction with Shuttle Materials at Orbital Altitudes", NASA, 58246, pp. 37-47.

12. Dauphin, J., "European Contamination Concerns", AFML-TR-190, NASA-CP-2039, pp. 1138-1154.

Chapter II. Sources and Prevention of Contamination

Debris from Spallation of Foam Insulation of Cryogenic Fuel Tanks in Space Launch Systems

E. P. del Casal[*]
Energy Incorporated, Idaho Falls, Idaho

Abstract

Because of low tensile strength, pressure buildup due to aerodynamic heating during ascent could cause local spallation of foam insulators. Results of an analytical investigation of debris formation from the spallation of foam insulation in cryogenic fuel tanks for space launch systems are presented. It is shown that foam spallation can be a significant source of debris, possible spacecraft contamination, and insulation loss.

Nomenclature

C_{lj} = heat capacity of condensing species j
C_{pj} = heat capacity at constant pressure for species j
C_{ps} = heat capacity of porous solid
C_{vi} = heat capacity at constant volume for species j
E_{vj} = internal energy per unit mass of species j in vapor phase
E_{lj} = internal energy per unit mass of species j in liquid phase
h_j = specific enthalpy of species j in vapor phase
ΔH_{vj} = heat of vaporization of gaseous species j

Presented as Paper 83-1457 at AIAA 18th Thermophysics Conference, Montreal, Canada, June 1-3, 1983. Copyright © American Institute of Aeronautics and Astronautics, Inc., 1983. All rights reserved.
[*]Principal Scientist. Presently Specialist Engineer, Boeing Aerospace Company, Seattle, Washington.

ΔH_{AB} = heat of ablation of porous matrix
k_s = thermal conductivity of porous solid
k_f = thermal conductivity of gas (bulk)
L = thickness of slab
m = mass flux
M_j = molecular weight of species j
R = gas constant
P = pressure
P_j = partial pressure of species j
ΔP_T = tensile limit of porous matrix
T = temperature
T_{AB} = temperature of ablation of insulator material
T_j = permeability of species j into porous matrix
t = time
x = distance along thickness of insulator layer
ρ_j = mass per unit volume (vapor + liquid + solid) of species j
ρ_s = mass per unit volume of porous solid
ρ_{vj} = mass per unit volume of species j in vapor phase
$\rho_{\ell j}$ = mass per unit volume of species j in condensed phase
ξ = porosity of insulator matrix
$(\#)^*$ = saturated (equilibrium) conditions

Introduction

The results of an analytical investigation dealing with possible debris formation resulting from the spallation of foam insulation of cryogenic fuel tanks in space launch systems are presented. In view of the low tensile strength of foam insulators, pressure buildup within the insulation layer due to aerodynamic heating during ascent could cause local foam spallation. An immediate case in point is the spray-on-foam insulation (SOFI) used for the Space Shuttle external fuel tank. Although foam insulation had been previously used in other launch systems utilizing cryogenic fuels (e.g., Apollo-Saturn upper stage), the consequences of debris formation due to spallation were of less importance to the mission. This may not be the case with the STS since foam debris of sufficient size may pose potential spacecraft contamination problems.

The problem investigated here is intuitively obvious and is generally recognized as a possible debris-generating mechanism in spray-on-foam insulators. It has been variously labeled as the "popcorn effect," the "trapped gas" problem, and various other sundry names. However, there appears to be a lack of a systematic analytical model that quantifies the problem.

Mathematical Formulation and Method of Solution

The physical model for the problem under consideration is illustrated in Fig. 1. A porous permeable insulator slab of uniform thickness is exposed to cryogenic temperatures on one side and a warm environment on the other. The insulator may contain some resident gases (e.g., Freon) in addition to external gases that diffuse into it. During the cryogenic process, some or all of the resident and ingested external gas species condense along the colder portions of the layer creating a density (mass per unit volume of insulator) gradient in the vapor phase of each condensible species along the slab thickness. This creates a driving potential for mass transfer from the warmer to the colder portions of the layer and in turn from the external environment into the insulator.

During launch and ascent, the cryogenic as well as external sides are heated up by hot gas pressurization and aerodynamic heating, respectively. Localized buildups in pressure can occur within the insulator layer because of the change of phase of the trapped condensates. The low permeability of the foam material impedes the rapid relief of the pressure. When the pressure exceeds the tensile limit of the material, spallation can take place.

A one-dimensional mathematical model was formulated with allowances made for the following:
1) Vapor condensation within the porous slab (any number of species);
2) Gas ingestion into the insulator slab;
3) Pressure buildup and outgassing due to insulator warmup;
4) Surface recession due to ablation;

5) Surface recession due to spallation using the tensile limit of the material as criterion for spallation;
6) Multilayer insulation;
7) Variable physical and thermodynamic properties of the porous material and ingested as well as resident gases.

The equations governing the transient flow of a vapor through a porous media are continuity:

$$\xi \frac{\partial \rho}{\partial t} + \nabla \cdot \vec{m} = 0 \qquad (1)$$

momentum:

$$\vec{m} = \sum_{1}^{Ns} \vec{m}_j = - \sum_{1}^{Ns} \Gamma_{jj} \Delta P_j \qquad (2)$$

energy:

$$\frac{\partial}{\partial t} \left[\int_0^{Ts} \rho_s C_{ps} dTs + \xi \sum_i^{Ns} (\rho_{vj} E_{vj} + \rho_{\ell j} E_{\ell j}) \right] - \xi \sum_1^{Ns} \Delta H_{vj} \frac{\partial \rho_{\ell j}}{\partial t} +$$

$$\sum_1^{Ns} \nabla \cdot h_j \vec{m}_j = \nabla \cdot (k_s \nabla T_s + k_f \nabla T_f) \qquad (3)$$

$$\rho = \sum_i^{Ns} \rho_j = \sum_i^{Ns} (\rho_{vj} + \rho_{\ell j} + \rho_{sj}) \qquad (4)$$

$$\rho_{vj} = \frac{P_{vj} M_j}{RT_j} \qquad (5)$$

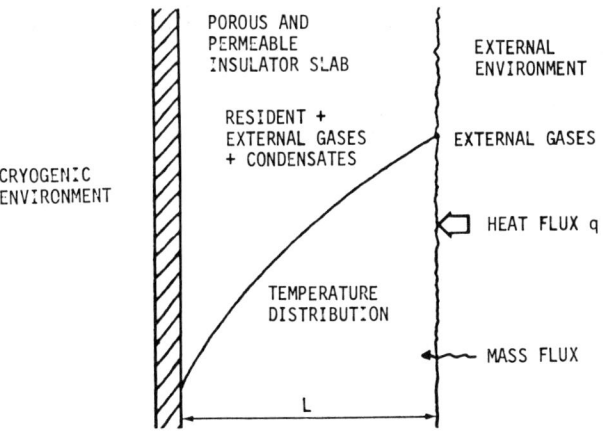

Fig. 1 Physical model.

$$\frac{\partial}{\partial t}\left[\int_0^{T_s} \rho_s C_{ps} dT_s + \xi \sum_i^{N_s} (\rho_{vj} E_{vj} + \rho_{\ell j} E_{\ell j})\right.$$

$$+ \xi \sum_i^{N_s} \Delta H_{vj} \frac{\partial \rho_{vj}}{\partial T_j} \frac{\partial T_j}{\partial t} +$$

$$\sum_i^{N_s} \nabla \cdot (h_j + \Delta H_{vj} \Psi_j) \vec{m}_j$$

$$= \nabla \cdot (k_s \nabla T_s + k_f \nabla T_f) \qquad (6)$$

where $\Psi_j = 0$ pure vapor and $\Psi_j = 1$ multiphase flow.

To make the problem tractable and yet still close to physical reality, the model of the insulator layer was formulated on the basis of the following simplifications and assumptions:

1) The fluid and porous matrix temperatures are equal. This assumes that the internal heat-transfer coefficient is sufficiently large

to keep the temperature difference between fluid and porous matrix to within a few degrees; otherwise, the internal heat-transfer coefficient must be known.

2) Only one spatial dimension is considered (this is consistent with the assumption of a thin slab), thereby reducing Eqs. (1) and (3) to the forms,

continuity:

$$\xi \frac{\partial \rho}{\partial t} + \frac{\partial}{\partial x} m = 0 \qquad (7)$$

momentum:

$$m = \sum_i^{Ns} m_j = - \sum_i^{Ns} \Gamma_{jj} \frac{\partial \rho_{vj}}{\partial x} \qquad (8)$$

energy:

$$\left[\rho_s C_{ps} + \xi \sum_i^{Ns} (\rho_{vj} C_{vj} + \rho_{\ell j} C_{\ell j} + \frac{\partial \rho_{vj}^*}{\partial T} \Delta H_{vj}) \right] \frac{\partial T}{\partial t} + \sum_i^{Ns} \frac{\partial}{\partial x} (C_{pj} T + \Delta H_{vj} \Psi_j) m_j = \frac{\partial}{\partial x} \left(k_s + k_f \right) \frac{\partial T}{\partial x} \qquad (9)$$

3) The porous matrix is chemically inert to both external and resident gases (vapors).

4) Equilibrium exists between phases (a direct analogy with assumption 1 can be made here-it is assumed that the mass transfer coefficient is sufficiently large).

5) The vapor phase obeys the ideal gas law,

i.e.,
$$\rho_{vj} = \frac{R_{vj} M_j}{RT} \tag{10}$$

so that the Clausius-Clapeyron equation can be used

$$\frac{dP^*_{vj}}{dT} = \frac{\rho^*_{vj} \Delta H_{jv}}{T^2} \tag{11}$$

6) As a consequence of the above assumption, the criteria for equilibrium condensation becomes:

a) If $\rho^*_{vj} > \rho_j RT/M_j$ no condensation occurs

b) If $\rho^*_{vj} < \rho_j RT/M_j$ condensation occurs

The liquid - (+ solid) - phase density is thus simply deduced from

$$\rho_{\ell j} = \rho_j - \rho^*_{vj} M_j/RT \tag{12}$$

7) The volume of the vapor phase is very much greater than that occupied by the condensate; i.e., the porous cavities are far from being saturated by liquid/solid condensates.

The initial and boundary conditions for the insulator slab are as follows:

Initial Conditions

1) Temperature distribution:

$$T(x,o) = T_1(x)$$

$$0 \leq x \leq L$$

2) Species vapor pressure distribution:

$$P_{vk}(x,o) = P_{1j}(x)$$

$$0 \leq x \leq L, \; j \leq 10$$

Boundary Conditions

1) Cryogenic side temperature history:

$$T(o,t) = f_1(t)$$

2) Outer side heat-transfer coefficient (from boundary layer data) and recovery temperature history:

$$C_h = C_h(t)$$

$$T_r = T_r(t)$$

The outer temperature may thus be computed from a simple heat balance for the cases with and without ablation:

$$q = C_h(T_r - T_w) = k \frac{\partial T}{\partial x} + aT_w^4 \quad \text{without ablation}$$

$$C_h(T_r - T_{ABL}) = k \frac{\partial T}{\partial x} + \varepsilon a T_{ABL}^4 - \dot{L} \, \rho s \, \Delta H_{ABL} \quad \text{with ablation}$$

4) Partial pressures of resident as well as external gaseous species at the outer surface (freestream values are used-diffusion boundary layer is ignored): same as freestream.

Equations (7-9) are numerically solved using an implicit predictor-corrector time integration scheme. Spatial derivatives are approximated by

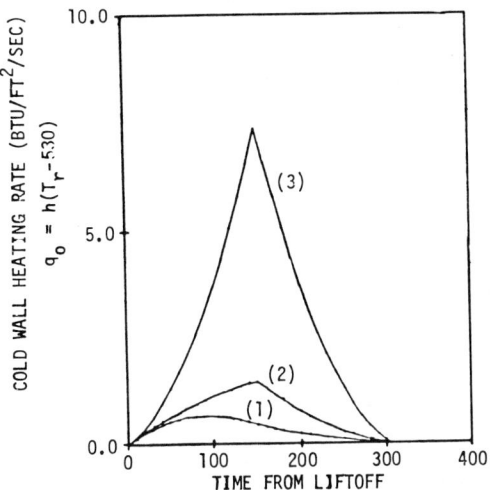

Fig. 2 Aerodynamic heating signatures.

finite differences. Spallation is assumed to occur when the difference between the local and external (boundary-layer) pressure exceeds the tensile limit of the foam material. The amount of material that breaks off is determined by the location where the pressure difference exceeds the tensile limit.

To account for surface recession due to ablation, the model Eqs. (7-9) were transformed from the (x,t) coordinated system into the (η,t) system where

$$t = t$$

$$\eta = \frac{x}{L}$$

so that

$$\frac{\partial}{\partial t} = \frac{\partial}{\partial t} - \eta \frac{\dot{L}}{L} \frac{\partial}{\partial \eta}$$

$$\frac{\partial}{\partial x} = \frac{1}{L} \frac{\partial}{\partial \eta}$$

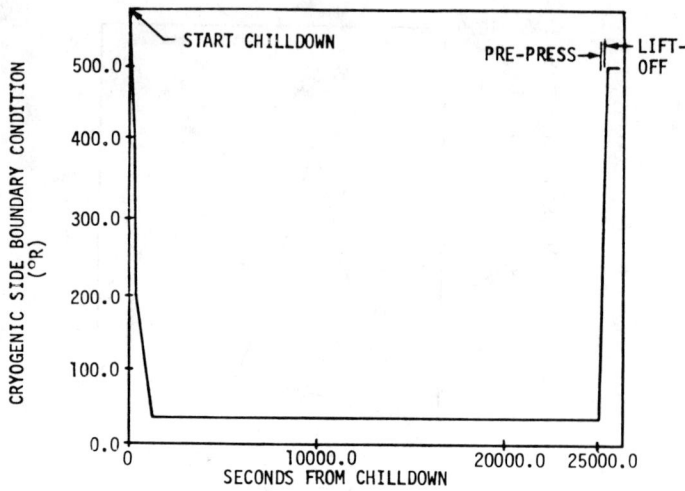

Fig. 3 Cryogenic side boundary conditions.

Results and Discussion

Three aerodynamic heating time signatures were considered in this study. These are presented in terms of a cold wall surface heat fluxes defined here as:

$$q_{cw} = h(T_r - T_0)$$

T_0 = sea level ambient temperature.

The heating signatures are shown in Fig. 2.

The thermal properties of the SOFI were taken from Ref. 6. The tensile limit of the SOFI was set at a nominal value of 35 psi. Ablation was assumed to occur at 600 degrees Fahrenheit and heat of ablation set at 200 Btu/lb. Four condensible gas species were considered in the simulation, namely, water vapor, air (taken as a single species), carbon dioxide, and Freon 12. The permeability of the SOFI to water vapor was estimated using the conservative value of 2% gain by weight of a 1-in-thick specimen within a period of 16 hrs from bone dry conditions in 90% humid air at 90 degrees Fahrenheit. The permeabilities of the SOFI to other gas (vapor) species were estimated using data from Ref. 6 on the relative permeabilities of water vapor, carbon dioxide,

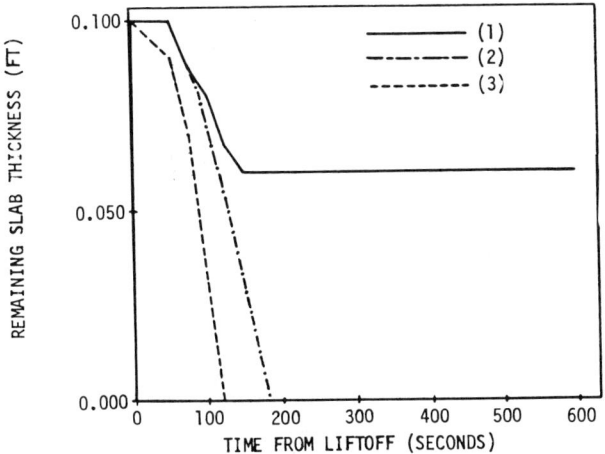

Fig. 4 Remaining slab thickness vs time.

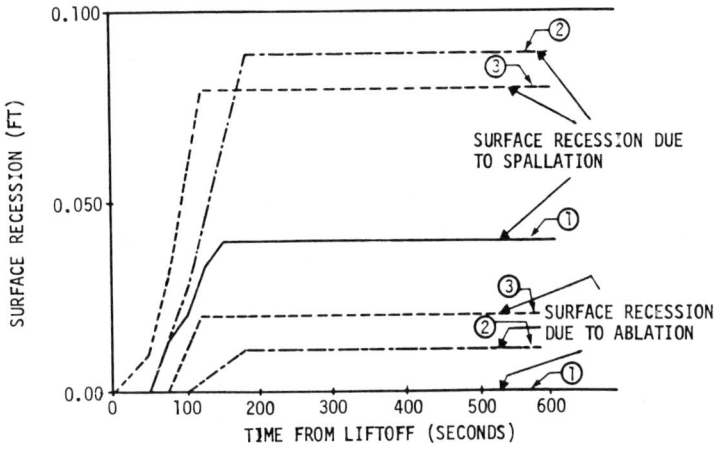

Fig. 5 Causes of surface recession.

air, and Freon in POLYURETHANE foams. The physical and thermodynamic properties of the gas species were obtained from Ref. 2.

The cryogenic side temperature history (supplied here as a boundary condition) is shown in Fig. 3. The calculations were started using conditions at the beginning of chilldown and terminated at MECO conditions. A total of 25,800 s of simulated time span was involved. The initial partial pressure distribution within the slab was set to correspond to a foam

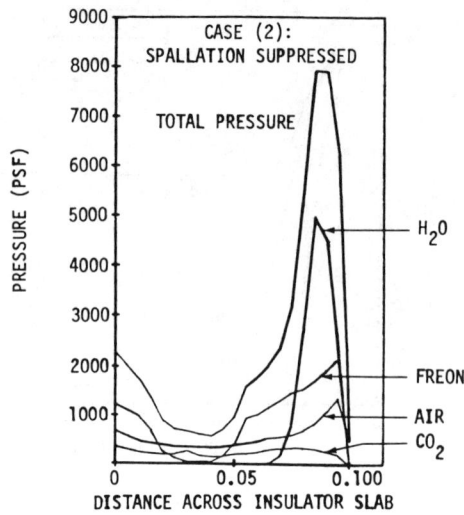

Fig. 6 Total and partial pressure distribution.

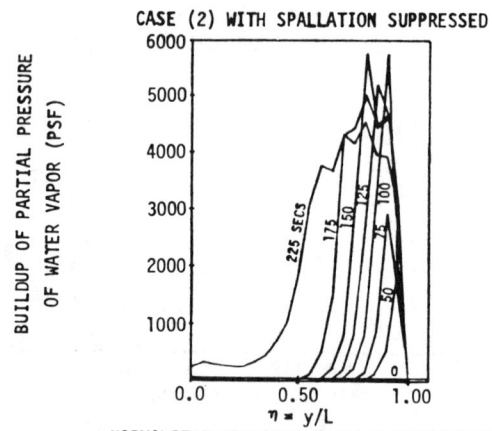

Fig. 7 Water vapor pressure distribution vs time.

that has been cured for six months. The partial pressures were thus set as follows: Freon 12 - 1,200 psf, carbon dioxide - 300 psf, air - 552 psf, and water vapor - 60 psf. The ambient external temperature was assumed to be 80 degrees Fahrenheit with a relative humidity of 50%. A residual moisture content of 1% by weight was also assumed present in the SOFI. The initial thickness of the foam was set at 0.1 ft.

DEBRIS FROM SPALLATION OF FOAM INSULATION 51

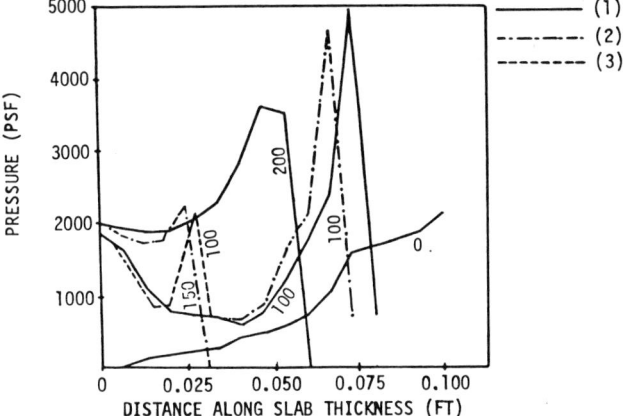

Fig. 8 Total pressure distribution vs time.

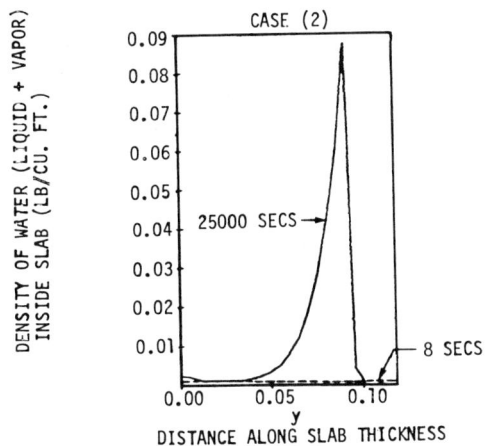

Fig. 9 Moisture density distribution.

The results of the numerical simulation are ummarized below.

1) Surface recession: The surface recession of the SOFI for the three aerodynamic heating signatures was calculated. The remaining thickness as a function of time for the three signatures is shown in Fig. 4. A dissection of the causes of surface recession is shown in Fig. 5. In all instances, spallation was a dominant cause.

2) Pressure buildup: To determine the amount of pressure buildup on the slab, the

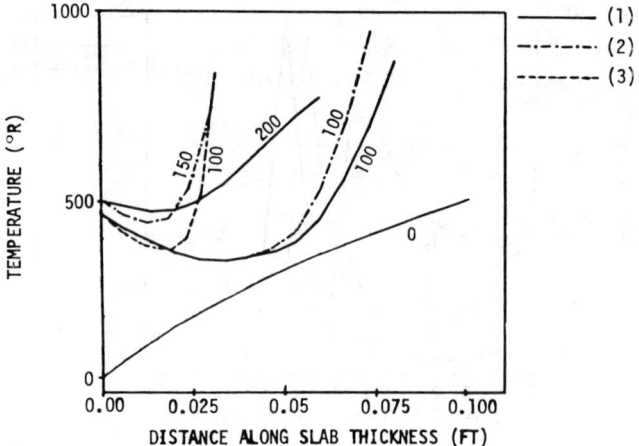

Fig. 10 Temperature distribution.

spallation mode was suppressed by assuming a much larger tensile limit for the material. The distribution of total pressure and its constituent partial pressures across the slab are shown in Fig. 6. The main contributor to the total pressure is water vapor followed by Freon, air, and carbon dioxide. The buildup of the partial pressure of water vapor as a function of time is shown in Fig. 7. The main culprit is water vapor since its partial pressure alone exceeds the tensile limit of the SOFI which is about 500 psf. The pressure distribution for each of the three aerodynamic heating signatures is shown in Fig. 8. In these cases spallation was allowed.

3) Gas ingestion into the insulator: The amount of water vapor ingested into the insulator slab is shown in Fig. 9 when the density distribution of the water vapor and its condensate are plotted along the slab thickness for prechilldown and at the start of prepressurization. Although the total amount of moisture (vapor plus liquid) amounts to only a few percent by weight, most of it piles up close to the surface. A bone dry foam was assumed for this purpose. Air ingestion is not very significant.

4) Temperature distribution: The temperature distribution inside the porous slab for each of the three aerodynamic heating signatures

is shown in Fig. 10. The surface temperatures are on the average significantly lower than the assumed ablation temperature of the SOFI (600 degrees Fahrenheit). This result is attributable to the spallation process taking place.

Conclusions

In summary, the following conclusions can be made:

1) A mathematical formulation and computational procedure has been developed to consider the problem of spallation due to pressure buildup in cryogenic insulators.

2) The calulations show that spallation is a significant cause of surface recession in SOFI.

3) As a consequence SOFI spallation is a a major debris generating mechnism in cryogenic fuel systems. This may pose spacecraft contamination problems for the STS and similar missions.

4) The sizing of the SOFI insulation must take into account spallation. Surface recession models based on ablation alone may not be conservative.

References

[1] Cunningham, R. E. and Williams, R. J. J., Diffusion in Gases and Porous Media, Plenum Publishing Corporation, New York, 1980.

[2] Perry, R., Chemical Engineer's Handbook, McGraw-Hill, Fourth Edition, New York, 1970.

[3] Scheidagger, A. E., The Physics of Flow Through Porous Media, MacMillan Company, New York, 1957.

[4] Forsythe, G. E. and Wasow, W. R., Finite-Difference Methods for Partial Differential Equations, John Wiley and Sons, New York, 1960.

[5] Collins, R. E., Flow of Fluids Through Porous Materials Reinhold Publishing Company, New York, 1961.

[6] Saunders, J. H. and Frisch, K. C., Polyurethanes, Chemistry and Technology, Part II: Technology, Interscience Publishers, New York, 1964.

[7] Williams, W. O., "Constitutive Equations for Flow of an Incompressible Viscous Fluid Through a Porous Medium," Quarterly of Applied Mathematics, Vol. 36, No. 3, pp. 255-267, Oct. 1978

Particle Dispersion around a Spacecraft

Aleck L. Lee*
Lockheed Missiles & Space Company, Inc., Sunnyvale, California

Abstract

Particulate material released from an orbiting spacecraft may return to the spacecraft and interfere with its operation. The equations of motion of a particle relative to the spacecraft are derived from Euler-Lagrange equation. The spacecraft may be in a circular, elliptic, ascending, or a specifically described orbit. Aerodynamic drag is also considered. The equations are solved numerically. Results of single-particle trajectories and dispersion of particles ejected from a spacecraft are presented.

Nomenclature

A	= radius of the Earth (= 6387 km)
Alt	= altitude of the spacecraft
C_d	= drag coefficient of the particle
D	= combined drag term = $[3/8\ (C_d \rho_a / r_p \rho_p)]$; distance between the ejected particle and the spacecraft
e	= eccentricity of an ellipse
g	= gravity acceleration at altitude $[= g_o\ (r_e^2/r^2)]$
g_o	= gravity acceleration at the surface of the Earth, (9.807 m/s²)
L	= Lagrangian
ℓ	= semilatus rectum of an ellipse

Paper 83-0243 presented at AIAA 21st Aerospace Sciences Meeting, Reno, Nev., Jan. 10-13, 1983. Copyright © American Institute of Aeronautics and Astronautics, Inc., 1983. All rights reserved.
*Research Specialist, Space Systems Division.

Q_i'	=	generalized forces
q_i	=	generalized coordinates
R	=	distance from spacecraft to the ejected particle $(=\sqrt{X^2+Y^2+Z^2})$
r	=	distance from center of the Earth to the particle $(=\sqrt{X^2+(r_s+Y)^2+Z^2})$
r_s	=	distance from center of the Earth to the spacecraft
r_e	=	radius of the Earth (= 6378 km)
r_p	=	radius of the particle
T_1, T_2	=	particle ejection launch angle and azimuth angle
$X,Y,Z;$ x,y,z	=	particle location with respect to the spacecraft coordinates
X_o, Y_o, Z_o	=	initial particle location
α	=	angle between velocity vector and local horizon; heading angle
μ	=	gravity constant
θ	=	angular displacement; true anomaly
$\dot{\theta}$	=	angular velocity
ρ_a	=	density of air around the spacecraft
ρ_p	=	density of the ejected particle
$\vec{\Omega}$	=	rotation vector of the frame of reference

Introduction

Particulate material is often released from a spacecraft during various phases of its operation. The trajectories of these particles are of vital interest if their potential contamination threat is to be predicted. The following questions are the main concerns:

1) Among the released particulates, how many will return to the spacecraft?
2) How is the particulate cloud around the spacecraft formed and/or dispersed?
3) How many of the particles will appear in a given field of view?

Answers to these questions are based on the accurate prediction of the motion of a single particle. Eggleston and Beck[1] give the derivation of equations of motion and calculate the satellites rendezvous velocities at several locations of both circular and elliptic orbits under drag-free conditions. Naumann[2] linearizes the equations derived from orbital mechanics and gives solutions for circular orbits. Similar results have been obtained[3] by computer simulation of the dynamic behavior of the three-body problem. Other approximate solutions are available in the literature for circular orbits[4,5] and noncircular orbits.[6]

The equations of motion are based on the Euler-Lagrange equations of the particles relative to the moving reference frame fixed on the spacecraft. The motion of the spacecraft is a part of the model. The orbit can be circular, elliptic, ascending, or any other specified one. The drag force plays an important role in the motion of the particle and it is included in the model. The model solves the set of exact equations numerically.

A particle cloud can be represented by a finite number of sample particles. Assuming no interaction among the particles, the model keeps track of the motion of each representative particle. The results are displayed as the streaks of the particles over a period of time or as the snapshots of the cloud at instants of time.

This paper presents the model and the results of examples.

Method of Solution

The trajectory of a particle separated from spacecraft is best described by using rectangular coordinates. The coordinate system, as shown in Fig. 1, is fixed on the spacecraft. The X direction is the local horizon along the spacecraft velocity vector; the Y direction is opposite to the center of the Earth. The coordinate system is rotating about the Earth at angular velocity $\vec{\Omega}$. The Lagrangian of the particle in a noninertial frame of reference without translation can be written as,[7]

$$L = \tfrac{1}{2} m \vec{V}^2 + m \vec{V} \cdot \vec{\Omega} \times \vec{r} + \tfrac{1}{2} m (\vec{\Omega} \times \vec{r})^2 - \mu / |\vec{r}| \quad (1)$$

In the case of rotating coordinates, it is assumed

$$\vec{\Omega} = \vec{k} \dot{\theta} \quad (2)$$

and that

$$\vec{r} = \vec{i} X + \vec{j} (Y + r_s) + \vec{k} Z \quad (3)$$

Also, for rectangular coordinates, the relative velocity is

$$\vec{V} = \vec{i} \dot{X} + \vec{j} (\dot{Y} + \dot{r}_s) + \vec{k} \dot{Z} \quad (4)$$

PARTICLE DISPERSION AROUND A SPACECRAFT

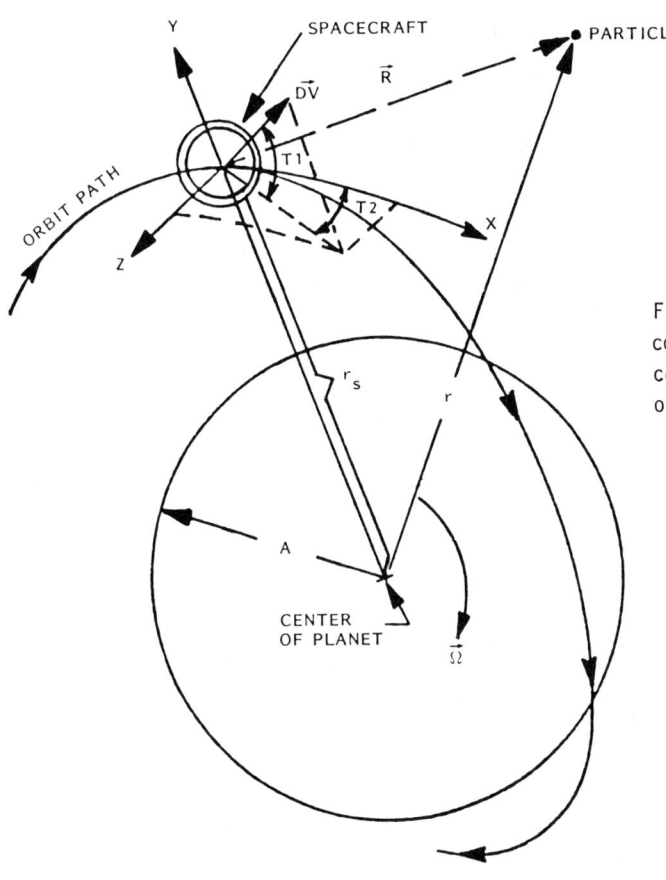

Fig. 1 Rectangular coordinate system centered on a body in orbit.

and the last term of Eq. (1) is given by

$$\frac{\mu}{|\vec{r}|} = -\frac{m \, g_e \, r_e^2}{(X^2 + (Y + r_s)^2 + Z^2)^{\frac{1}{2}}} \quad (5)$$

where g_e is the acceleration of gravity at the surface of the Earth and r_e the radius of the Earth. Hence, the Lagrangian becomes,

$$L = \tfrac{1}{2}m \, [\dot{X}^2 + (\dot{Y} + \dot{r}_s)^2 + \dot{Z}^2] + m\dot{\theta} \, [X(\dot{Y}+\dot{r}_s) - \dot{X}(Y+r_s)]$$

$$+ \tfrac{1}{2}m\dot{\theta}^2[(Y+r_s)^2 + X^2] + \frac{m \, g_e \, r_e^2}{\sqrt{X^2 + (Y+r_s)^2 + Z^2}} \quad (6)$$

The aerodynamic drag force of the particle can be included in the Lagrangian by considering the term

$$\vec{F} = \frac{3}{8} C_d \frac{\rho_a}{r_p \rho_p} \left[(r_s\dot{\theta}+ \dot{X})^2 \vec{q}_1 + (\dot{r}_s+\dot{Y})^2 \vec{q}_2 + \dot{Z}^2 \vec{q}_3 \right] \quad (7)$$

where ρ_a is the density of air, ρ_p the density of the particle, and r_p, the radius of the particle. The drag coefficient C_d is assumed to be constant in the flow regime.

In order to simplify the notations, let

$$D = (3/8)(C_d/r_p)(\rho_a/\rho_p) \quad (8)$$

$$r = \sqrt{X^2 + (Y + r_s)^2 + Z^2} \quad (9)$$

$$g = g_e (r_e^2/r^2) \quad (10)$$

The exact equations of motion in rectangular coordinates are obtained by evaluating the terms in Lagrange's equation. The term on the right-hand side represents the generalized force Q'_i, which is the aerodynamic drag in this case,

$$\frac{d}{dt}\left(\frac{\partial L}{\partial \dot{q}_i}\right) - \frac{\partial L}{\partial q_i} = Q'_i \quad (i = 1,2,3) \quad (11)$$

The generalized coordinates q_i are x, y, or z, depending on i = 1, 2, or 3, respectively. The aerodynamic drag per unit mass, assuming the particle is in spherical shape, is

$$Q'_i = D V_i^2$$

$\vec{V}$ is the relative velocity of the particle with respect to the surrounding air,

$$V_1 = r_s \dot{\theta} + \dot{x}$$
$$V_2 = \dot{r}_s + \dot{y}$$
$$V_3 = \dot{z}$$

Substituting the Lagrangian equation (6) and the generalized forces into the Euler-Lagrange differential equations (11), the set of differential equations of motion is as follows:

$$\ddot{x} + (y + r_s)\ddot{\theta} + 2(\dot{y} + \dot{r}_s)\dot{\theta} - x\dot{\theta}^2 + g\frac{x}{r} + D(r_s\dot{\theta} + \dot{x})^2 = 0$$

$$\ddot{y} - x\ddot{\theta} - 2\dot{x}\dot{\theta} + \ddot{r}_s - (y+r_s)\dot{\theta}^2 + g\frac{y}{r} + D(\dot{r}_s + \dot{y})^2 = 0$$

$$\ddot{z} + g\frac{z}{r} + D\dot{z}^2 = 0 \qquad (12)$$

Solutions of Eqs. (12) give the relative position of the particle, $\vec{R}(x,y,z)$, with respect to the spacecraft. The position and motion of the spacecraft are described by r_s, θ, and their derivatives.

The set of Eqs. (12) is solved numerically by the modified Runge-Kutta method as outlined in Ref. 8. The Runge-Kutta method is extended to solve the equations of mth order with n variables. A subroutine of this method, known as MRKDE,[9] has been adopted in the model. Clohessy and Wiltshire[4] linearized the equations by expanding the nonlinear gravitational force terms and retaining the significant linear terms. The linearized equations of motion are

$$\ddot{x} + 2\dot{\theta}\dot{y} = 0$$

$$\ddot{y} - 2\dot{\theta}\dot{x} - 3\dot{\theta}^2 y = 0$$

$$\ddot{z} + \dot{\theta}^2 z = 0 \qquad (13)$$

A first integral of the first of Eqs. (13) can be obtained directly since both terms are time derivatives. Then $\dot{x}$ from the solution of this equation can be used in the $(2\dot{\theta}\dot{x})$ term of the second equation to obtain a complete solution in terms of Y. Finally, this solution can be applied in the first integral of the first equation to solve it completely. The third equation, being uncoupled from x and y, is just simple harmonic motion and can be solved directly. The solutions are as follows:

$$x = -(4\frac{\dot{x}_o}{\dot{\theta}} + 6y_o)\sin\dot{\theta}t + 2\frac{\dot{y}_o}{\dot{\theta}}\cos\dot{\theta}t - (6y_o + 3\frac{\dot{x}_o}{\dot{\theta}})\dot{\theta}t - 2\frac{\dot{y}_o}{\dot{\theta}} + x_o$$

$$y = -(2\frac{\dot{x}_o}{\dot{\theta}} + 3y_o)\cos\dot{\theta}t + \frac{\dot{y}_o}{\dot{\theta}}\sin\dot{\theta}t + 4y_o + 2\frac{\dot{x}_o}{\dot{\theta}}$$

$$z = z_o\cos\dot{\theta}t + \frac{\dot{z}_o}{\dot{\theta}}\sin\dot{\theta}t \qquad (14)$$

Equations (14) are referred to as Clohessy-Wiltshire equations. As it is obvious from the derivation, the solutions in Eqs. (14) are valid only for circular orbits under drag-free conditions. The solutions may seem restrictive; however, they offer quick and accurate answers for short durations. These equations were actually incorporated in the Gemini spacecraft rendezvous computer in shell-coordinate form for the first successful attempts at Earth orbital rendezvous.[7]

Spacecraft Trajectory and Eject Velocity

The trajectory of the spacecraft is described by r_s, θ, and their derivatives. In the mathematical model, three common trajectories (i.e., circular, elliptic, and ascent orbits) are provided. Provision is also made for any other specified orbit.

When the spacecraft moves in a circular orbit, the orbit is determined by the altitude, Alt, alone. The equations of motion are simplified considerably by the conditions.

$$r_s = Alt + r_e$$

$$\dot{r}_s = \ddot{r}_s = 0$$

$$\dot{\theta} = const = |\vec{\Omega}|$$

$$\ddot{\theta} = 0 \qquad (15)$$

Most of the orbiting satellites fly in elliptic orbits with the center of the Earth at one of the foci. The orbit is defined by giving the altitude Alt, the velocity V, and the angle α between the velocity and local horizon at a given time. The other parameters of the ellipse can be computed with relations in orbital mechanics.[10] An elliptic orbit around the Earth is shown in Fig. 2. The orbit is described by

$$r_s = \ell/(1 + e \cos \theta) \qquad (16)$$

where e is the eccentricity and ℓ the semilatus rectum. The angle θ is true anomaly and is measured from perigee of the orbit. When the orbit parameters r_s, α, and V_o are given at instant t_o, the angular velocity $\dot{\theta}$, and radial velocity $\dot{r}_s$ can be calculated.

$$\dot{r}_s = V_0 \sin \alpha$$
$$\dot{\theta} = V_0 \cos \alpha / r_s \qquad (17)$$

Let the angular momentum h and gravity constant K be defined as

$$h = r_s^2 \dot{\theta} = \text{const}$$

$$K = g_0 r_0^2 = g\, r_s^2 = \text{const}$$

The true anomaly θ_0 at $t = t_0$ is given by[11]

$$\theta_0 = \tan^{-1}\left(\frac{(r_s V_0^2/K)\sin\alpha\cos\alpha}{(r_s V_0^2/K)\cos^2\alpha - 1}\right) \qquad (18)$$

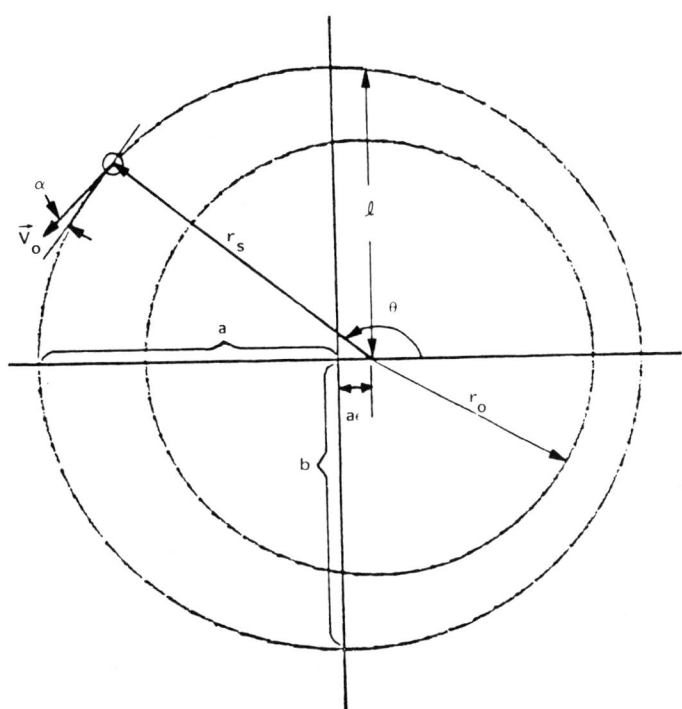

Fig. 2 An elliptic orbit around the planet.

Eccentricity e and semilatus rectum ℓ are given as

$$e^2 = \left(\frac{r_s V_o^2}{K} - 1\right)^2 \cos^2\alpha + \sin^2\alpha \qquad (19)$$

$$\ell = r_s (1 + e \cos \alpha) \qquad (20)$$

The derivatives of r_s and θ can be calculated for any time instant,

$$\dot{\theta} = h/r_s^2 = r_o V_o \cos\alpha/r_s^2 \qquad (21)$$

Because the angular momentum h is constant,

$$\frac{dh}{dt} = 2r_s \dot{r}_s \dot{\theta} + r_s^2 \ddot{\theta} = 0$$

Therefore,

$$\ddot{\theta} = - (2\dot{r}_s \dot{\theta}/r_s) \qquad (22)$$

From the definition of the ellipse,

$$r_s = \ell/(1 + e \cos\theta)$$

the derivatives of r can be computed simply by taking the time derivative. The results are

$$\dot{r}_s = (r_s^2 \dot{\theta}/\ell) \, e \sin\theta \qquad (23)$$

$$\ddot{r}_s = (r_s^2 \dot{\theta}^2/\ell) \, e \cos\theta \qquad (24)$$

These quantities as given in Eqs. (21-24) are continuously computed so that the updated information will be available in solving Eqs. (12).

In an ascending orbit, as shown in Fig. 3, the orbit can be established by given altitude r_s, velocity V_o, and the ascent angle α. The radial velocity V_r and the angular velocity $\dot{\theta}$ are given by

$$V_r = \dot{r}_s = V_o \sin \alpha$$

$$\dot{\theta} = (V_o/r_s) \cos \alpha \qquad (25)$$

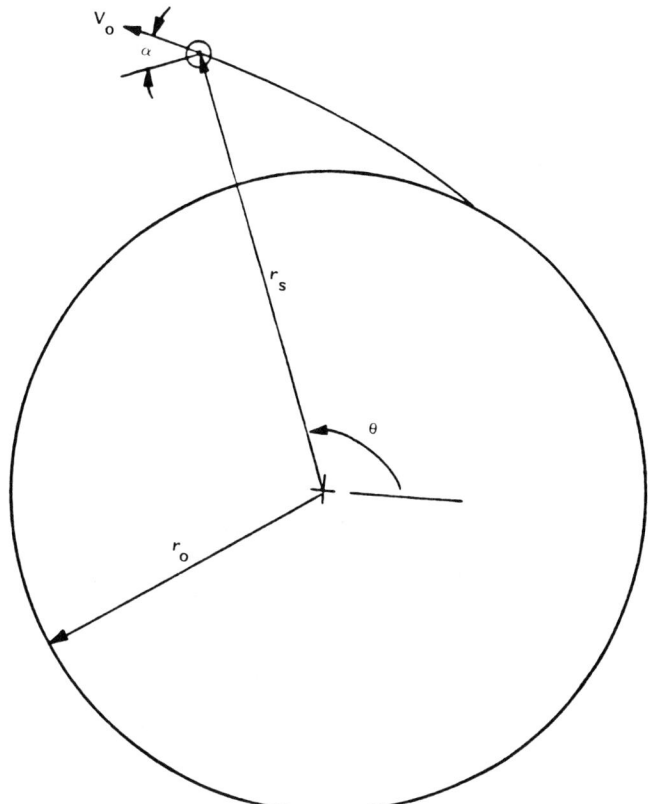

Fig. 3 An ascending orbit.

Furthermore, assuming the velocity of the spacecraft is not changed during the time interval of interest, the second derivatives of r_s and θ can be easily obtained,

$$\ddot{r} = 0$$
$$\ddot{\theta} = - (V_0 \cos\alpha/r_s^2)\dot{r}_s \qquad (26)$$

Any specially specified orbits may be considered in the model, so long as the quantities r, $\dot{r}$, $\ddot{r}$, $\dot{\theta}$, and $\ddot{\theta}$ are calculated at each time interval. This class of orbits may include either a parabolic orbit, hyperbolic orbit, or a segment of powered flight. The user will provide the logic in the subroutine to compute the orbit positions and derivatives as functions of time.

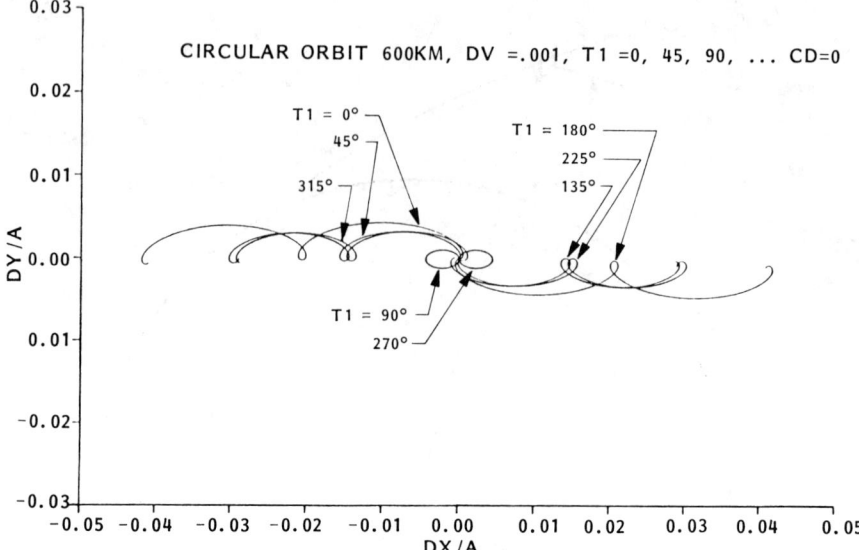

Fig. 4 Trajectories of particles ejected from spacecraft in circular orbit.

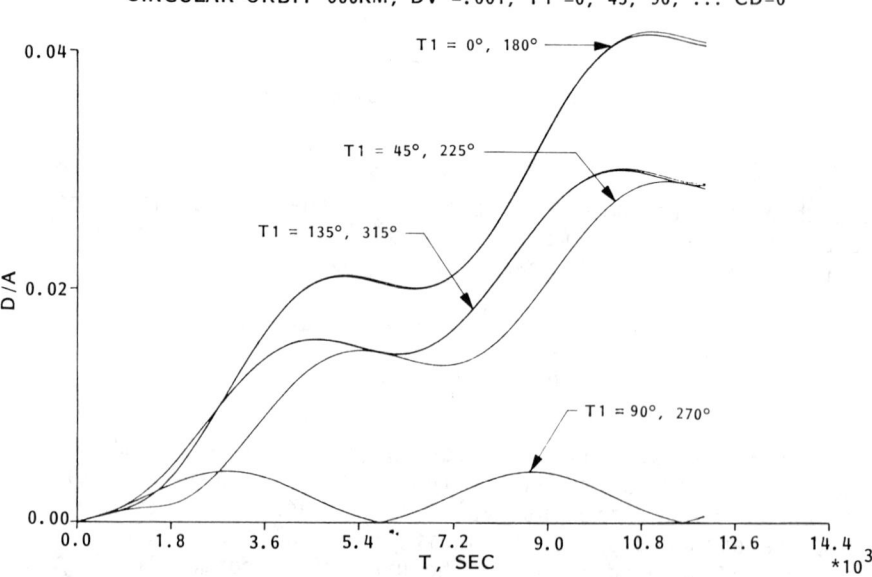

Fig. 5 Separation distances of the ejected particles.

PARTICLE DISPERSION AROUND A SPACECRAFT 65

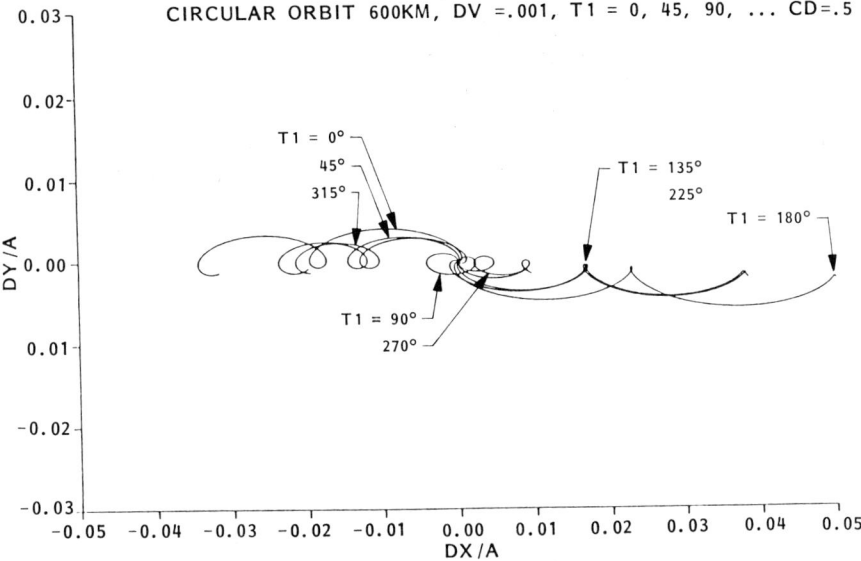

Fig. 6   Trajectories of particles with aerodynamic drag.

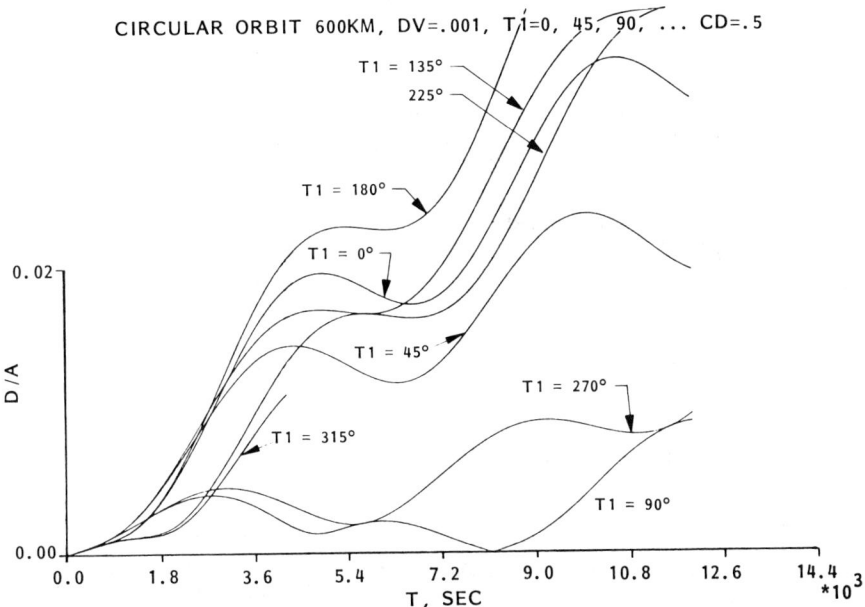

Fig. 7   Separation distances of the ejected particles with aerodynamic drag.

The particle ejection velocity relative to the spacecraft is denoted by $\vec{DV}$. It can be expressed either in meters per second or as a fraction of the orbit speed. Its initial direction is defined by two angles, T1 and T2. T1 is the local elevation angle, measured from the X axis toward the Y axis, while T2 is the local azimuth angle, measured from the X axis toward the Z axis. The initial velocity is, therefore

$$V_1 = DV \cdot \cos(T1) \cdot \cos(T2)$$

$$V_2 = DV \cdot \sin(T1)$$

$$V_3 = DV \cdot \cos(T1) \cdot \sin(T2) \qquad (27)$$

Examples and Results

The trajectories of a particle ejected at various angles from a spacecraft in circular orbit are shown in Figs. 4-7. The altitude is 600 km.

The ejection velocity is 0.001 of the orbit velocity. The ejection angles start at 0 deg and increase by 45-deg increments. The time covered is two orbit periods (~12,000 s). Figure 4 shows the trajectories when no drag is considered (CD = 0). The X and Y distances are normalized on the Earth radius and are denoted by DX/A and DY/A, respectively. Symmetry of the trajectories is obvious. Figure 5 shows the normalized distances of the particle (D/A) relative to the orbiting spacecraft time. Curve 1 in the figure shows the distances of the particles ejected at T1 = 90 and 270 deg. Note that at the end of one orbit period the particles return to the originating spacecraft. Similar conditions are rerun with aerodynamic drag considered. The symmetry is destroyed (Fig. 6) and the particles are swept away from the spacecraft (Fig. 7). The drag force, although small at this altitude, is not negligible as demonstrated in the example.

The Clohessy-Wiltshire (C-W) solutions for circular orbits are compared with the numerical solution of the exact equations under no-drag conditions. The results are shown in Fig. 8. The altitude is 600 km. The total time duration is 13,000 s or 2.24 periods. The C-W solution is periodic, while the exact solution is influenced by the nonlinear terms in the equations. The deviation between the two solutions becomes noticeable after half of an orbit period.

The trajectories of a particle ejected at various angles from a spacecraft in elliptic orbit are shown in

PARTICLE DISPERSION AROUND A SPACECRAFT

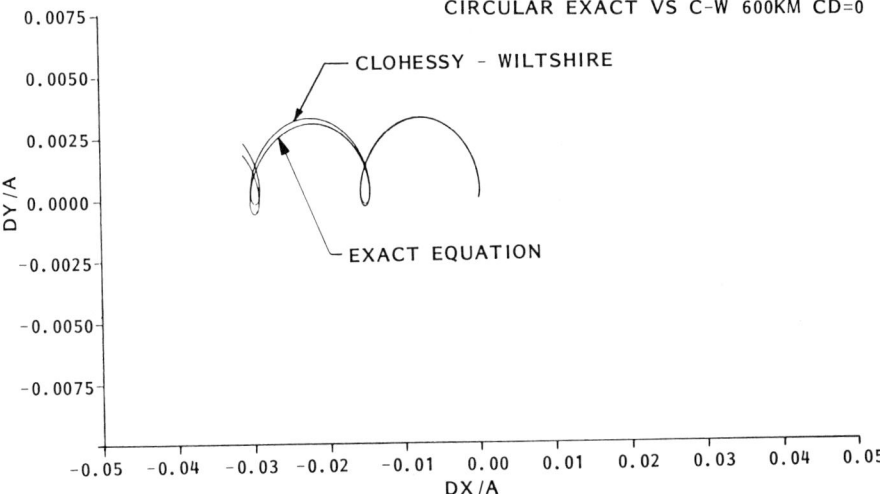

Fig. 8 Trajectories obtained by the exact solution and by the Clohessy-Wiltshire solutions.

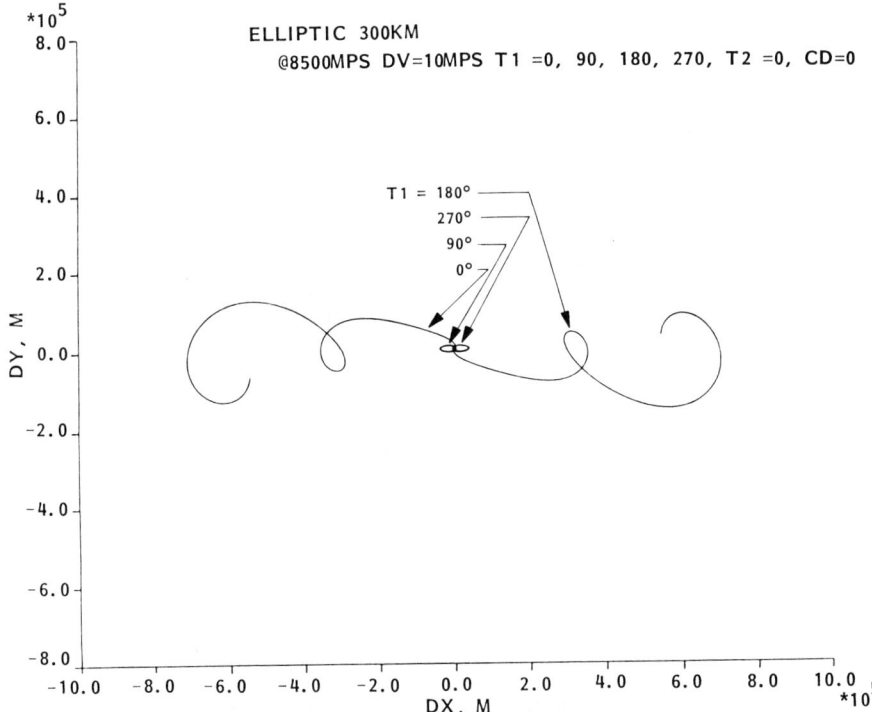

Fig. 9 Trajectories of particles ejected from a spacecraft in elliptic orbit, drag neglected.

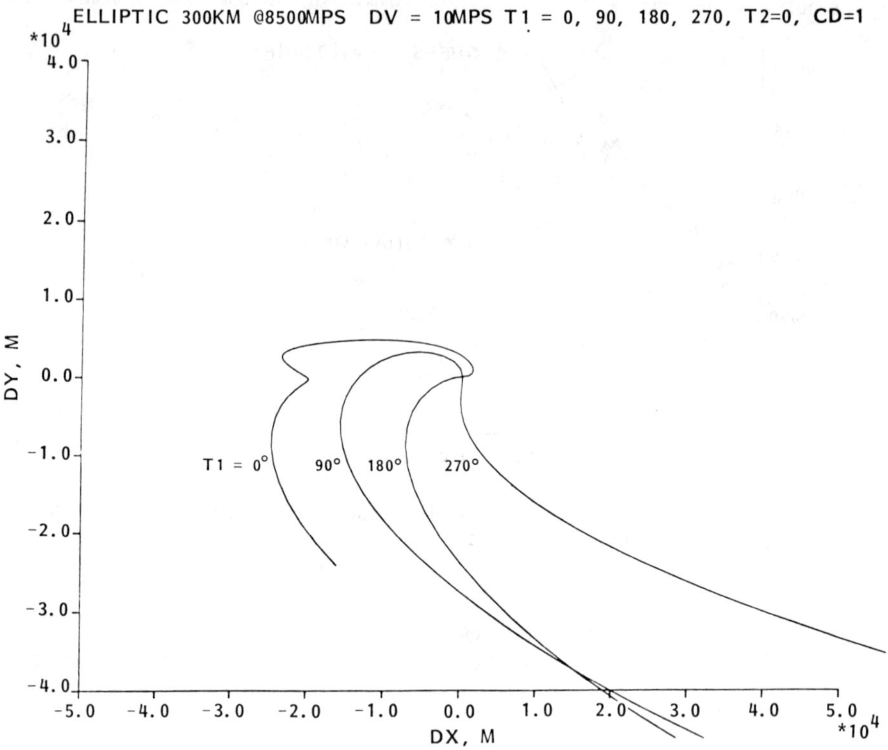

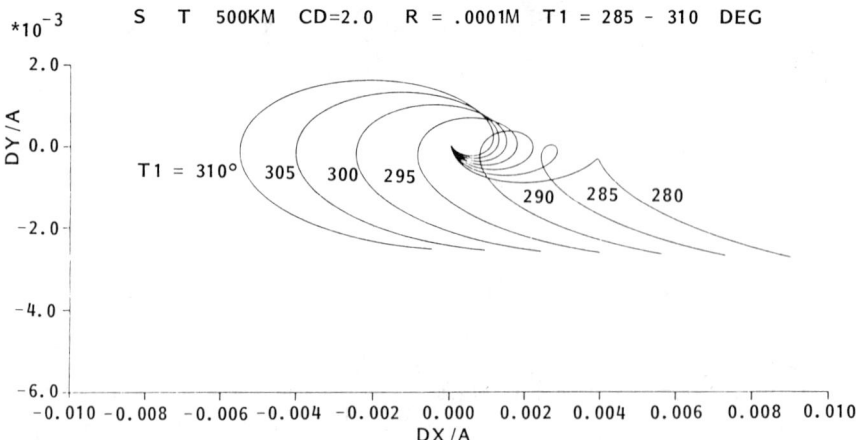

Fig. 10 Trajectories of particles ejected from a spacecraft in elliptic orbit with drag effects.

Fig. 11 Prediction of returning particles.

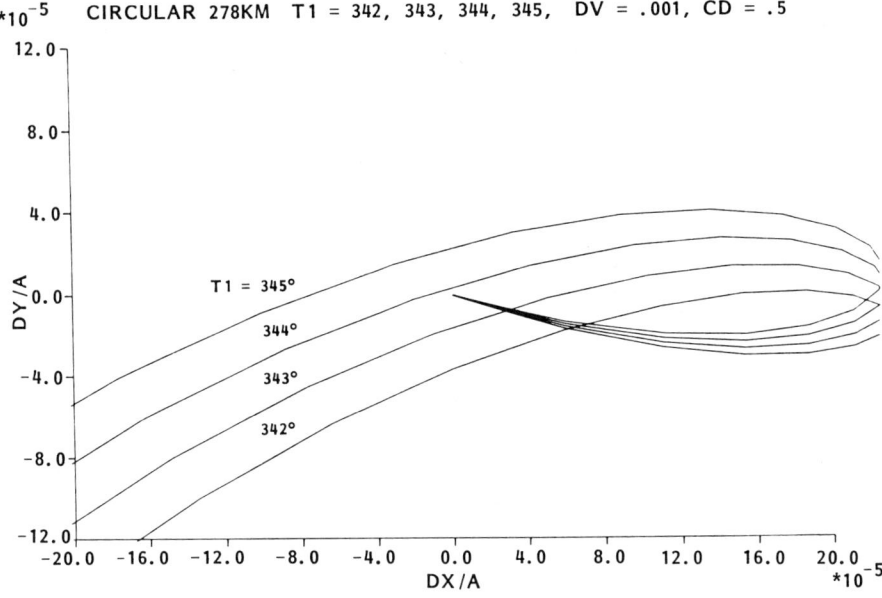

Fig. 12   Magnified particle trajectories.

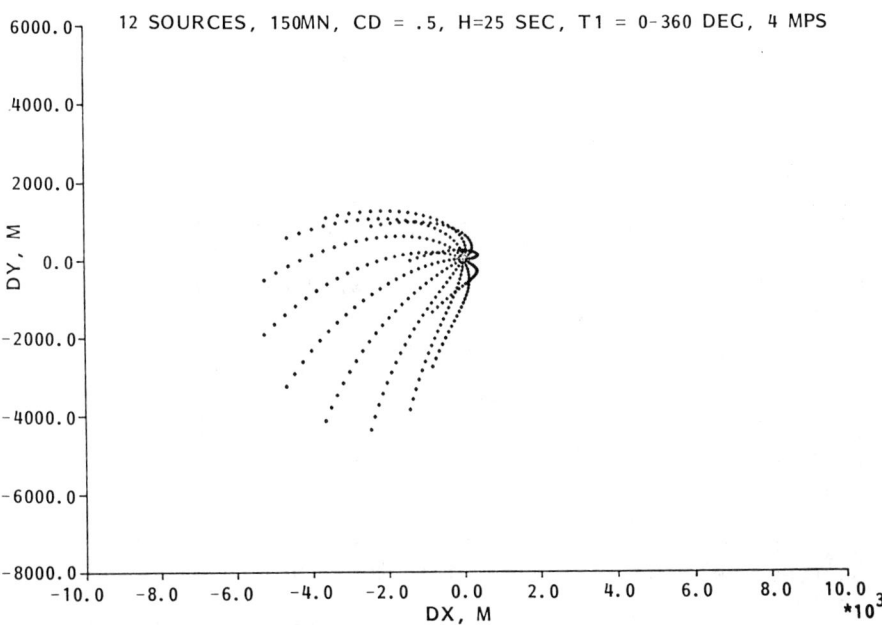

Fig. 13   Particle dispersion at low altitude to 600 s at 25-s intervals.

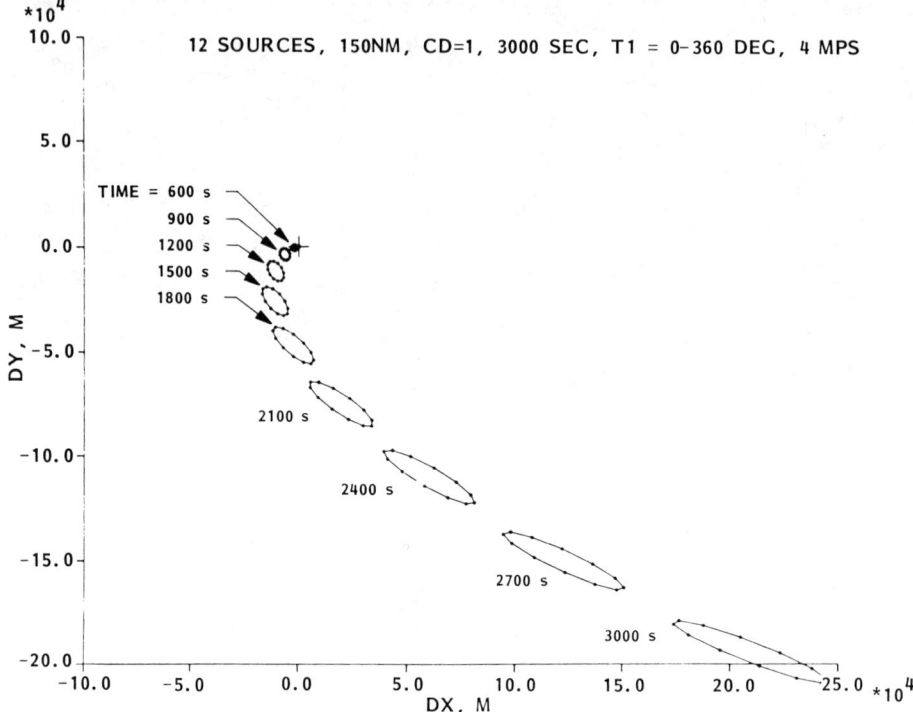

Fig. 14 Particle dispersion at low altitude to 3000 s at 300-s intervals.

Figs. 9 and 10. The spacecraft was moving at 8500 m/s at the perigee point of 300 km. The particles were ejected with a speed of 10 m/s at T1 = 0, 90, 180, and 270 deg and T2 = 0. Figure 9 shows the trajectories for two orbital periods with no drag. Figure 10 shows the same case with C_d = 1 for the particles. The time span was about one-tenth of the orbit period.

An interesting application of the model is to predict if a particle ejected from a spacecraft could return to the spacecraft. A parametric study is shown in Fig. 11, where the trajectories with the launch angles of 280-310 deg are plotted. The particles launched between T1 = 290 and 295 deg are likely to return. A magnified parametric study is shown in Fig. 12, where the trajectories of T1 = 342-345 deg are traced. When the characteristics of the particle sources are given, it is possible to conduct a quantitative analysis.

A particle cloud can be defined by its representative sample particle points. Figure 13 shows the dispersion of a particle cloud ejected uniformly at 4 m/s from a space-

PARTICLE DISPERSION AROUND A SPACECRAFT

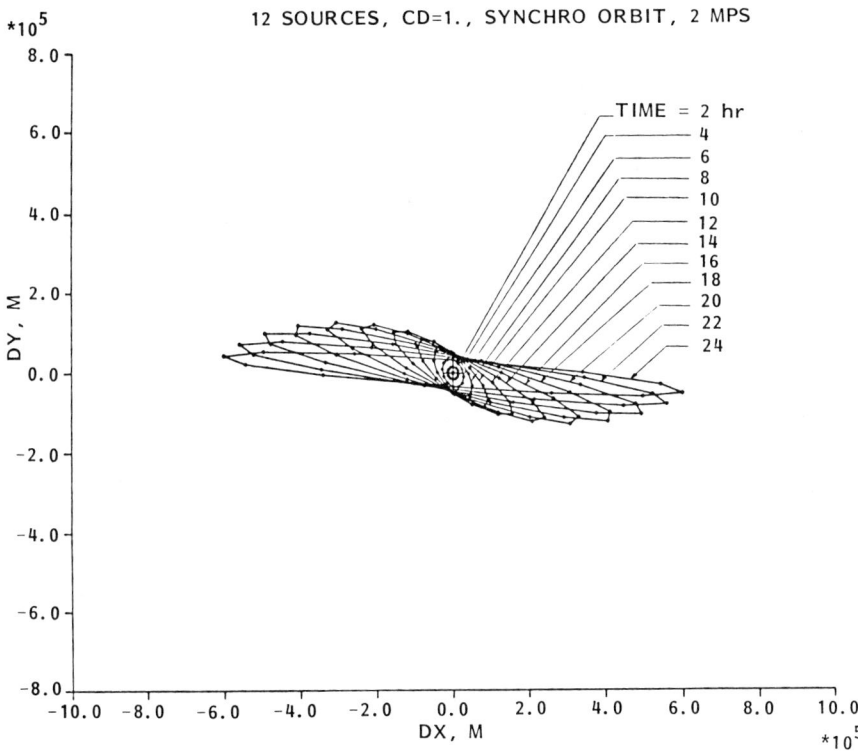

Fig. 15 Particle dispersion at high altitude.

craft at 150 nmi. The cloud is represented by 12 particles. The time span is 600 s. The locations of the particles are plotted at every 25 s. Due to the drag force acting on the particles, the particle cloud is moving downstream and downward initially. It moves forward and further downward as the tracking continues. Figure 14 shows the results to 3000 s at 300-s intervals. Under the influence of drag, all particles evolve into the forward and downward trail, regardless of the launch velocities.

Dispersion of the particle cloud can be visualized by connecting the sample particles of the same time instant. Figure 15 shows a group of particles at geosynchronous orbit. The drag is negligible here. The particles disperse into a layer forward and behind the spacecraft.

Snapshots of the particles can also be recorded on movie film. With properly timed intervals and repetition of frames, the dispersion of particles can be animated for special visual effects.

Conclusions

A mathematical model to predict the trajectories of particles ejected from orbiting spacecraft is developed based on the concept of satellite rendezvous computations. The model includes the drag effects and covers all types of orbits. The trajectories of particles with respect to the spacecraft can be precisely tracked with this model. Characteristics of particle dispersion around a spacecraft can be thus described.

References

[1] Eggleston, J. M. and Beck, H. D., "A Study of the Positions and Velocities of a Space Station and a Ferry Vehicle during Rendezvous and Return," NASA TR R-87, 1961.

[2] Naumann, R. J., "Dynamics and Column Densities of Small Particles Ejected from Spacecraft," NASA TN D-7590, Feb., 1974.

[3] Fong, M. C. and Lee, A. L., "P80-2 Contamination Control - Design Analysis Report," Lockheed Missiles & Space Co., Inc., Sunnyvale, Calif., Rept. LMSC/D616248, Dec. 1978.

[4] Clohessy, W. H. and Wiltshire, R. W., "Terminal Guidance System for Satellite Rendezvous," Journal of the Aerospace Sciences, Vol. 27, Sept. 1960, pp. 653-658.

[5] London, H. S., "Second Approximation to the Solution of the Rendezvous Equations," AIAA Journal, Vol. 1, July 1963, pp. 1691-1693.

[6] Anthony, M. L. and Sasaki, F. T., "The Rendezvous Problem for Nearly Circular Orbits," AIAA Paper 65-32, 1965.

[7] Dunning, R. S., "The Orbital Mechanics of Flight Mechanics," NASA SP-325, 1973.

[8] Ralston, A. and Wilf, H. S., Mathematical Methods for Digital Computers, John Wiley & Sons, New York, 1960.

[9] "Manual of Scientific Program Digest," Lockheed Missiles & Space Company, Inc., Sunnyvale, Calif., 1972.

[10] Greenwood, D. T., Principles of Dynamics, Prentice-Hall, Englewood Cliffs, N.J., 1965.

[11] Thomson, W. T., Introduction to Space Dynamics, John Wiley & Sons, New York, 1961.

Impact of the STS Ground/Launch Particle Contamination Environment on an Optical Sensor

L. E. Bareiss* and F. J. Jarossy†
Martin Marietta Denver Aerospace, Denver, Colorado

Abstract

A satellite payload analysis has been performed as a first attempt to develop an integrated analytical approach to predict the effect of the particle contamination environment encountered during payload final assembly, STS integration, and launch. The objective of the analysis was to determine if the particle contamination deposited on the exposed surfaces of an optical sensor during ground and launch operations would significantly affect sensor performance. Based on a simplified scattering model the minimum solar offset angle for the uncontaminated sensor is 2.0 deg. Assuming a Class 100-K air cleanliness level for all ground facilities results in a minimum solar offset angle prediction of 2.3 deg and a corresponding surface obscuration of 0.5%. The surface cleanliness level predicted for a Class 1000-K assembly area/100-K STS integration facilities flow equates to level 800 per MIL-STD-1246A. This results in a surface obscuration ratio of 3.8%. This effect is considered unacceptable for an optical element. The STS 2/4 payload bay environment measured during launch was severe. The predicted worst case surface obscuration ratio of 8% resulting from this environment would not only degrade optical components but thermal

---

Presented as Paper 83-1566 at the AIAA 18th Thermophysics Conference, Montreal, Canada, June 1-3, 1983. Copyright © American Institute of Aeronautics and Astronautics, Inc., 1983. All rights Reserved.
  *Unit Head, Contamination Effects Technology.
  †Senior Staff Engineer, Contamination Effects Technology.

control surfaces and solar arrays as well. The severity
of this prediction resulted primarily from the initial
assumption of payload bay surface cleanliness prior to
launch ("visibly clean level 1" which is the standard STS
service baseline) and a worst-case particle transport and
deposition model. While the validity of these predictions
has not yet been established, the potential for severe
particle contamination effects must be considered.

Introduction

The impact of the induced particulate contamination
environment to which sensitive optical systems will be
exposed during Space Shuttle ground and launch operations
has long been a concern of potential payloads. With the
acquisition of recent particulate environment data for the
Kennedy Space Center (KSC) ground processing facilities
and during the Shuttle launch phase in conjunction with a
systems level analytical approach, the capability exists
to estimate the ultimate effects of the ground and launch
environment on optical systems. To demonstrate this, an
example analysis of a typical Shuttle-launched optical
sensor has been conducted.

The objective of this analysis was to determine if the
particle contamination deposited on an optical sensor filter during ground and launch operations would significantly
affect the performance of the system. The performance
criteria was based on the end-of-life (EOL) maximum allowable scattered background power which is 6.0×10^{-7} W. For
the sensor system, this power corresponds to a minimum
solar offset angle of 1.9 deg. If the particle deposition
resulted in an increase in this angle of 0.2 deg, then
the sensor performance would be considered unacceptable
and modifications to the baseline system and/or operations
would be recommended.

Prediction of Particle Deposition

The following predictions of particle deposition are
based on the anticipated ground and launch environments.
The ground environment to be seen at the launch site is
assumed to be equivalent to the existing facilities for
Shuttle payloads at Cape Canaveral Air Force Station
(CCAFS) and Kennedy Space Center (KSC).

Ground Environments

After an eighteen month assembly period, the payload
sensitive surfaces and external surfaces will be cleaned.

It is expected that the exposed optical element will be repeatedly cleaned reaching a surface cleanliness level of 100 per MIL-STD-1246A (Ref. 1).

Figure 1, extracted from Ref. 1, defines the particle size distributions corresponding to the various surface cleanliness levels. The general equation for these relationships is:

$$\log n = 0.9260 (\log{}^2 x_1 - \log{}^2 x) \qquad (1)$$

where n is the number of particles/ft^2 greater than the given size (x), x_1 the cleanliness level, and x the particle size in μm. Also shown in Fig. 1 are surface cleanliness level ranges estimated for various contamination conditions. These will be discussed later in the paper.

After cleaning, the payload will undergo testing for an additional 90 days; however, the sensor is no longer accessible. It is assumed that the room environment is not a cleanroom, but a controlled area and will be a Class 1,000,000 air environment per FED-STD-209B (Ref. 2) as shown in Fig. 2. The payload will then be double bagged and shipped to the launch site.

The initial receipt location will consist of an aircraft landing strip or a truck loading dock, and it represents the first facility which interfaces with the spacecraft. At this point in the payload ground flow, the entire unit will remain in its protective covers; the

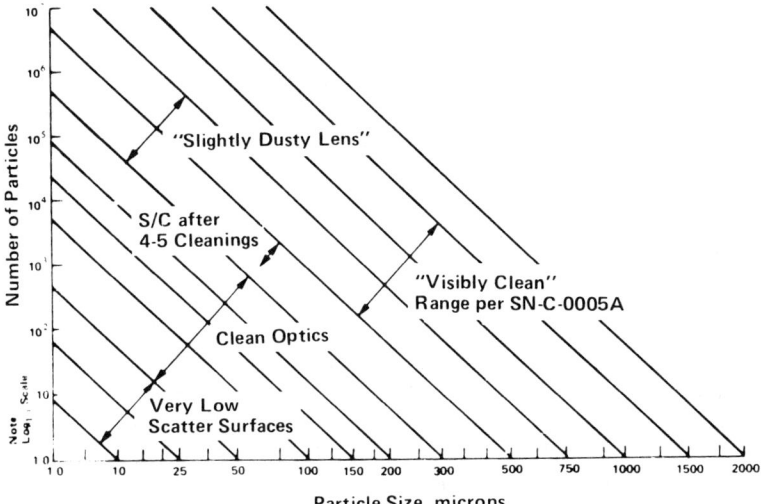

Fig. 1 Surface cleanliness levels definition.

facility cleanliness is not a major consideration. The facility/environment exposure flow is summarized in Table 1.

The vertical processing facility (VPF) will accommodate assembly, test, and checkout operations for Shuttle payloads. In general, this type of facility is designed to meet Class 100,000 and operate as a clean area. About 40 days is a typical period of exposure for a payload in the VPF.

Payloads will then be transported to the launch site in the payload canister to the payload changeout room (PCR) where final checkout functions will take place. Then the Shuttle Orbiter will mate with the PCR, and the payload will be inserted. The PCR will be supplied with input air meeting nominally Class 100, guaranteed Class 5000. The payload is assumed to be in the PCR for 6 days.

The payload bay (PLB) inlet air will nominally meet Class 100, guaranteed Class 5000, and the PLB will maintain

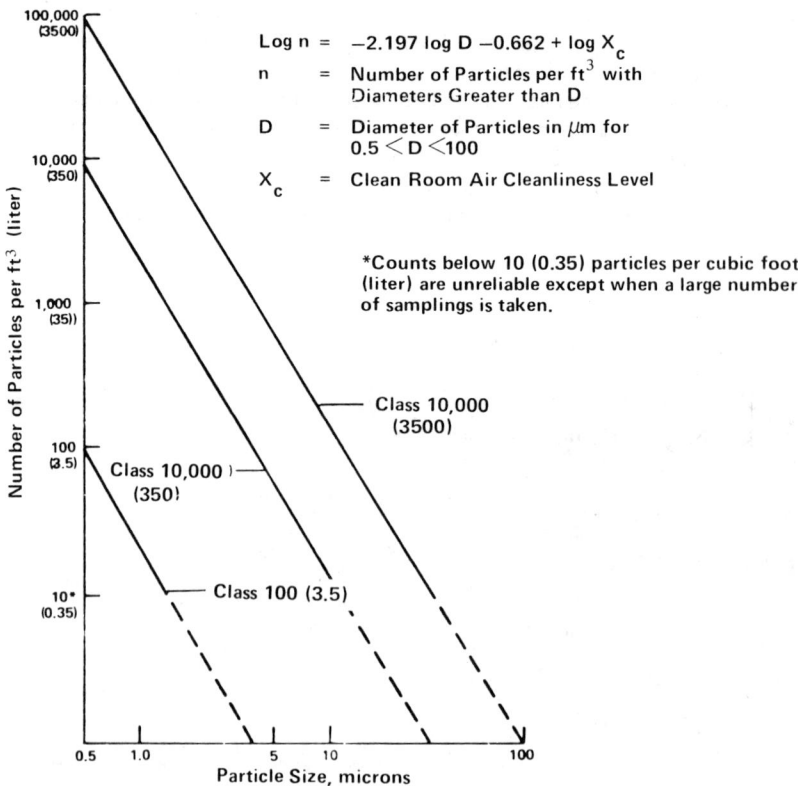

Fig. 2 Air cleanliness level definition.

Table 1 Ground flow timeline

Location	Air cleanliness level	Exposure time (days)
Assembly/test location	N/A[a]	90
Shipping container	100,000	...
VPF	100,000	38
Transport canister	100,000	2
PCR	100,000	6
PLB (prelaunch)	100,000	10
Total		146

[a]This area is not a clean room.

a positive pressure relative to uncontrolled areas. The PLB will have been cleaned to visibly clean (level 1) per NASA-SN-C-0005 (Ref. 3). The payload is assumed to be in the PLB for 10 days. Both the PCR and the PLB are assumed to be equivalent to a Class 100,000 clean area.

The model used to predict the deposited particle surface density is based on a compilation of test data acquired by various groups. This model, developed by Hamberg (Ref. 4), curve fits the data in terms of particle fallout in 24 h vs the environment air cleanliness level. Predicted fallout rates shown in Fig. 3 account for average cleanroom activity over a 24-h period.

The mean rate is expressed by:

$$n = 2.851 \times 10^3 \, N_c^{0.773}$$

where n is the fallout rate (particles/ft^2 >5 μm in 24 h) and N_c the air cleanliness level (particles/ft^3 >5 μm).

In applying the model, Hamberg suggests the following ground rules:

1) For conventional type cleanrooms with 15 to 20 air changes/h, use the mean fallout rate.

2) For cleanrooms with considerably less than 15 to 20 air changes/h approaching still air, use the maximum 95% confidence limits to obtain conservatively high fallout rates.

3) For cleanrooms with directional velocities near 90 ft/min approaching laminar flow type rooms, use the minimum 68% confidence limits to obtain conservatively high fallout rates.

For 100-K conventional cleanrooms, the fallout rate becomes ($N_c = 634$):

$$n = 4.0 \times 10^5 \text{ particles/ft}^2/\text{day} > 5 \text{ μm}$$

For the assembly/test location the 95% confidence limit is used to determine the fallout rate of:

$$n = 4.5 \times 10^6 \text{ particles/ft}^2/\text{day} > 5 \text{ μm}$$

Multiplying the exposure duration by the proper fallout rates results in the fallout for the total ground flow of:

$$n = 4.3 \times 10^8 \text{ particles/ft}^2 > 5 \text{ μm}$$

Measurements of particle fallout in KSC ground facilities (Refs. 5 and 6) have provided data for model correlation. Plots of the particle data show that the size distributions of the fallout do not follow the log-normal distributions defined in MIL-STD-1264A (Fig. 1) that are skewed towards the larger sizes. It is suggested that the skewness is caused by local activity and is not characteristic of the cleanroom performance during quiescent periods (no activity). However, since the sensor will be

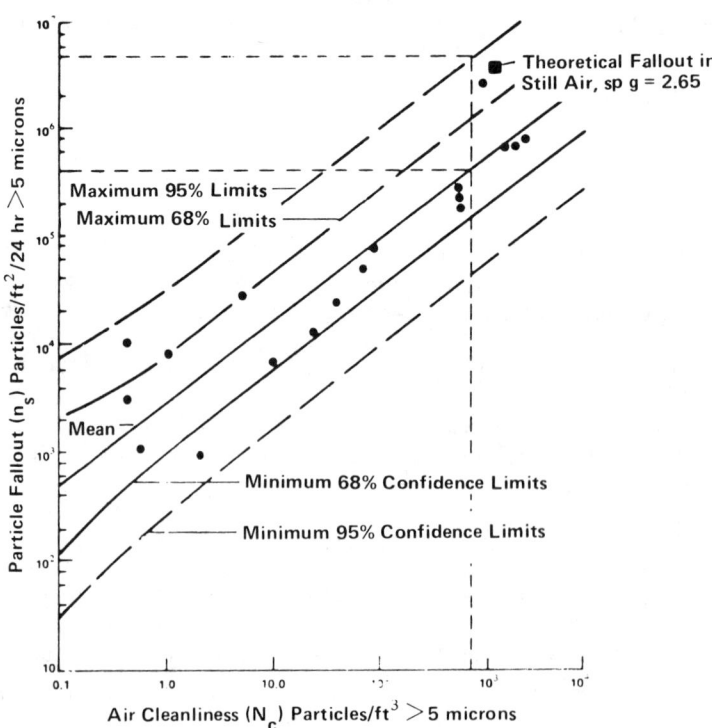

Fig. 3 Particle fallout rates.

exposed to the local environment, the resultant total distribution must be considered. In order to compare the overall effect of the log-normal and non-log-normal (skewed) size distributions in terms of equivalent surface cleanliness levels, the obscuration ratio was plotted as a function of surface cleanliness level for the log-normal distributions in 1246A. (The obscuration ratio is the cumulative cross-sectional area of deposited particles per unit area of substrate). The obscuration ratios were also computed for the KSC facility data. The results, plotted in Fig. 4, allow a comparison of fallout effect in terms of obscuration ratio and equivalent surface cleanliness level. Based on: 1) the exposure times shown in Table 1; 2) assuming mean obscuration values for the VPF (8.7×10^{-4}/day) and PCR/PLB (4.9×10^4) and; 3) assuming that the assembly/test location is equivalent to the mean orbiter processing facility (OPF) value (4.4×10^{-3}/day); the resulting obscuration for the entire period is 0.44 or an equivalent surface cleanliness level of ~1300.

If it is assumed that the Hamberg fallout model corresponds to the log-normal size distribution, then solving

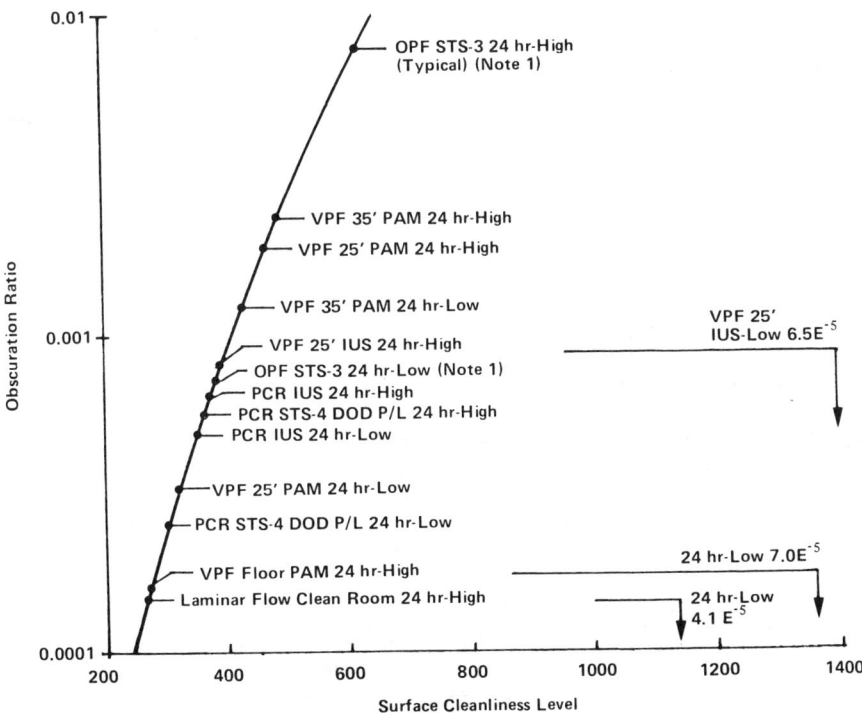

Fig. 4 KSC facilities data.

equation (1) for surface cleanliness level (X_1) by entering $n = 4.3 \times 10^8$ particles/ft^2 and $X = 5$, the equivalent surface cleanliness level is ~1300. Therefore, as a first approximation, the fallout model appears valid, although the assumptions and variations in facility cleanliness levels are quite gross.

The fallout data represented by the Hamberg function and the KSC measurements were all obtained for horizontal surfaces while the sensor front element is stowed in a near vertical orientation. Test data from Arnold Engineering Development Center (AEDC) (Ref. 7) and TRW (Ref. 8) indicate that the ratio of particle deposition on vertical surfaces to deposition on horizontal surfaces can range from 0.0023-0.10.

The corrected fallout for vertical surfaces, n_v, then becomes:

$$9.9 \times 10^5 < n_v < 4.3 \times 10^7 \text{ particles/ft}^2 > 5 \text{ μm}$$

From Fig. 1, this is equivalent to a surface cleanliness level range per MIL-STD-1246A of from 430-890 with a mean of 780 (based on an average vertical to horizontal deposition ratio of 0.05).

If a more optimistic estimate of the air cleanliness level in the assembly and test facility is assumed (100-K), then the fallout range, corrected for a vertical surface, becomes:

$$1.3 \times 10^5 < n_v < 5.8 \times 10^6 \text{ particles/ft}^2 > 5 \text{ μm}$$

which is equivalent to a surface cleanliness level of between 280 and 620 with a mean of 540.

Launch Environment

The approach used to evaluate the PLB launch environment for particle redistribution was developed by Hamberg (Ref. 9). The validity of this model is uncertain at this time; however, preliminary data are available from the Space Transportation System (STS) induced environment contamination monitor (IECM) measurements and are compared to the model predictions in a very preliminary assessment.

Redistribution Model Prediction. The Hamberg redistribution model assumes that during periods of large vibrations and gas flow (launch environment) all particles in a closed environment are dislodged and uniformly redistributed over all surfaces. This phenomenon is described by:

$$N/A = \frac{(N/A)_1 A_1 + (N/A_2) A_2 + \text{-------} + (N/A)_n A_n}{A_1 + A_2 + \text{-------} + A_n}$$

where N/A are the final uniformly redistributed particles >5 μm/ft^2, $(N/A)_{1,2,n}$ are the original particles >5 μm/ft^2 on surfaces A_1, A_2, A_n, and $A_{1,2,n}$ is the area of surfaces A_1, A_2, A_n.

Hamberg indicates that predictions based on this approach are probably conservative. Considering the geometry of the payload and the sensor exposure to the PLB, this is most likely true for the case being analyzed. However, from a worst-case aspect, a preliminary assessment of the launch environment will be made using this approach.

Spacecraft particle surface cleanliness level specifications typically vary from level 100 (MIL-STD-1246A) for systems with sensitive optical devices to visibly clean[3] for nonoptical systems. The visibly clean criteria are currently estimated to be equivalent to a surface cleanliness level range of between 500-1500.

Figure 5 (Ref. 9) is a plot of the surface cleanliness level achieved vs the number of cleaning operations based

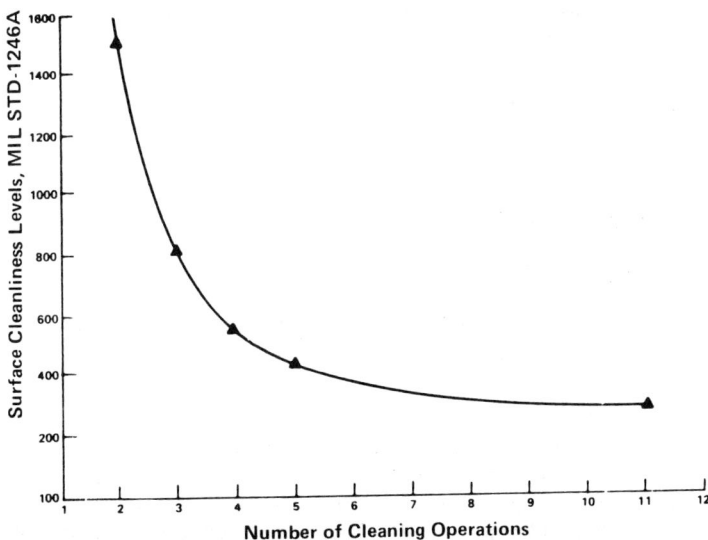

Fig. 5 Spacecraft cleaning effectiveness.

Table 2 Cascade impactor data

Diameter particles (μ)	Measured mass per unit volume ($\mu g/m^3$)		
	Flight	Ascent	Descent
>5	STS-2	30	10
	STS-3[a]	10	10
	STS-4	Nonfunctional	20
1–5	STS-2	500	250
	STS-3[a]	10	10
	STS-4	300	10
0.3–1	STS-2	250	125
	STS-3[a]	10	10
	STS-4	90	Nonfunctional

[a]Minimum instrument sensitivity.

on test data on a thermal vacuum test chamber using standard cleaning techniques. The plot shows that a practical cleaning effort (4 to 5 cleanings) will result in a surface cleanliness of 400 to 500. A level 450 will be assumed for this payload after cleaning operations.

The Space Shuttle Orbiter PLB standard cleanliness level (Ref. 3) is specified as visibly clean, level 1 which is assumed for this analysis to be equivalent to a surface cleanliness level of 1000. Better cleanliness levels are available to the payload at additional cost.

Estimating the surface area of the payload (A_P) and PLB (A_{PLB}) (neglecting cracks and crevices) and assuming that no other payloads are in the PLB, results in a surface cleanliness level of:

$$A_P = 650 \text{ ft}^2$$

$$A_{PLB} = 3180 \text{ ft}^2$$

$$(N/A)_{PL} = 1.2 \times 10^6$$

$$(N/A)_{PLB} = 7.6 \times 10^7$$

$$N/A = \frac{(1.2 \times 10^6)(650) + (7.6 \times 10^7)(3180)}{650 + 3180}$$

$$= 6.3 \times 10^7,$$

which is equivalent to a surface cleanliness level of ~950.

This indicates that per the Hamberg model, payload cleanliness levels will be driven by the Shuttle Orbiter PLB surface cleanliness; and if protective covers, shrouds, purges, etc. are not provided by the payload, surface cleanliness levels would be relatively poor.

IECM Flight Data Predictions. The induced environment contamination monitor (IECM) was flown on several early Shuttle flights to monitor the PLB contamination environment. The IECM cascade impactor instrument collected particles during Shuttle ascent and descent on a quartz crystal microbalance. Apiezon grease was used to capture the impacting particles. Table 2 summarizes the currently available Cascade Impactor data (Refs. 10-12). Since the instrument can only operate in a continuum flow regime (nonvacuum), the data shown represent conditions existing only during the initial 80 s after launch.

The magnitude of the measured particle environment can be put in perspective by comparing the mass concentration for that size range found in a Class 100,000 cleanroom. The comparison is weak in that the mechanisms for particle deposition differ. Varying acceleration and centrifugal forces, reduced buoyancy by low pressures, etc., all contribute some variation. However, some insight can be gained by the comparison.

The standard definition of a Class 100,000 environment includes all particles greater than 0.>5μm. Assuming a particle mass density of 2 g/cm^3 and integrating the functions plotted in Fig. 2 over the size ranges of 1-5- and 0.3-1- μm results in the curves shown in Fig. 6. Comparing the cascade impactor data to these curves indicates

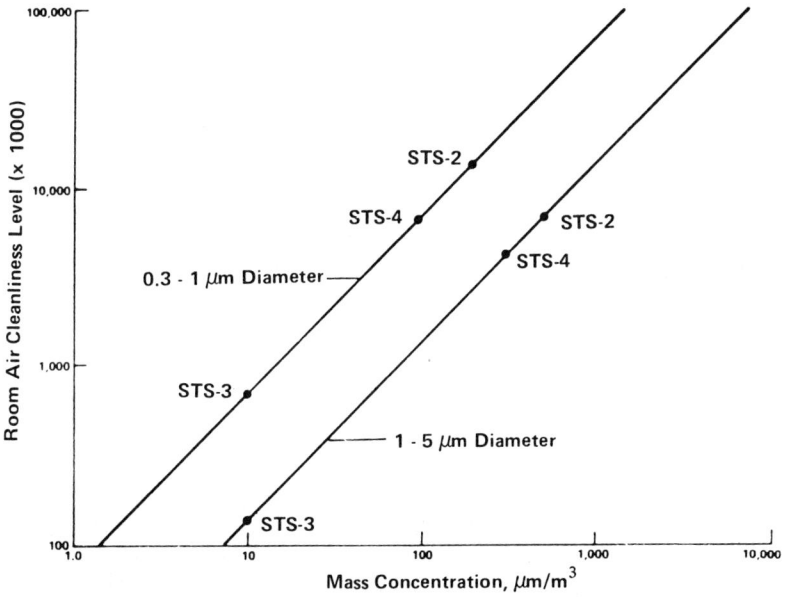

Fig. 6 STS launch environment predictions.

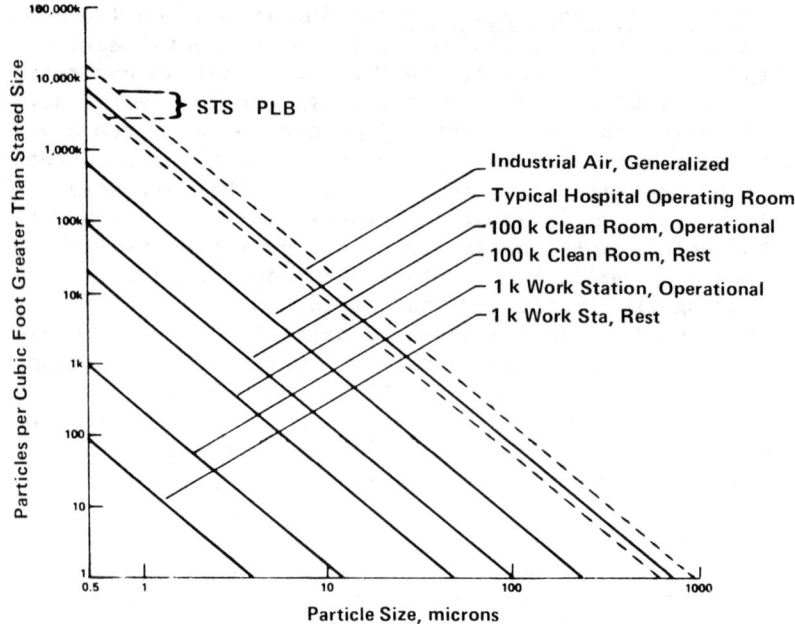

Fig. 7 Typical room air cleanliness levels.

that the measured environment exceeds a Class 100,000 air cleanliness level for all cases, and except for the STS-3 data, implies an equivalent air cleanliness level range of Class 4,000,000 to 14,000,000.

Quantitative assessments of cleanliness levels of these magnitudes would appear to be questionable. However, Fig. 7 (Ref. 13) shows a comparison between various types of environments and the STS PLB environment. The figure indicates that based on the IECM measurements, the PLB environment is equivalent to worse than "generalized industrial air" during launch. The STS-3 data corresponds to a relatively benign environment indicating the potential significance of payload mix considerations since STS-2 and -4 contained payload elements with possibly severe particle sources.

A comparison between the cascade impactor measurements and the Hamberg redistribution model predictions can be made considering the total particle population trapped in the PLB at launch. Based on the predicted surface cleanliness level and assuming that all particles become detached and uniformly distributed throughout the PLB, ($\sim 1 \times 10^4$ ft^3) an equivalent particle number density distribution can be plotted. Figure 8 shows this plot overlayed on the air cleanliness specification curve. The baseline curve shows that the resulting volumetric particle distribution does not follow the air cleanliness specifications functions. If, however, a conservative assumption is made that the actual volumetric distribution of surface particles is weighted heavily towards larger particles (test data show that smaller particles are more difficult to remove from the surface) and a curve is drawn through the intersection of the baseline distribution and the unit number density axis parallel to the air cleanliness functions (modified PLB particle distribution), the PLB suspended particle distribution becomes equivalent to a Class 10,000,000 environment. This level is consistent with the equivalent predicted air cleanliness levels based

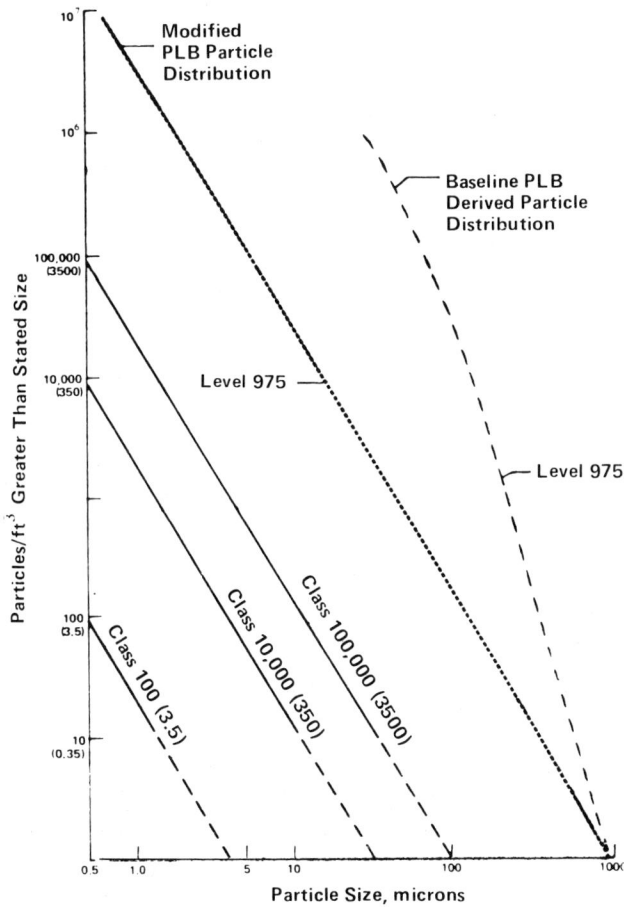

Fig. 8 Effective PLB air cleanliness levels.

on the cascade impactor measurements (Fig. 6); and therefore adds some credibility to the predicted levels.

Computing incident flux/surface cleanliness levels from air/PLB volume cleanliness levels predicted from the IECM measurements requires a new approach since the Hamberg fallout model (Eq. 2) applies to a viscous flow environment. Studies and testing to determine adhesive forces and release velocities for spacecraft environments (vibration/shock) are only now beginning to be performed (Ref. 13). Therefore, particle released velocities must be evaluated parametrically.

One can assume that during launch the velocities of all size particles are equal. The release velocity will probably be determined by a momentum exchange phenomenon and will vary with particle mass and attachment forces. However, since the intent of this assessment is only to scope the launch environment effects, constant velocities were assumed.

For a Class 10,000,000 environment, the volumetric density for particles >5 μm is ~6×10^4 particles/ft^3. For velocities of 0.1 and 1.0 ft/s (estimated) the flux impinging on a surface is 6×10^3 and 6×10^4 particles/ft^2 s, respectively. For a 3-h exposure (maximum Shuttle Orbiter PLB doors closed period), the resulting surface particle density becomes 6.5×10^7 and 6.5×10^8 particles/ft^2. This is equivalent to a surface cleanliness level of between 950 and 1400 if all impinging particles were to stick, which exceeds the mean level (950 max).

Comparing these predictions to the Hamberg redistribution model prediction (950) shows reasonable correlation. However, because of the assumptions and extrapolations required to obtain these results and the conservatism of the Hamberg model, care must be exercised in applying them to establishing system requirements.

<u>Launch Environments Analysis Summary</u>. The results of the launch environments analysis are summarized in Table 3. The redistribution model predicts an average surface cleanliness level existing on all PLB and payload surfaces after launch based on initial surface cleanliness levels.

Table 3 PLB launch environments summary

Prediction approach	Volume cleanliness (particles/ft^3 >0.5 μm)	Surface cleanliness level (MIL-STD-1246A)
Redistribution model	1×10^7	950
IECM flight data (preliminary)	$4 \times 10^6 - 1.4 \times 10^7$	950 (max)

The model is simplistic since it assumes that all particles will be detached and redeposited during launch due to the vibration and shock environment and the blowdown of the PLB. However, the model is also conservative and is useful for estimating only a worst-case condition.

The IECM results reported here are preliminary as the majority of the data collected have not yet been analyzed in detail. Until correlation between all data sets (cascade impactor, camera photometer, astronaut comments, etc.) is obtained, a quantitative definition of the PLB environment cannot be established with high confidence.

A comparison between the redistribution model predictions and the preliminary IECM data indicates that the

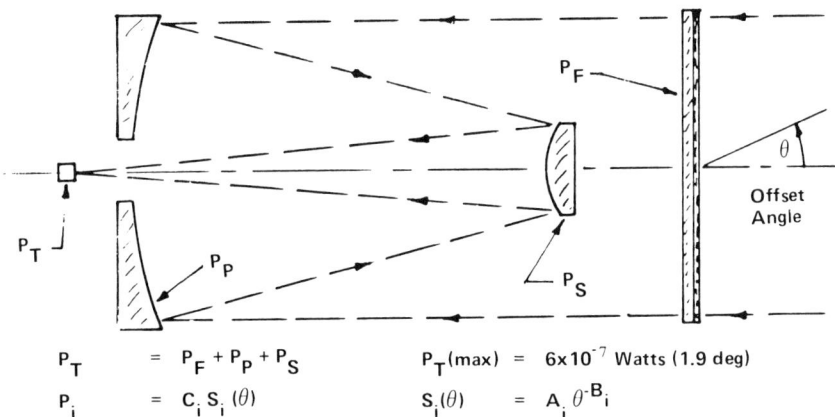

$P_T = P_F + P_P + P_S$ $P_T(\text{max}) = 6 \times 10^{-7}$ Watts (1.9 deg)
$P_i = C_i S_i(\theta)$ $S_i(\theta) = A_i \theta^{-B_i}$

Fig. 9 Sensor schematic.

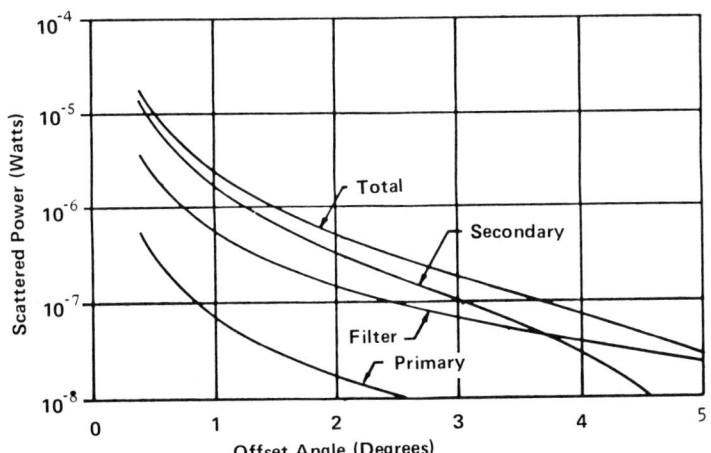

Fig. 10 Sensor baseline performance.

predicted environments are at least consistent. Both approaches predict that a severe particle environment exists in the Shuttle Orbiter during launch. However, it must be kept in mind that this correlation is based on several gross assumptions. For example, recent discussions (Ref. 20) with MSFC indicate that the particle environment in the vicinity of the Shuttle Orbiter immediately after the first door opening is extremely severe requiring 12-15 h to dissipate. The source of these particles is unknown. If a substantial fraction of these particles emanate from the PLB, this would imply that all particles do not redeposit as the model assumes. Until additional data are obtained with respect to the particle deposition process, these predictions should only be used as worst case guidelines.

Optical Effects

The ground and launch environment degradation effects were addressed separately since the contamination control techniques eventually implemented may differ. For example, if a cover is installed around the payload at the time of integration with the Shuttle Orbiter, contamination would still have accumulated during ground operations and may still impact sensor performance.

Baseline Sensor Performance

A functional schematic of the sensor is shown in Fig. 9. The simplified sensor scattering functions are

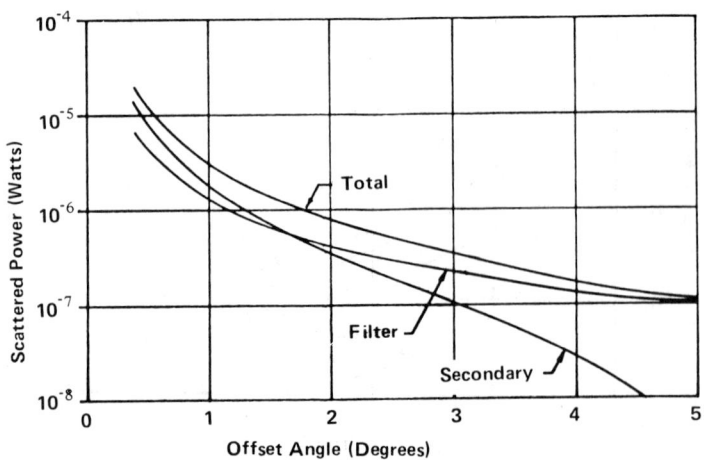

Fig. 11 Contaminated sensor performance.

shown below (Ref. 15):

$$S_F(\theta) = 0.17\, \theta^{-2.0}$$
$$S_P(\theta) = 0.021\, \theta^{-2.15}$$
$$S_S(\theta) = 0.021(6\theta)^{-2.15}$$

$\lambda = 0.82\ \mu m$ (sensor wavelength)

where $S_F(\theta)$, $S_P(\theta)$, $S_S(\theta)$ = scattering functions (SR^{-1})

This approach was followed for the sensor performance analysis described in this paper. Alternative more sophisticated stray radiation analysis techniques are available (Ref. 16 and 17, for example). However, they were not used for this study. Regardless of the analytical technique finally used to evaluate sensor performance, final certification and acceptance will be based upon test.

The scattered power as a function of solar offset angle for the baseline sensor is shown in Fig. 10. The offset angle corresponding to the EOL background level (6×10^{-7} W) is 1.9 deg.

Ground Contamination Effects

The only test data relating to the scattering functions of realistic particles deposited on optical surfaces were found in Refs. 7, 18, and 19. The Hughes data applies to mirrors at $\lambda = 10.6\ \mu m$. Wavelength scaling could be performed to 0.8 μm; however, since insufficient data

Table 4 Environments analysis results summary

Environment	Cleanliness levels	
	Air	Surface
Ground operations		
1) Facilities environments		
a) baseline	100 K/>100 K	
b) optional	100 K	
2) Analysis results		
a) baseline		780(430–890)
b) optional		540(280–620)
Launch		
1) Initial conditions		
a) PLB visibly clean, Level 1		1000(500–1500)
b) payload		450(400–500)
c) sensor		100
2) Analysis results		
a) redistribution model		
1) model output		950
2) estimated air equivalent	10,000 K	
b) IECM data analysis		
1) analysis results	4000–14,000 K	
2) estimates surface equivalent		950 (max)

are presented and could not be verified via contact with the author, the data were not used. The AEDC data were not used because as stated by the author the particle spatial distribution was not always uniform, and the resultant measurements were ambiguous.

The contaminated lens function from the Leinert paper applies to "about 50 particles/cm^2 visible on a dark field photograph." This equates to 4.6×10^4 particles/ft^2 and was judged to be equivalent to a surface cleanliness level of 300-700. However, the contaminated lens scattering function actually includes scattering functions for the deposited particles and the substrate. Extending the superposition approach (implicit in the scattering model) to deposited particles and subtracting the substrate scattering function (also from the Leinert paper) results in the scattering function for the deposited particles.

$$S_C(\theta) = 0.08 \; \theta^{-1.86} - 0.02 \; \theta^{-2.12}$$

S_C = particle scattering function (SR^{-1})

The scattered power as a function of angle for the particle scattering function added to the stack scattering function is shown in Fig. 11. Therefore the minimum solar offset angle for a surface cleanliness level of 300-700 is ~2.3 deg.

The surface cleanliness level for the deposited particles (300-700) corresponds to the level predicted for the optional (100-K) ground flow and therefore the minimum solar offset angle (2.3 deg) is also applicable. However, the cleanliness level for the baseline ground flow (>100 K/100 K) exceeds (780 mean value) the Leinert data point and although measurements are not available, the anticipated minimum solar offset angle is expected to greatly exceed 2.3 deg and is therefore unacceptable.

Launch Environment Effects

A surface cleanliness level, as shown in Table 3, of 950 results in an obscuration ratio of ~6%. Scattering calculations are not required to establish that this condition is unacceptable.

Assuming that the PLB could be cleaned similarly to a spacecraft, then a surface cleanliness level of 400 to 500 would be achievable. This would result in scattering levels and minimum solar offset angles similar to those predicted for the ground environment (2.3 deg). However, the ability to clean the PLB has not yet been demonstrated.

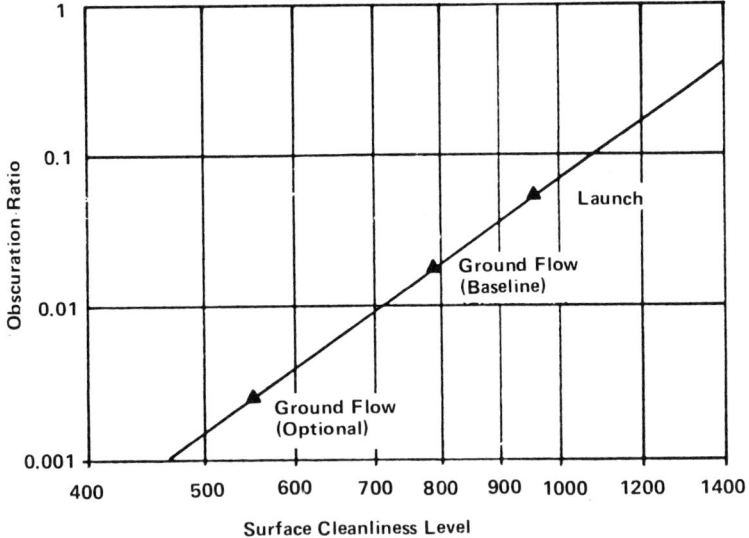

Fig. 12 Particle deposition effects summary.

In fact, flight data and astronaut comments indicate a very severe particle environment. This is due in part to the basic design of the Orbiter PLB which is not conducive to cleaning to levels compatible with optical system requirements. The visibly clean criteria imposed on PLB surfaces has yet to be accurately quantified due to the somewhat subjective nature of the criteria and the influence of various surface characteristics on the ability to detect visible contamination (i.e., visibly clean for the silver teflon PLB radiator surfaces may be as good as level 50, while for the diffuse porous PLB liner, it may be as high as level 1500). Based on the capability to clean the bay and liner, the level of activity required to integrate a payload and the severity of the launch environment make it impossible at this time to estimate with any degree of confidence the minimum cleanliness level achievable on surfaces in the PLB after orbit insertion considering the current data base.

Results Summary

The results of this analysis are summarized in Table 4. The baseline ground operations air cleanliness level was assumed to be 100 K for all facilities except the assembly and test facility which is not a cleanroom. The surface cleanliness levels were based on the Hamberg fallout model.

Table 5 Sensor performance effects summary

Summary	Surface cleanliness	Obscuration ratio	Minimum offset angle
1) Baseline sensor (clean)	100	...	1.90 deg
2) Ground contamination			
Optional flow (100 K)	540	0.005	2.3 deg
Baseline flow (>100 K/100 K)	780	0.028	≫2.3 deg
3) PLB launch contamination	950	~0.08	

An optional ground flow assuming all 100-K facilities was also addressed.

The launch environment was computed assuming: 1) a PLB standard service baseline surface cleanliness level (visibly clean level 1 per SN-C-0005A), and 2) a nominal payload cleanliness level based on 4 to 5 cleanings (400-500). The cleanliness levels predicted by the Hamberg redistribution model showed some correlation with the preliminary IECM data which indicates a potential for a severe particle environment in the PLB during launch. However, these results are worst-case and should only be used as a guideline until more quantitative results based on additional flight data become available. An additional consideration is the anticipated improvement in Shuttle processing facilities cleanliness levels due to operational and hardware modifications and controls.

The effects of the predicted contamination environment on sensor performance is summarized in Table 5. Based on the simplified scattering model and applying scattering functions scaled by wavelength, the baseline uncontaminated sensor will meet the 2-deg offset angle requirement.

Superimposing the particle scattering function of the Leinert test data on the sensor filter results in a minimum offset angle of 2.3 deg. This value corresponds to the predicted effect of the optional ground flow environment. The baseline ground flow is predicted to result in a much more severe deposition level. No scattering data were obtained for surface cleanliness levels of 780. However, the 2.8% obscuration ratio (ratio of projected area of the particles to surface area of the substrate) for the 780 level compared to 0.5% for the 550 level (2.3 deg) would indicate a very severe increase in scattering and consequently the minimum offset angle. The relationship between obscuration and surface cleanliness is shown in Fig. 12.

The sensitivity of the predicted offset angles to the particle scattering functions can be assessed by considering the Leinert data. The Leinert "slightly dusty" condition was estimated to correspond to an equivalent surface cleanliness level of 300-700. This range spans a particle number density range of a factor of 100 (see Fig. 1). If, for example, the actual scattering function were x2 higher, then the offset angle would increase to 3 deg. If the scattering function were x10 higher, the offset angle would increase to ~5 deg. Therefore the uncertainty in the Leinert "slightly dusty" cleanliness level results in a substantial uncertainty in the prediction of the offset angle and must be considered when applying the results of this analysis.

The OFT 2/4 PLB environment measured during launch was severe. The predicted 8% surface obscuration would not only degrade optical system performance but thermal control and solar power subsystems as well. Although the source(s) of the particles is unknown (PLB/payload mix) the potential for severe payload degradation exists.

Conclusions

1) The existing analytical tools and data base are insufficient for developing accurate particle contamination effect predictions. Additional analytical tool development and verification are required using validated test data.

2) Preliminary ground environment predictions indicate that sensor background scattering due to particle deposition for the baseline (>100-K/100-K) ground flow is extremely severe. Predictions for the optional ground flow (100-K) are marginal. The predictions are based on: 1) the 2.1-deg maximum offset angle requirement; 2) nominal ground flow timeline and facility air cleanliness levels; 3) a simplified scattering model; 4) a preliminary fallout model; and 5) minimum optical effects data.

3) Preliminary IECM flight data for STS-2 and STS-4 indicate a very severe particle environment may exist in the PLB during launch. Predictions using the worst-case Hamberg redistribution model tend to support this conclusion. However, the flight data analysis is preliminary and until all sources of data have been evaluated and correlated these results should be viewed as an upper bound.

Acknowledgment

This work was supported in part by AFSD contract F0701-81-C-0122.

References

[1] MIL-STD-1246A, "Product Cleanliness Levels and Contamination Control Program," Aug. 1967.

[2] FED-STD-209B, "Clean Room and Work Station Requirements in Controlled Environments," April 1973.

[3] NASA-SN-C-005, "Specification, Contamination Control Requirements for STS Payloads," March 1978.

[4] Hamberg, O., "Particulate Fallout Predictions for Clean Rooms," The Journal of Environmental Sciences, Vol 25, No. 3, May/June 1982.

[5] Whitehead, V., "Relationship Between Air and Surface Cleanliness Levels at KSC Facilities," Proceedings of the Seventh Contamination Control Seminar, G&N Community, Palo Alto, Calif., Oct. 1982.

[6] Borson, E., "Evaluation of the Ground Contamination Environment for STS Payloads," Shuttle Environment Workshop, Washington, D.C., Oct. 1982.

[7] Young, R. P., "Degradation of Low Scatter Mirrors by Particle Contamination," AEDC Arnold Air Force Station, Tennessee, AD-A-0040103, Jan. 1975.

[8] Ruel, R. P. et al, "A Forecasting Technique for Accumulated Particulate Contamination on Spacecraft Assemblies," TRW Systems, Redondo Beach, Calif., Dec. 1977.

[9] Hamberg, O., "Prelaunch and Orbiter Bay Contamination Control at KSC," Aerospace Corp., El Segundo, Calif., 78-5124.17-15, Nov. 1978.

[10] NASA TM-82457, "STS-2 Induced Environment Contamination Monitor (IECM) Quick Look Report," Jan. 1982.

[11] NASA TM-82489, "STS-3 IECM Quick Look Report," Jan. 1982.

[12] Private communication, B. J. Duncan, NASA-MSFC, Huntsville, Ala., Sept. 1982.

[13] NASA Handbook NHB-53003, "Handbook for Contamination Control on the Apollo Program," Sept. 1966.

[14] Glassford, A. P., "Design of an Experiment to Measure Particle Detachment Forces from Surfaces," Proceedings of the Seventh Inertial Guidance Community Contamination Control Seminar, Palo Alto, Calif., Aug. 1982.

[15] Informal Memo, Aug. 1982. Subject: Sensor Performance Model (Additional Data Not Available).

[16] Breault, R. P., "APART/PADE Version: A Deterministic Computer Program Used to Calculate Scattered and Diffracted Energy," *Proceedings of the SPIE Symposium*, Vol 257, 1980.

[17] Freniere, E. R., "Simulation of Stray Light in Optical Systems with GUERAP III," *Proceedings of the SPIE Symposium*, Vol 257, 1980.

[18] Leinert, C., "Stray Light Suppression in Optical Space Experiments," *Applied Optics*, Vol 13, March 1974.

[19] ANON, "Informal Communication & Data--Mirror BRDF Plot," Hughes Aircraft Co., Culver City, Calif., Not Published.

[20] Private communication, G. Owens, NASA-MSFC, Huntsville, Ala., March 1983.

Analysis of Contamination Degradation of Thermal Control Surfaces on Operational Satellites

J. E. Ahern,* R. L. Belcher,* and R. D. Rufft†
Aerojet ElectroSystems Company, Azusa, California

Abstract

The change in performance of a passive radiator, when vent paths are modified to avert contamination flow, is presented. Solar absorptance increases for the radiator were calculated and compared for three different vent path configurations, on a series of synchronous orbiting satellites. The improved design on the latest flight has drastically reduced the degradation rate.

Computer programs have also been developed which model the contamination process using a line-of-sight assumption. This analysis predicted a contaminant distribution that is similar to flight data results. The information can be used to direct the analysis of new satellite designs to minimize the contamination of thermal control surfaces.

Introduction

The analysis of data from orbiting satellites has shown a general warming trend that can shorten the operational life of electronics and optical systems (Refs. 1-4). To extend spacecraft life, the cause of this temperature rise

---

Presented as Paper 83-1449 at the AIAA 18th Thermophysics Conference, Montreal, Canada, June 1-3, 1983. Copyright © American Institute of Aeronautics and Astronautics, Inc., 1983. All rights reserved.

* Senior Technical Specialist, Space Surveillance Division.
† Technical Staff, Space Surveillance Division.

ANALYSIS OF CONTAMINATION DEGRADATION

must be ascertained. Previous reports (Refs. 1 and 2) have disclosed that two basic types of degradation exist for thermal control surfaces. One of these involves the change in the material as a result of exposure to the solar and space environment. The behavior of several materials under these conditions was reported in Ref. 2, including fused silica second surface mirrors, silvered Teflon, zinc orthotitanate white paint, silica cloth, and silver-alumina-silica. This paper is concerned with the other cause of degradation, which involves the effect of contamination on thermal control surface performance.

Solar absorptance (α_s) increases on a cylindrical radiator have been calculated for seven different synchronous orbiting satellites. For the purposes of this paper they will be labelled satellites A, B, C, D, E, F and G in the order of their launch dates.

The degradation of satellites A and B was first reported in Ref. 1. Solar absorptance increases on the cylindrical radiator were found to be nonuniform and much higher than expected. Many possible explanations for the phenomenon were proposed and investigated (Ref. 5). Quartz crystal microbalance experiments on a later flight indicated that mass was accumulating on the high degradation surfaces, but the source of the contamination was uncertain. Surfaces exposed to vent openings from an electronics compartment showed the highest degradation rates, indicating that volatile materials from this region might be outgassing and condensing on the cold radiator.

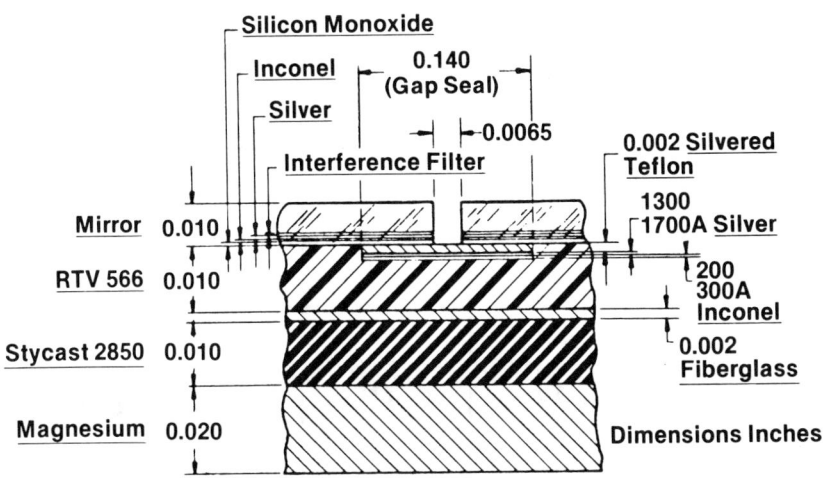

Fig. 1 Radiator thermal control surface composition.

On more recent flights the vent paths have been modified. This paper presents the impact of these changes by comparing solar absorptance increases on the radiator for three different venting configurations. Included is a description of the radiator modeling and analysis procedure.

In addition, a series of computer programs have been developed to analyze the deposition of outgassing materials on the radiator, using a line-of-sight assumption. Runs were made simulating the different venting arrangements, and the results are compared to flight measurements. A brief description of the programs is included.

Radiator Description

The radiator being studied is basically cylindrical with 16 flat sides. A magnesium shell support structure is covered with 1-in. square silver backed fused silica mirrors that are 0.010 in. thick. Satellites subsequent to A also have silvered teflon seals in the gaps between mirrors. The surface composition is shown in Fig. 1.

For computer modeling, the radiator is divided into three horizontal zones with four sections circumferentially in each zone for a total of twelve nodes. A projected layout of the radiator is shown in Fig. 2. Temperature monitors are located at the center of the nodes where possible.

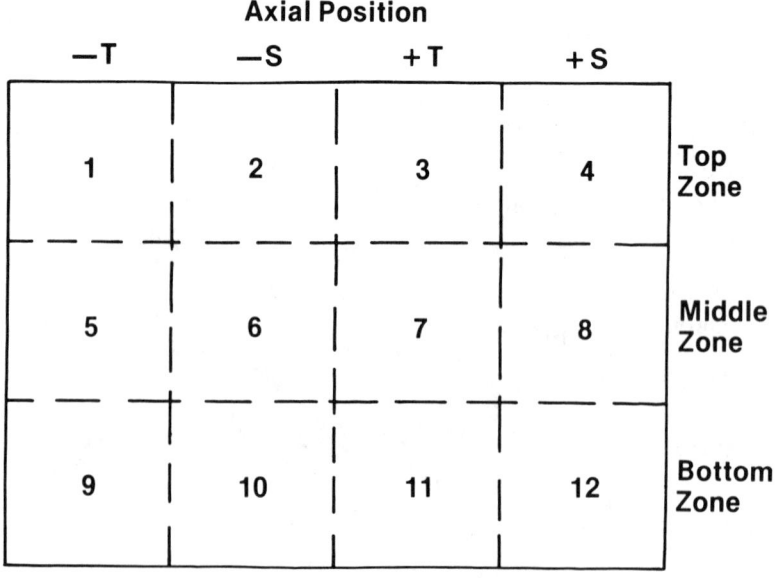

Fig. 2 Radiator node and zone designations.

Except for satellites A and B, which had a calorimeter on each node in the top zone, all flights have a single calorimeter on node 1. Flight data on node 1 was judged to be unsuitable on flights following B due to this nonuniformity and the inability to model it accurately. Usable flight data is also not available for nodes 6, 8, and 11 due to the presence of other satellite hardware.

A typical diurnal temperature range for the radiator surface is -220 to -120°F.

Electronics Region Vent Paths

The top of the electronics compartment is located just below the bottom zone of the radiator. There are five sunshades which protrude from this region in front of nodes 10 and 11 as shown in Fig. 3. The original vent paths from the electronics area are located at the base of the sunshades, the interior of sunshades d and e, and at the base of the radiator as shown in Fig. 4. The relative venting areas are given in Table 1. Note that 64% of the total venting area is through the interior of sunshades d and e.

A typical diurnal temperature range for the electronics baseplate is 10 to 30°F.

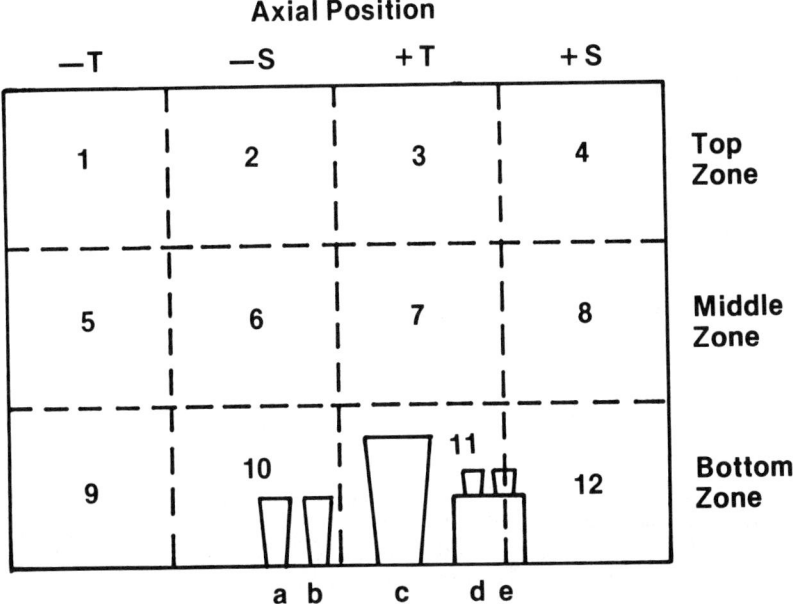

Fig. 3 Sunshade location and designation.

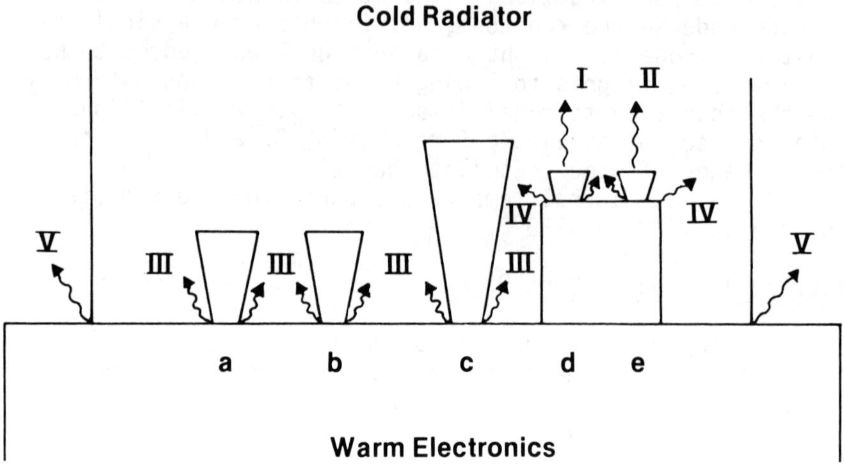

Cold Radiator

Fig. 4 Original vent paths (satellites A-D).

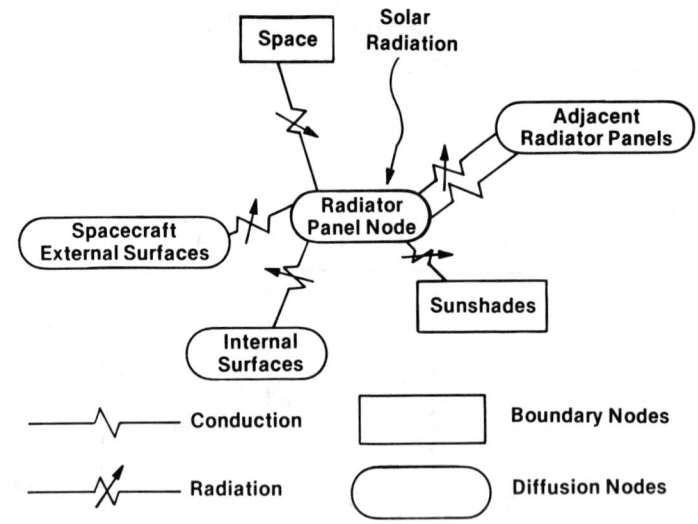

Fig. 5 Thermal connections to a typical radiator panel node.

Flight Data Analysis

The thermal model for the satellites contains 350 nodes with over 4000 conduction and radiation connectors. Since the radiator is thermally connected to the spacecraft, accurate α_s calculations require that the full 350 node model be used to establish a heat balance each time a data point is taken. The typical thermal connections to a radiator node are shown in Fig. 5. Even though there are numerous connections, most of the nodes have small heat flows to the rest of the spacecraft.

A flow chart describing the procedure for calculating α_s is shown in Fig. 6. The calculation for a particular flight is normally made four times per year. Even though this introduces major variations in satellite orientation and solar input, the computed α_s values are consistent. An analysis of data trends has shown the rms deviation in α_s to be less than 0.003.

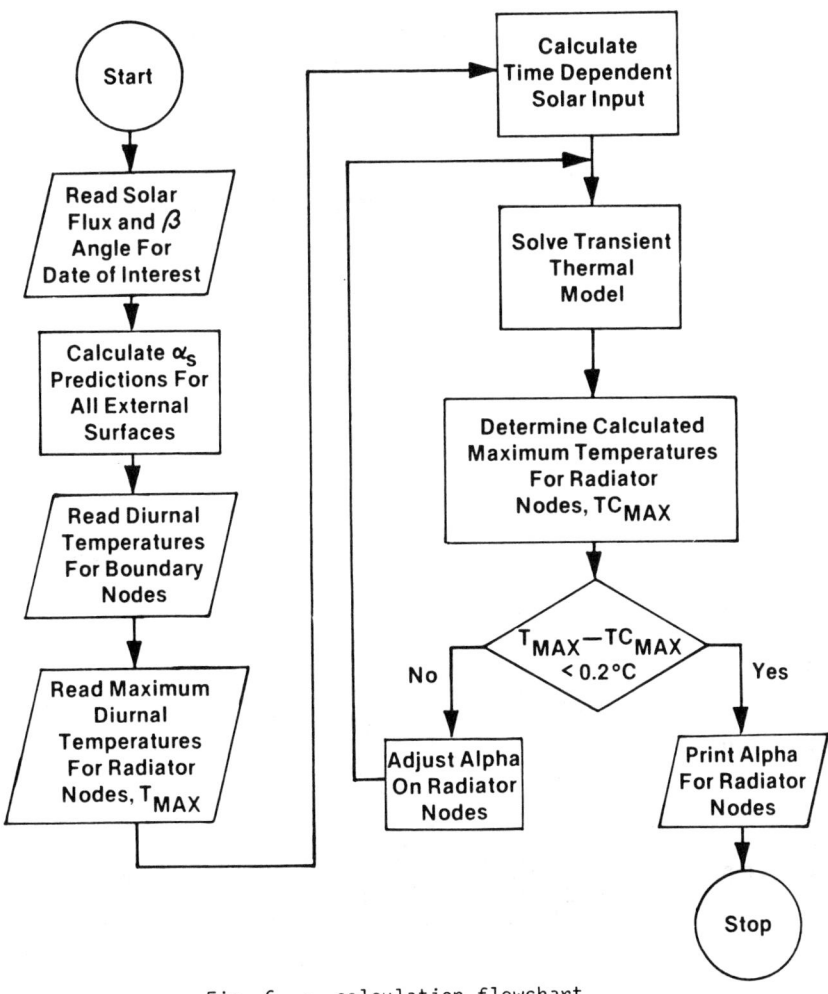

Fig. 6 α_s calculation flowchart.

As was previously noted, four of the twelve nodes do not have usable temperature monitor data. Solar absorptance increases for these locations are estimated based on the trends of adjacent monitored nodes and expected degradation patterns. Even though the estimates could be in error, the effect on monitored nodes is small due to low conductances and small temperature differences between panels.

Flight Data Results

The original vent path configuration described earlier was flown on satellites A, B, C, and D. A circumferential plot of solar absorptance increases and distribution typical of these flights is shown in Fig. 7. The data shows degradation that is quite pronounced in the sunshade area compared to the opposite side of the radiator which is relatively clean.

Table 1 Relative venting areas

Vent	Percent of total venting area
I	25.3
II	38.8
III	19.4
IV	11.7
V	4.8

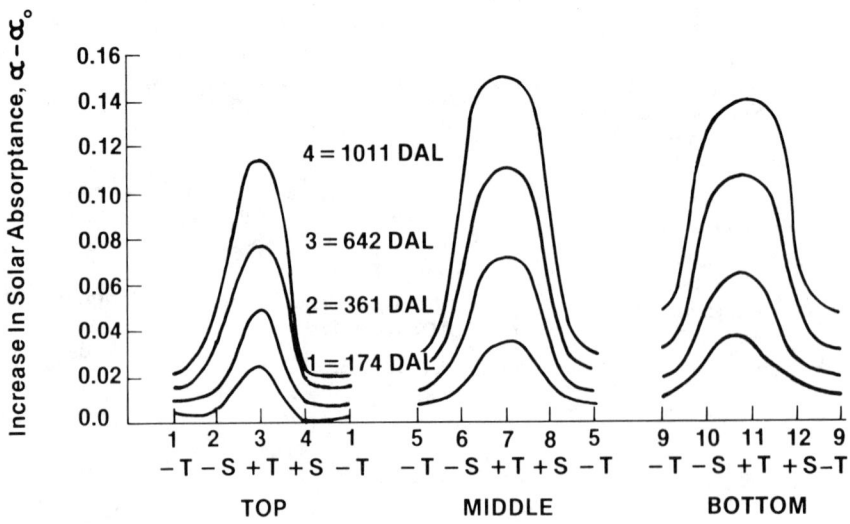

Fig. 7 Satellite D circumferential radiator degradation.

The opportunity occurred to seal some of the vent openings on satellites E and F. New vents designed to direct contaminants away from the satellite were installed on the side of the electronics compartment. The remaining vent paths opened to the radiator are shown in Fig. 8. Unfortunately, the major openings through the interior of sunshades d and e could not be closed off, so the original vent area was decreased by only 31%. Fig. 9 shows the effect of this modification on satellite E by comparing it to satellite D. The reduction in degradation is small but encouraging.

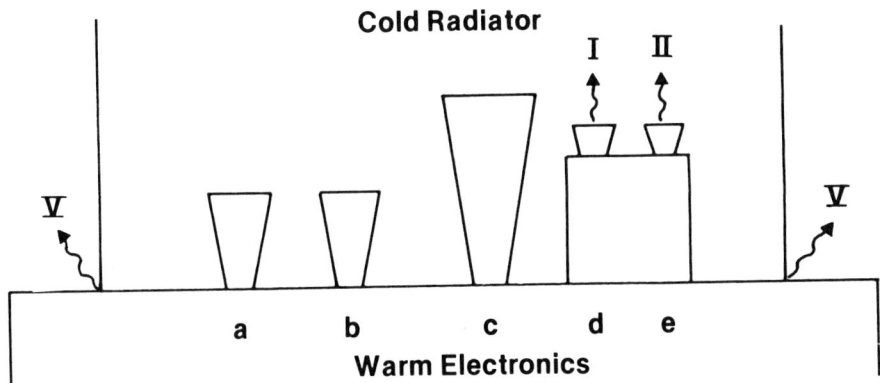

Fig. 8 Vent paths on satellites E and F.

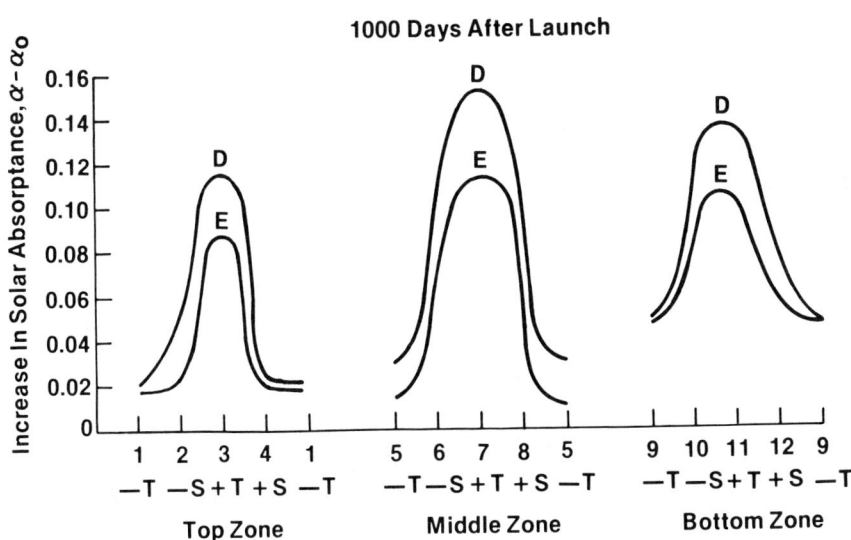

Fig. 9 Comparison of radiator degradation on satellites D and E.

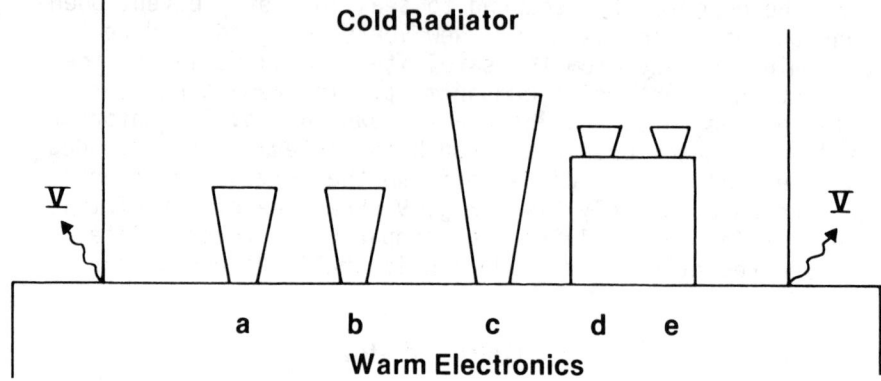

Fig. 10 Vent paths on satellite G.

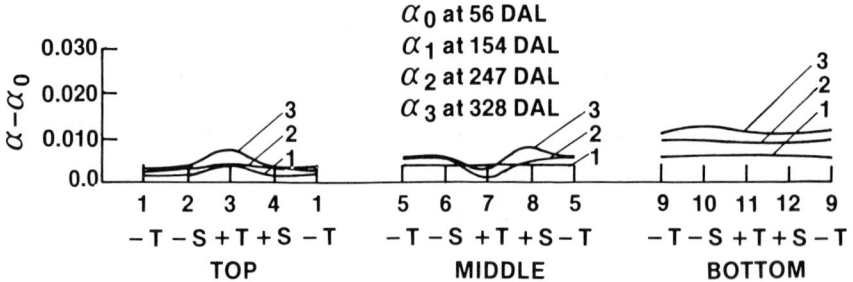

Fig. 11 Satellite G radiator degradation.

It indicates that further improvements could be expected by sealing the rest of the vents.

On satellite G the sunshades are completely sealed. The only remaining vent path opened to the radiator is the annular space between the radiator and the top of the electronics enclosures as shown in Fig. 10. This is less than 5% of the original vent area. As a result of the modifications, increases in α_s on satellite G have been small and nearly uniform around the circumference of the radiator, as shown in Fig. 11.

The most recent analysis of flight data on satellite G was conducted at 328 days after launch. A comparison of α_s on flights D, E, and G for the same time period is shown in Fig. 12. Sealing the vent paths has nearly eliminated contamination of the radiator, except for the bottom zone which may still be affected by the remaining vent opening.

The reduced degradation should have a marked effect on long range satellite performance. In fact, the potential

ANALYSIS OF CONTAMINATION DEGRADATION

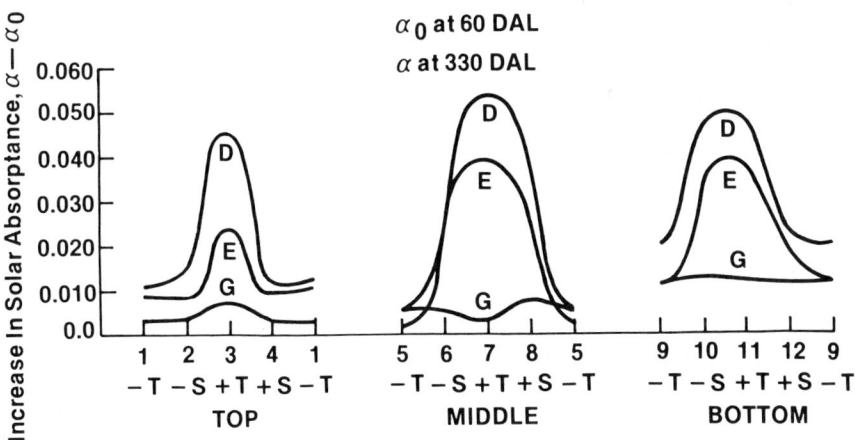

Fig. 12 Comparison of radiator degradation on satellites D, E, and G.

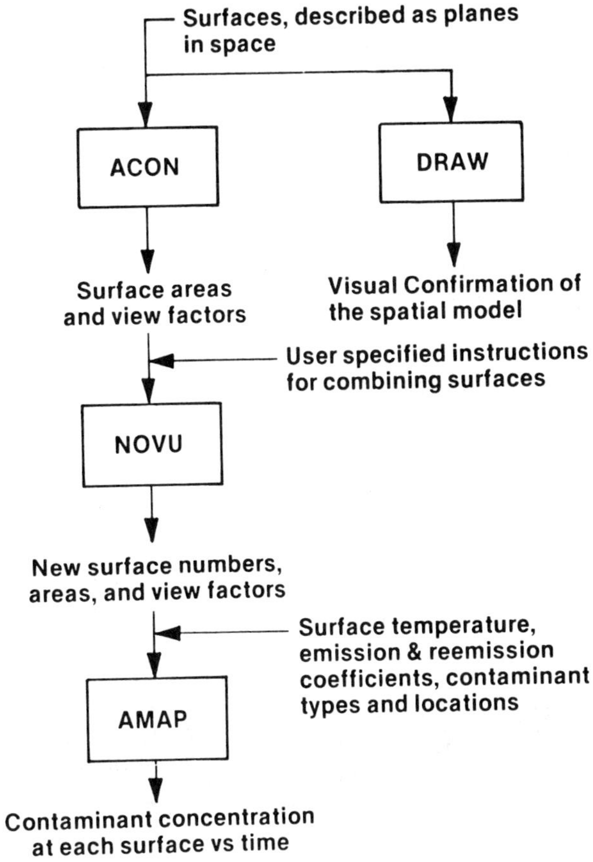

Fig. 13 Data flow in Aerojet mass analyzer programs.

useful life of satellite G is expected to be twice as long as satellite D's.

Contamination Computer Analysis

The Aerojet mass analyzer program (AMAP) is an analytical tool designed to determine mass deposition on surfaces of a vehicle due to material outgassing in a space environment. A line-of-sight assumption is used. A more rigorous analysis would include the effects of particle collision and charging.

The analysis is made up of a number of distinct steps as shown in Fig. 13. A model of the body to be analyzed is constructed of plane surfaces, and the surface-to-surface view factors are determined. A plot program to aid in model verification is available. The original model, and its view factor matrix may be modified to produce a model containing fewer nodes. Finally, the area and view factor data, along with surface temperatures and contaminant emission and re-emission coefficients are used to determine the amount of contaminant deposited on each surface as a function of time.

Computer calculations simulating the vent configurations of satellites D, E, and G were made. The results are shown in Fig. 14. The contaminant distribution calculated is similar to the observed surface degradation. The deposition results are qualitative at present, because the data pertaining to emission and re-emission of various contaminants are incomplete. More research needs to be done to determine these characteristics before accurate, quantitative, contamination predictions can be made.

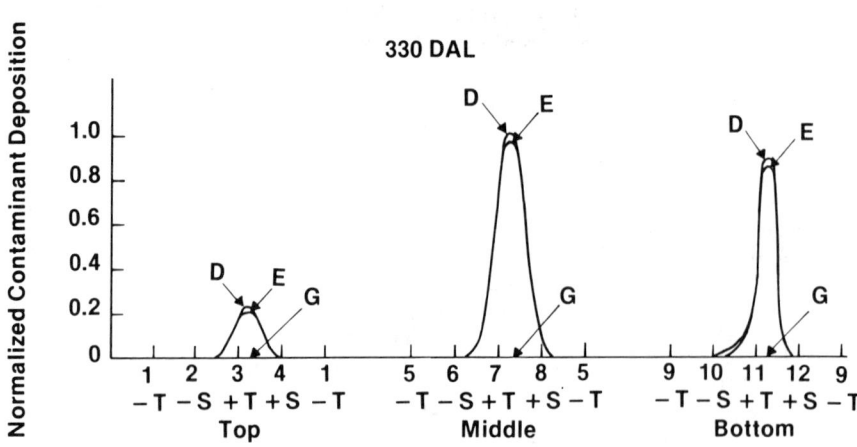

Fig. 14 Radiator contaminant distribution on satellites D, E, and G using the Aerojet mass analyzer program.

Conclusion

The analysis of satellites presented in this paper has demonstrated the importance of isolating thermal control surfaces from outgassing contaminants. Even though an excellent surface material may be selected, contamination can cause it to degrade at many times the clean, uncontaminated rate. As a consequence, spacecraft performance can be seriously impaired and operational life drastically shortened.

When designing a satellite, the various mechanisms of contamination should be analyzed, to determine the quantitative effects on thermal control surfaces. Unfortunately, the technology needed to predict these contamination rates is relatively new and incomplete. Until better modeling is possible, the experiences of earlier flights and experiments can be used to help direct the design of new satellites.

As shown in this study, the selective arrangement of vent paths can lead to substantial improvements in thermal performance.

References

[1] Curran, D. G. T. and Millard, J. M., "Results of Contamination/Degradation Measurements on Thermal Control Surfaces of an Operational Satellite," AIAA Paper 77-740, 12th Thermophysics Conference, Albuquerque, N. Mex., June 1977.

[2] Ahern, J. E. and Karperos, K., "Calorimetric Measurements of Thermal Control Surfaces on Operational Satellites," AIAA Paper 83-0075, 21st Aerospace Sciences Meeting, Reno, Nev., Jan. 1983.

[3] Rajagopalan, R. and Willson, V. J., "Thermal Performance of Anik-B Satellite in Orbit," AIAA Paper 80-1498, 15th Thermophysics Conference, Snowmass, Colo., July 1980.

[4] Bouchez, J. P. and Howle, D., "The Orbital Test Satellite (OTS) Thermal Experience After 3.5 Years in Orbit," AIAA Paper 82-0831, AIAA/ASME 3rd Joint Thermophysics Fluids, Plasma and Heat Transfer Conference, St. Louis, Mo., June 1982.

[5] Anonymous, "Evolution of AESC Satellite Contamination Studies," Report No. 4724, Aerojet ElectroSystems Company, Azusa, Calif., Feb. 1973.

Abatement of Gaseous and Particulate Contamination in a Space Instrument

John J. Scialdone*
NASA Goddard Space Flight Center, Greenbelt, Maryland

Abstract

Contaminants from sources external or internal to an instrument may degrade its critical surfaces and/or create an environment which limits the instrument's intended performance. Analyses have been carried out to investigate methods to prevent the ingestion of external contaminants into the instrument and to limit the effect of the self-generated contaminants during ground, launch, orbiting and landing phases of flight. It is proposed that a positive pressure and purging flow of clean gas inside the instrument be maintained while on the ground, during launch, and for a period of time in orbit. The pressure to be maintained and the required purging flow are examined in terms of the effectiveness in preventing gaseous and particulate contaminants ingestion and the abatement of the self-generated contaminants. Considerations have been given to the venting requirements for the structural integrity of the instrument during launch, the limitations on the volume and the pressure of the purging gas to be carried along in orbit, and the requirements to operate the instrument after a reasonable time in orbit. The required venting area is established based on the internal volume of the instrument, the allowable pressure differential, and the rate of external pressure change during launch. Relationships have been developed for the time dependence of the diffusion of external contaminants into the volume in terms of the volume, vent area, depth of vent passage, volume characteristic dimensions, molecular mass of contaminants, and internal purging pressure. The exclusion of particulates is based

---

Presented as Paper 83-1567 at the AIAA 18th Thermophysics Conference, Montreal, Canada, June 1-3, 1983. This paper is declared a work of the U.S. Government and therefore is in the public domain.
*Aerospace Technologist, Instrument System Branch, Instrument Division.

on drag forces induced by the density and velocity of the purging gas. The volumetric flow requirements are established for the indicated vent area and purging pressure. The abatement of self-induced contamination is predicated on the partial suppression of the materials outgassing and the scattering and transport of its products out of the volume by the purging density and flow. Full internal outgassing occurs when the outside environment has improved, the purging is stopped, and the instrument is fully open. The results of the analysis have been applied to a telescope which is very sensitive to contaminant deposits on energy reflecting surfaces and to other equally contaminant sensitive systems.

Introduction

An instrument or a system may be degraded by gaseous and particulate materials deposited on its internal or external surfaces. The degradation results from changes in thermo-optical properties of the contaminated surfaces. The contaminant modifies the absorption, reflectance, and scattering of the incoming radiations. The instrument performance may be degraded also because of absorption, reemission, and scattering of certain radiation by the molecular and particulate medium in its field of view. The contaminants may originate from several sources: residues of machining and assembly operation, handling; exposure to polluted environment; release of weakly bound molecules from materials under vacuum conditions; exhaust products of nearby propulsive or discharge systems; micrometeorites impacts on materials; abrasion of moving parts; oxidation/degradation of materials and others. Many precautions have been employed to reduce the generation of those contaminants and their effects in space flights. However, several have depended on the availability of time to allow the environment to become acceptable and the instrument to dissipate its own outgassing and clean itself. With the advent of the Shuttle, the environment of an instrument located in its bay is affected by other nearby systems and by the many service functions to be performed by the Shuttle. In addition, the time available for an instrument attached to the Shuttle to accomplish its mission has been curtailed to the order of weeks at least for now. The delay tactic cannot be employed. On the other hand, the Shuttle has provided many other benefits: reflights with refurbishment of instruments and systems; hands-on in orbit for corrective actions and operation; and, more important, it can carry many payloads and ancillary devices and systems.

It is proposed to utilize gaseous purging available from containers attached to payloads or from service facilities on the Shuttle to prevent contamination of an instrument. The purge can disperse and prevent the entrance of external contaminants to critical locations of the instrument while it is on the ground and during launch, re-entry and the early hours of the flight when contamination hazards are at their maximum. It can also prevent self-contamination of the instrument by reducing the outgassing of the internal surfaces, and by sweeping out those outgassed molecules and residual contaminants from surfaces. The benefits, limitations, and operational parameters have been discussed in the paper. An application of the purging is carried out for a large telescope which will be making solar observation from the Shuttle bay. Several factors which influence the choice of the purging parameters have been examined, namely: 1) the venting requirements necessary to limit the pressure differentials caused by the rapid pressure drop occurring during the transonic region of the launch; 2) the operational constraints on the length of time during which the purging and the natural self-outgassing of the instrument can be accomplished; 3) the quantity, volume, and pressure of the purging gas which can be carried in orbit; 4) the gas flow which would not create in the bay a high pressure while the bay doors are closed, and would not be a source of propulsive effect. The analysis examines the following parameters: 1) the vent area requirements for the launch conditions, 2) the diffusion of external gases into a pressurized volume via a vent passage, 3) the partial pressure of external gases which can build up in a volume under the purging conditions, 4) the outgassing of materials in the volume under pressure, 5) the deposit of contaminant and its effect on the absorptive property of the surface, 6) the flow requirements to maintain a pressure in the volume, and finally 7) the sizes and velocities of particulates which can be stopped and blown away by the purge gas velocity and density. The results of the purging analysis as applied to the telescope protection during ground launch, orbit and deorbiting phases of the flight are reported. A summary and recommendation completes the paper.

Venting Area

The flowfields surrounding a structure during atmospheric flights produce pressure distributions which determine the rates of energy and mass transfer through vents. The flow rates through the vents are sensitive to the pressure variations at the vents. Insufficient venting and improper locations may subject a compartment to high pressure differentials and structural failures. The differential pressures can be limited by using favorable vent paths, vent areas, orifice configurations and locations. A suitable approach to the vent of a compartment is to design the compartment to have a flow response time comparable or faster than the external flowfield disturbance time. The author employs the following to estimate the pressure differential at a volume produced by an external pressure drop rate dP/dt occurring at a base pressure P_0. Assuming isothermal flow conditions and small pressure differentials, the conservation of mass applied to the gas in volume V venting through an orifice of area A and discharge coefficient C_D is $d(\rho V)/dt = \rho v A C_D$, where ρ is the gas density and v its velocity at the orifice. The density in terms of pressure and temperature using the gas law is $\rho = P/RT$ and its derivative is $d\rho/dt = 1/RT \cdot dP/dt$. The gas velocity is $v = \sqrt{2gh} = \sqrt{2g(\Delta P/\rho)} = \sqrt{2gRT(\Delta P/P)}$. The substitution of these expressions in the above relation where V, A, and C_D are constants, give

$$\frac{dP}{dt} = \frac{A\ C_D\ P}{V} \sqrt{2gRT(\Delta P/P)} \qquad (1)$$

Solving for ΔP, one has

$$\Delta P = \frac{1}{2gRT} \left(\frac{V}{A\ C_D}\right)^2 \frac{(dP/dt)^2}{P} \qquad (2)$$

or the ratio of $V/A\ C_D$ which produces a pressure differential when the external pressure drop rate is $dP/dt = \dot{P}$ and the pressure is $P = P_0$, is given by

$$\left(\frac{V}{A\ C_D}\right) = \sqrt{2gRT(P_0 \Delta P/\dot{P}^2)} \quad (m) \qquad (3)$$

In this expression $g = 9.81$ (m/s^2) is the acceleration of gravity, $R = 29.2$ (m/K) is the gas constant for air and T(K) is the gas temperature. The orifice coefficient C_D can be taken conservatively to be about 0.6. It can be shown that these relationships relate the time constant of the volume-orifice system with the time involved in the external flowfield change. The evaluation of the required vent area requires a knowledge of $\dot{P}$ and P_0 and the specification of an acceptable ΔP. Experimental measurements of pressures vs time on a spacecraft have been used to specify $\dot{P}$ and P_0. Alternatively, flowfield analyses based on configuration and trajectory of the vehicle have provided these parameters.

The above equation provides the size of the vent area which allows a certain pressure differential to occur in a volume during the launch phase of the flight. Unless the vent area can be changed during other phases of flight and on the ground, the area size so established, becomes basic to the purging flow requirements and the protection against contaminants in a system.

Penetration and Diffusion of an External Gas into a Pressurized Volume

The mass increase with time into a volume $V (cm^3)$ via a film of thickness d (cm) and area A (cm^2) of a gas at density ρ_0 (g/cm^3) can be expressed by the equation

$$V \frac{d\rho}{dt} = \frac{DA}{d}(\rho_0 - \rho) \qquad (4)$$

where D (cm^2/s) is the diffusion coefficient of the entering gas through another gas representing the film. The solution of this equation for $\rho = \rho_0$ at $t = 0$, which can be verified by substitution, is

$$\rho = \rho_0(1 - e^{-t/\tau}) \qquad (5)$$

where $\tau_p = dV/DA$ (s) is the time constant which corresponds to the time when $\rho/\rho_0 = (1 - 1/e) = 0.632$. The diffusion coefficient from kinetic theory of gases, (Ref. 1) is $D = \lambda c/3$, where c (cm/s) is the average velocity of the gas, and λ (cm) is the mean free path of the gas. The velocity is given by $c = \sqrt{8kT/\pi m}$ and the m. f. p. by $\lambda = kT/\pi\sqrt{2}\delta^2 P$, where k is the Boltzman constant and the characteristics of the gas are given by the temperature $T(K)$, the molecular mass m (g/molec), the molecular diameter δ (cm), and P (Torr) is the pressure. The expression for the diffusion coefficient with these substitutions becomes

$$D = \frac{2}{3} \frac{1}{\delta^2 P} \sqrt{\frac{k^3 T^3 N}{\pi^3 M}} \quad (cm^2/s) \qquad (6)$$

where N the Avogadro number and M the mole mass have replaced $m = M/N$. An equivalent value or reduced mass of M and δ is needed to calculate the diffusion coefficient of two gases. The diameter δ must reflect combined forces between the two molecules in addition to the molecular sizes. Experimental data are available in the literature (Refs. 2

and 3) giving the coefficient D_0 for several combinations of gases at a temperature T_0 and pressure P_0. An approximate value of the coefficient at different temperature T and press P and for the gases having an equivalent molecular mass M can be calculated by modifying the theoretical expression for D, i.e.,

$$D = D_0 \frac{P_0}{P} \left(\frac{M}{M_0}\right)^{1/2} \left(\frac{T_0}{T}\right)^{3/2} \quad (7)$$

The experimental value of D_0 for H_2O diffusing into air atmospheric pressure (P_0 = 760 Torr) and at T = 273 + 20 = 293 K is D_0 = 0.24 cm²/s.

The time constant τ_p = dV/AD applicable to the diffusion of a gas through a layer, d of air into a volume V can be written with the above substitution as

$$\tau_p = \frac{dV}{AD} = \frac{P}{D_0 P_0} \left(\frac{M_0}{M}\right)^{1/2} \left(\frac{T}{T_0}\right)^{3/2} \frac{dV}{A} \quad (s) \quad (8)$$

and for water vapor diffusing into air at 20°C, the time constant is τ_p = (P/0.24P_0)(dV/A). This expression which reflects the pressure, the opening area, the passage thickness, and the volume has been plotted in Fig. 1 in terms of V/A and pressure P. The plot is for d = 1 cm so that for other values of d, one should multiply by d (cm) and for

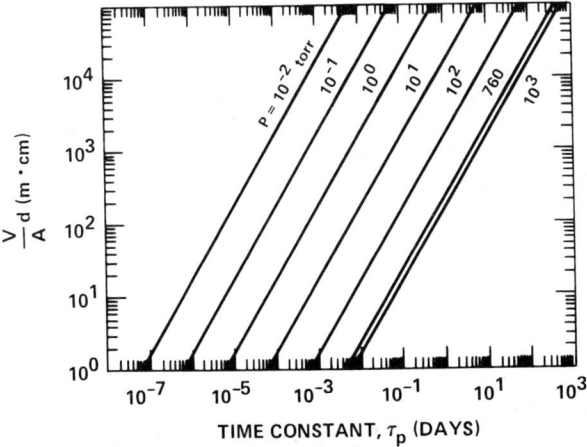

Fig.1 Time constant for penetration of H_2O in a volume V (m³) with air at pressure (Torr) via an opening of area A (m²) and depth d (cm).

other temperature and masses by the ratio shown by the theoretical equation for D.

The previous time constant reflects the penetration through the entrance to the volume. It may be necessary for a large volume to consider the variation of the density ρ at a location ℓ from the surface of the volume or in the present case from the surface immediately after the entrance. This can be estimated using the time-dependent, one-dimensional diffusion equation $d\rho/dt = Dd^2\rho/dx^2$, where D is the diffusion coefficient. The solution of this equation for the uniform diffusion of the density ρ_0 from a large surface to a surface 2ℓ distant from it is (Ref. 1, page 581).

$$\frac{\rho}{\rho_0} = 1 - \frac{8}{\pi^2} \left[e^{-(\pi^2 D/4\ell^2)t} + e^{-(9\pi^2 D/4\ell^2)t} + e^{-(28\pi^2 D/4\ell^2)t} + \ldots \right] \quad (9)$$

It is shown that for $Dt/\ell^2 > 0.089$, the second term in bracket is less than 2% of the first term.

Hence for long time t and relatively small ℓ, the solution to a first approximation can be written as

$$\frac{\rho}{\rho_0} = 1 - \frac{8}{\pi^2} e^{-t/\tau_D} \quad (10)$$

where $\tau_D = 4\ell^2/\pi^2 D$ is a characteristic time. For $t = \tau_D$ the density is $\rho \approx 0.70 \rho_0$, so τ_D has not the same significance of the time constant defined as the time when $\rho \approx 0.64 \rho_0$. The error which will be made with the assumption that the characteristic time has the same meaning of the usual time constant, can be determined by its comparison to the penetration time constant. This pseudo-time constant can be expressed with the substitution for D as

$$\tau_D = \frac{4\ell^2}{\pi^2 D} = \frac{P}{P_0 D_0} \left(\frac{M_0}{M}\right)^{1/2} \left(\frac{T}{T_0}\right)^{3/2} \left[\frac{4\ell^2}{\pi^2}\right] (s) \quad (11)$$

The above has been plotted in Fig. 2 for the diffusion of water vapor into air at 20°C in terms of the pressure P and for ℓ varying from 0.2-6 m.

With the above assumptions, the total time constant τ for a gas to penetrate the entrance of the volume, fill it up, and diffuse to a distance ℓ is the sum of the two time constants since the flow resistances are in series, or

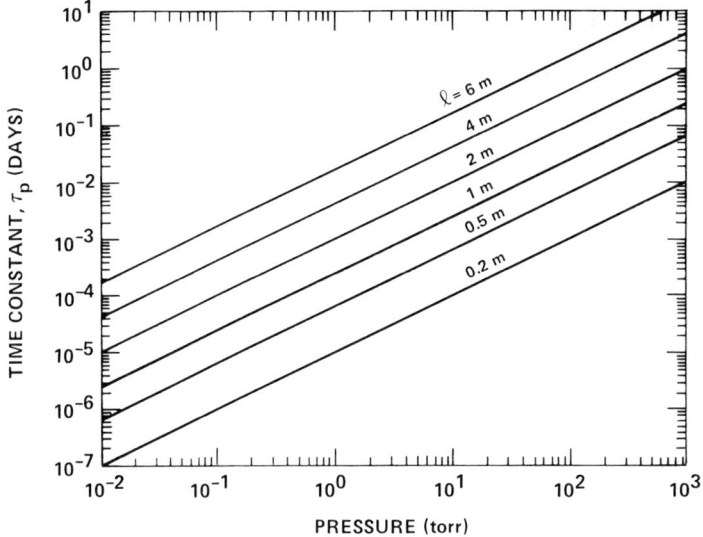

Fig.2 Pseudo-time constant for H_2O diffusing to a distance $\ell(m)$ in a large section volume containing air at pressure P (Torr).

$$\tau = \tau_p + \tau_D = \frac{Vd}{DA} + \frac{4\ell^2}{\pi^2 D} = \frac{P}{P_0 D}\left(\frac{M_0}{M}\right)^{1/2}\left(\frac{T}{T_0}\right)^{3/2}\left[\frac{Vd}{A} + \frac{4\ell^2}{\pi^2}\right] \quad (12)$$

A comparison of the values of τ_p and τ_D can be obtained by using these data: $V = 350$ m^3, $A = 0.5$ m^2, $d = 0.1$ m, $\ell = 2.5$ m, $P = 760$ Torr, and $D_0 = 0.24$ cm^2/s. These values are representative of the Shuttle bay and the time constants which may represent the time for the diffusion of water vapor in the Shuttle bay at the normal condition of pressure and temperature. It is found that τ_p is about 33.7 days and τ_D about 0.305 day indicating a total time constant of about 34 days. The time constant, disregarding the diffusion, would be about 3.6 days with $d = 10^{-2}$ m. This compares favorably to the time constants obtained from a series of experiments on humidity infiltration in a volume via thin orifices (Ref. 4).

Equilibrium Partial Pressure of a Contaminant Gas in a Volume Under Purging Condition

The partial pressure P of a gas at pressure P_0 infiltrating and diffusing in a volume, while a purging flow Q is opposing the infiltration, can be obtained from the

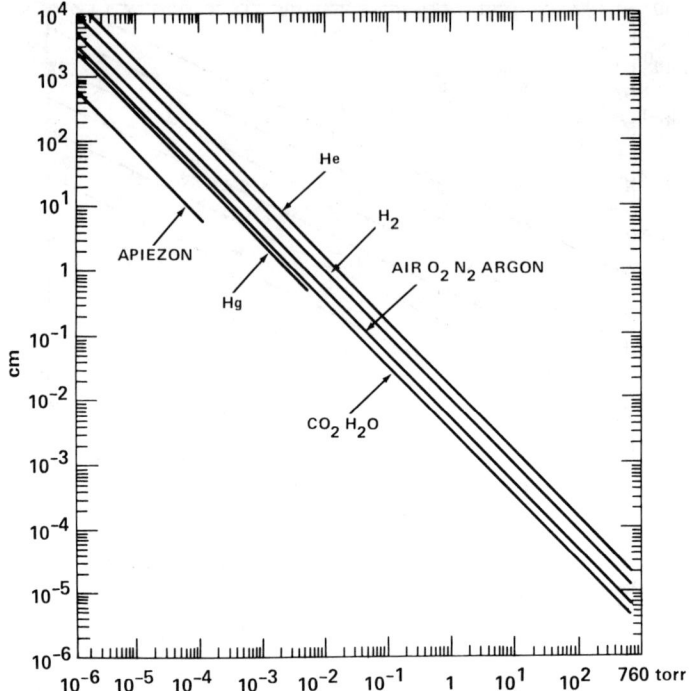

Fig.3 Mean free path as a function of pressure at 20°C.

equation expressing a mass balance in the volume under isothermal conditions

$$V \frac{dP}{dt} = C(P_0 - P) - QP \qquad (13)$$

where C is the conductance for the inflow of the contaminant. Rearranging the equation and noting that $C/V = 1/\tau$ is the infiltration and diffusion time constant and $Q/V = 1/t_p$ is the change of purged volume per unit time, the above equation can be written

$$\frac{dP}{dt} = \frac{P_0 - P}{\tau} - \frac{P}{t_p} \qquad (14)$$

This equation can be solved to provide the contaminant pressure in the volume as a function of time. However, the equilibrium maximum pressure, which can occur when $dP/dt = 0$, will be

$$P = t_p/(t_p + \tau) P_0 \qquad (15)$$

If $t_p \ll \tau$, and one ignores the diffusion portion of the infiltration time constant, the contaminant pressure in the volume as given by the above equation can be approximate by $P \sim (D/\rho vd)P_0$, where D is given by equation (7), v is the exit velocity of the purging gas at density ρ, and d is the depth of the passage.

Effect of Pressure on Outgassing of Materials

The outgassing of materials is characterized by:
1) The nature of the material; for a given material, the outgassing is a function of the surface treatment and the atmospheric composition to which it was exposed.
2) The outgassing is proportional to the surface area and increases exponentially with temperature and decreases slowly with time and apparently never becomes zero at ambient temperature.
3) The outgassing rate of a material is independent of the total gaseous pressure on the surface when this pressure is less than about 10^{-3} Torr. This behavior has not been established with complete certainty but it is valid as a first approximation.

The effect of the total pressure on the outgassing can be evaluated by considering the "survival relation" which gives the probability of the number of molecules N out of a total N_0 which will reach a distance x from their origin given that the mean free path of the gas is λ. The equation is $N = N_0 e^{-x/\lambda}$. The number of molecules which will have made a collision and are scattered within the distance x are, $N_0 - N = N_0(1 - e^{-x/\lambda})$ so that for $x = \lambda$, 64% of the molecules will be scattered. The mean free path or average distance per collision for a gas at density n (cm^{-3}), pressure P (Torr) and temperature T(K) is given by $\lambda = 1/\sqrt{2}\pi\sigma^2 n = kT/\sqrt{2}\pi\sigma^2 P$ (cm), where k is the Boltzmann constant and σ (cm) is the effective molecular diameter. A plot of λ as a function of pressure for several gases is shown in Fig. 3. A large size molecule or a chain of molecules as those released by polymeric materials, would have a smaller mean free path than that offered by air or N_2 to a similar molecule. The mean free path for a N_2 purging pressure at 1 Torr is about 4×10^{-3} cm and 64% of the outgassed molecules would be scattered within this distance. A large fraction of those scattered molecules would return to their surface of origin and the outgassing rate would be reduced. Few data are available to indicate the reduction of outgassing or of distillation-evaporation at high total pressures. Dushman (Ref. 1, page 21) reports data obtained by Hickman on the distillation of Octoil at temperatures of 368, 383, and 393 K with residual air pressures on the Octoil varying

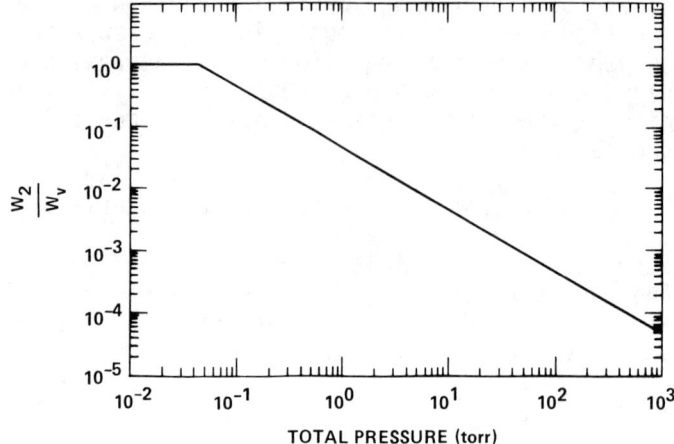

Fig.4 Estimated fraction of maximum material outgassing rates as a function of the total pressure on the surface.

from 3×10^{-4} to 5×10^{-2} Torr. The data show for example that when the saturated vapor pressure of the Octoil was 1×10^{-2} Torr, and the air pressure on the surface was 5×10^{-2} Torr, the rate of distillation was 20% of that with a residual gas pressure of 3×10^{-4} Torr. Another evidence of the effect of pressure on reducing outgassing can be obtained by comparing the rate of evaporation of water under vacuum and at atmospheric pressure. A semiempirical equation for the evaporation of water per unit area when the atmospheric pressure is P_o is Ref. (5).

$$W_a = (\alpha/P_o)(P_s - P_{va}) \quad (kg/h/m^2) \qquad (16)$$

where $\alpha = 24$ kg/h/m² for no wind conditions and $\alpha = 29$ when wind exists, P_s is the saturation vapor pressure of the water at temperature T, and P_{va} the vapor pressure in the air. This equation can be compared to the rate under vacuum given by the Langmuir equation

$$W_v = 5.833 \times 10^{-2} \sqrt{M/T} \; (P_s - P_{va}) \quad (g/cm^2/s)$$

$$= 5.18 \times 10^2 \; (P_s - P_{va}) \quad (kg/h/m^2) \qquad (17)$$

where M = 18 (g/mole), T = 293 K and P (Torr). If one lets $P_{va} = 0$ in both the equations, the ratio indicates that

$$W_a = (4.63 \times 10^{-2}/P_o) W_v \qquad (18)$$

where P_0 (Torr). This relation which has been plotted in Figure 4 for convenience, indicates that at $P_0 = 760$, $W_a = 6.09 \times 10^{-5}$ W_v, and the two rates are equal when $P_0 = 4.62 \times 10^{-2}$ Torr. According to Ref. (6) the rate of water evaporation at STP is probably 10^{-2}-10^{-3}; the rate under vacuum and according to Ref. (7) is about 10^{-4}. In any case, all these considerations show that the outgassing will be reduced considerably under a total pressure which provides a small mean free path. Two more effects of high pressure are important. The survival relation shows also that the flux $\phi = \phi_0 e^{-x/\lambda}$ at a distance x from the source when the m. f. p. is λ, so at $x = 10\lambda$, for example, the contaminant outgassing is a very small fraction of the surface flux and it has been mixed randomly within the gas. If a purging is present these scattered molecules will also be transported by the purging flow.

Particles Rejection by a Vent Gas

The velocities of particulates approaching an instrument may be:

1) Terminal velocities of particles falling in a gravity field through a medium at density ρ and viscosity η. As shown in Fig. 5, terminal velocities of large particles (~ 0.1 cm) may be 4-5 m/s at normal ambient conditions.

2) Velocities which the particles have acquired from a gaseous medium as for example by wind or by cleanroom air flows. Particles velocities in horizontal cleanroom facilities may be of the order of 1 m/s.

3) Velocities of particles produced by propulsive systems. These particles may approach sonic velocities or supersonic velocities ($v = \sqrt{2\gamma RT/\gamma-1}$) of gases at T(K) with specific heat ratios γ which can be of the order of 2-3 km/s.

4) Velocities of particles ejected from the impact of meteorites on surfaces. The meteorites may be traveling at average velocities of 15-20 km/s. Some ejecta may acquire these velocities if the meteorites have sufficient energy to overcome the various forces holding the particles (van der Walls, Coulombic, adhesion, etc.) and to impart a kinetic energy to the released particles. However, it is estimated that in general particles released in orbit by the Shuttle are a few m/s (Ref. 8).

5) Velocities of particles ingested via Shuttle filters during re-entry which could have values approaching re-entry velocities.

In order to evaluate the effectiveness of a purging gas in preventing particles from entering an instrument via a vent area, the following analysis has been carried out.

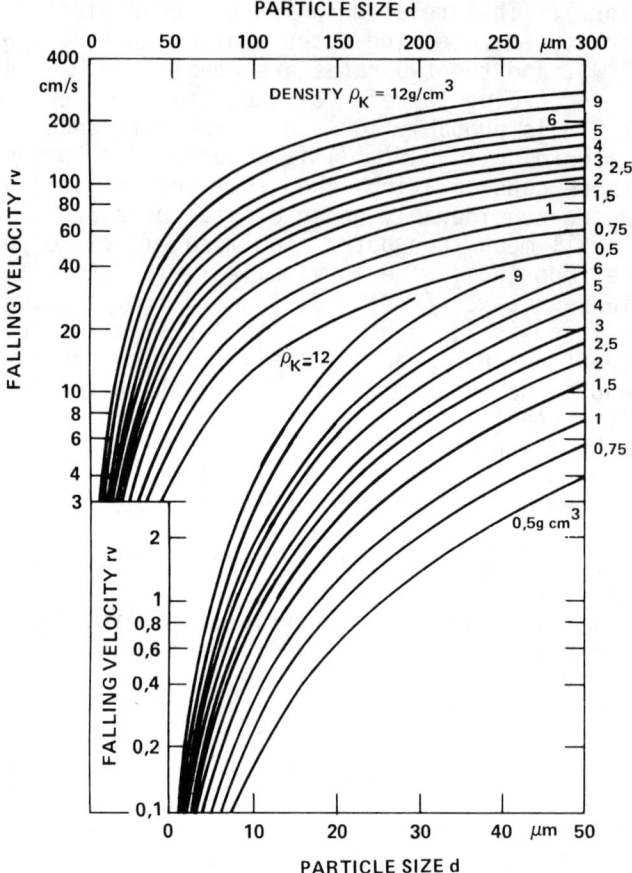

Fig.5 Falling velocity of spherical particles in air.

The terminal velocity of a falling particle is obtained when the particle weight equals the drag force on the particle, i.e., when $mg = F_D$. The drag force is a function of the Reynold's number $R = \rho v d/\eta$, where ρ (kg/m^3) and η (Pa·s) are the medium density and viscosity, respectively, v (m/s) and d (m) are the velocity and diameter of the particle. The dynamic viscosity varies with temperature and will vary with pressure when the particle dimensions are comparable to the mean free path in the medium. For a $R < 10$, the drag force is given by Stokes law, $F_D = 3\pi\eta dv$. For $10 \leq R \geq 10^5$, the drag is $F_D = C_D(\pi d^2/4)\rho v^2$, where C_D the drag coefficient is a function of R and the shape of the particle (Ref. 9). The terminal velocities are then given by

$$v_t = \rho_p g d^2/18\eta \quad (m/s) \tag{19}$$

if Stokes law applies or, in the other case by

$$v_t = \sqrt{\frac{4}{3} \frac{d}{\rho} \frac{\rho_p g}{C_D}} \quad (m/s) \qquad (20)$$

In these equations ρ_p (kg/m^3) is the density of the particle. If one assumes that a particle is moving at a velocity v_e through a gas moving in the opposite direction at velocity v_a, the drag force on the particle will be $F_D = k(v_a + v_e)$. The constant k is the constant in Stokes force and $(v_a + v_e)$ is the velocity of the particle relative to the gas stream. The equilibrium equation in that case will be $k(v_a + v_e) = mg$. If $v_a = 0$, then $v_e = v_t$, the terminal velocity. For $v_a \neq 0$, dividing $k(v_a + v_e) = mg$ by $kv_t = mg$, one obtains $v_a + v_e/v_t = 1$. This indicates that to stop the approaching particle ($v_t = 0$), one needs a gas velocity $v_a = -v_e$. For the purge gas to make the particle move away with a velocity equal and opposite the terminal velocity ($v_e = -v_t$), one needs a purge gas velocity ($v_a = 2v_t$), i.e., two times the terminal velocity.

For the calculations on the sizes of the particles which can be stopped by a purge flow, one may assume conservatively, that only those which can be given a velocity $v_a/2 = v_t$ will be prevented from entering the instrument. This assumption should insure that the arresting will occur far from the vent exit and compensate for the fact that the above analysis applies to the case of Stokes drag and not the case when the drag is proportional to the square of the particle velocity. With these assumptions, the size of the particles which can be prevented from entering can be estimated using the previous equations for terminal velocities with the substitutions $v_t = v_a/2$, i.e.,

$$d = \sqrt{\frac{18\eta}{\rho_p g} \frac{v_a}{2}} \quad (m) \qquad (21)$$

or

$$d = \frac{3}{4} \frac{C_D \rho}{\rho_p g} \left(\frac{v_a}{2}\right)^2 \quad (m) \qquad (22)$$

The appropriate equations to be used is determined by the Reynolds number. One may note that the maximum velocity of a gas at the throat of an orifice will be a sonic velocity ($\sim$310 m/s for Nitrogen gas at normal temperature). So, certain particles moving at these velocities can be stopped by the purging flow.

Rejection of Natural Atoms and Molecules at the Entrance of an Open Instrument

An instrument open to space such as a telescope is exposed to the flux of the natural atomic and molecular gases. The flux which is the product of the density and the relative velocity between the instrument and gas can be very high when the instrument is pointing into the velocity vector of the spacecraft. As recently reported in Ref. (10), fluxes of atomic oxygen at Shuttle orbital altitudes have been sufficiently high to cause oxidation and aging of several materials. The losses have been quite high considering the relatively short time ($\sim 3 \times 10^4$ s) exposure of the materials to the high oxygen fluxes.

The avoidance of these high fluxes into an instrument or over a surface can be prevented with a low purging non-contaminating gas. The purge gas must scatter and reflect the incoming gas within a convenient distance from a critical surface. This purging gas, as reasoned in Ref. (11), or by using a modified "survival equation" should have a mean free path

$$\lambda \leq (Lv/v_p) \qquad (23)$$

where L (cm) is the distance within which the scattering has to occur, v (cm/s) is the purging gas velocity, and v_p (cm/s) the relative velocity of the natural gas with respect to the instrument or surface ($v_p \sim 8$ km/s in the velocity vector). The pressure corresponding to λ is given by the relation indicated previously or obtained from Fig. 3.

Flow Requirements and Exit Flow Velocities

The contamination prevention against gaseous and particulate contaminants either external or internal to an instrument is predicated on the internal pressure and the exit velocity of the purging gas. The purging gas is either dry air or gaseous nitrogen. The quantity of purge gas is a function of the vent area, $A(m^2)$ the pressure to be maintained P (Torr) upstream of the vent area and the downstream pressure P_o (Torr). For air or nitrogen, when $P_o > 0.53$ P, the viscous flow of gas at standard pressure and temperature (SPT) from an orifice with small pressure differentials is given by

$$Q = \psi A v = \psi A \sqrt{2gRT_o(P-P_o/P_o)} \quad (m^3/s) \qquad (24)$$

which is obtained from the Torricelli equation. The coefficient ψ, which accounts for a velocity coefficient and vena contracta, may vary from 0.64 to 0.98 for an orifice. The other parameters are the temperature of the gas $T_0(K)$, the acceleration of gravity g (m/s^2), and the gas constant R = 29.26 (m/K) for air. The flow velocity at the exit for the same conditions of $P_0 > 0.53$ P is

$$v = \alpha\sqrt{2gRT_0(P-P_0/P_0)} = 24\alpha\sqrt{T_0(P-P_0/P_0)} \quad (m/s) \quad (25)$$

The velocity coefficient α is about 0.98 for an orifice. The continuum gas flow rate at SPT when $P_0 < 0.53$ P is from the "Fliegner equation" using A(m^2), T(K), and P (Torr)

$$Q = 4.34 \; C_D \; (AP/\sqrt{T_0}) \quad (m^3/s) \quad (26)$$

where C_D is the discharge coefficient. The above for T_0 = 293 K reduces to Q = 0.253 $C_D AP$(m^3/s) = 15.22 $C_D AP$(m^3/min). This equation normalized with the volume has been plotted in Fig. 6. The normalization provides the number of volume changes per unit time Q/V (m^3/m^3/min), or the time needed for one complete volume change t_p = V/Q (min) as a function of the purging pressures and the sizes of the orifice. The plot includes also the flow required when molecular flow conditions exists. This plot includes also the flow required when molecular flow conditions exist. This occurs when the gas m. f. p. is about 10 times the diameter of the orifice. The flow of atmospheric air at 20°C is then Q = 8.8 AP (m^3/min). The exit flow velocity for the continuum flow regime when $P_0 < 0.53$ P is sonic at the orifice and is given by

$$v = \alpha\sqrt{(2g\gamma/\gamma+1)RT_0} = 18.3\alpha\sqrt{T_0} \quad (m/s) \quad (27)$$

where $\gamma = c_p/c_v = 1.4$ is the ratio of the specific heats for air, and $\alpha \sim 0.98$ is the velocity coefficient. The other parameters are as indicated above.

Contaminant Deposits and its Optical Effect

The deposit on a surface by a contaminant with a vapor pressure less than the saturated vapor pressure corresponding to the temperature of the surface can be estimated using the Langmuir equation for adsorption or more accurately using the BET relation Ref. (12). The Langmuir adsorption isotherm is

$$\sigma = \gamma\phi\tau = \gamma(5.83\times10^{-2}\sqrt{M/TP})\tau_0 \; e^{-E/RT} \quad (g/cm^2) \quad (28)$$

where $\gamma \approx 1$ is a sticking coefficient, ϕ (g/cm²/s) is the flux of contaminant impinging on the surface, M (g/mole) is its molecular mass, T(K) its temperature, P (Torr) its partial pressure, and τ is a residence time of the molecule on the surface. The residence time is $\tau = \tau_0 \, e^{E/RT}$, where $\tau_0 \approx 10^{-13}$ (s) is an oscillation period of the molecule on the surface, E (cal/mole) is the activation or binding energy of the molecule on the surface at temperature T(K), and R (cal/mole K) is the gas constant. The activation energy may have a value approximately the same as the heat of evaporation if the adsorption involves physical forces and if many layers of deposits have been accumulated on the surface. Or, it may be several times the heat of evaporation if the adsorption involves chemical bonding.

For the case when the contaminant gas partial pressure is greater than its saturated vapor pressure at the surface

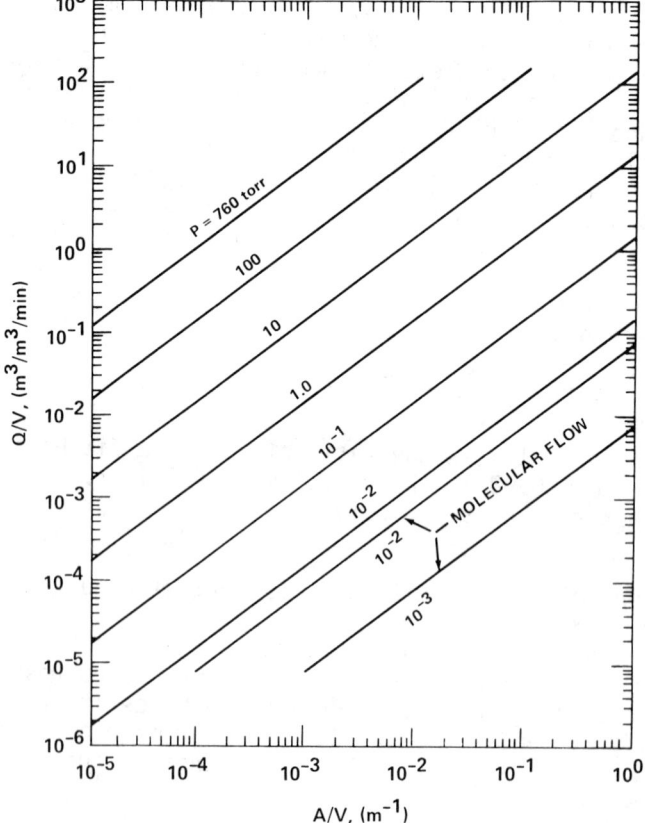

Fig.6 Volume air changes Q/V (min⁻¹) vs vent area volume ratio A/V (m⁻¹) as a function of internal pressure.

which exists, in general, when cryogenic surfaces are present, the accumulation on the surface is a condensation process and can be calculated by summing the net flux

$$\sigma = \int_{t_0}^{t} (\gamma\phi - \phi_s) dt = k \int_{t_0}^{t} (\gamma P - P_s) dt \quad (g/cm^2) \quad (29)$$

where ϕ is the incident flux and ϕ_s is the flux of contaminant departing the surface at its saturated conditions.

The optical effect of contaminant deposits on a surface is difficult to evaluate because one must know the complex index of refraction of the contaminant and substrate. The effect changes with the radiation wavelength, the contaminant thickness, and distribution on the surface. An estimate of the absorption at certain wavelengths by contaminants can be obtained by applying the Lambert relation. The loss of incident intensity through a thickness t (A) of material which has an absorption coefficient α (A^{-1}) at certain wavelengths is

$$(I - I_0)/I_0 = (1 - e^{-\alpha t}) \quad (30)$$

The absorption coefficient will change if the chemical nature of the deposit is modified by certain radiations or energetic particles.

Application

The results of the previous analysis have been applied to the contaminant protection of a solar telescope which will be carried in space by the Shuttle. The telescope will remain in orbit for about a week and will be making sun observations from the Shuttle bay. Solar energy in the telescope will be very high and part of it must be rejected via mirrors back into space. Contaminants on those surfaces can affect negatively the rejection requirements and introduce instrument loss of sensitivity in several regions of the spectrum, especially in the UV region. The telescope tube with communicating volumes has been estimated to have a volume of about 16 m³ and to be about 6-m long. The telescope closure door which for protection reasons will remain closed during ground, launch, initial orbital phase, and return has a diameter of 1.3 m.

The calculations and application of the theory which follow are preliminary. However, they indicate that the proposed purging is feasible and can protect the telescope against gaseous and particulate contaminants external and

Table 1 Summary of SOT Purging Requirements[a]

Parameter	Ground[b]	Launch	Orbit 0-1 hr	Orbit 1-24 hr	Deorbit Landing	Comments
Vent Area (cm^2)	159 (24 in^2)	159	159	1.3×10^4	159	Vent area required for $\Delta P < 0.25$ psi during launch phase
Purge Press (torr)	13 (0.25 psi)	$P_{AMB} + 1$	1	-	$P_{AMB} + 1$	P_{AMB} = variable press in bay
Purge Volume Rate (m^3/s)	0.53	$0.14 - 4.2 \times 10^{-3}$	4.2×10^{-3}	-	$4.2 \times 10^{-3} - 0.14$	Quantity for 1 hr purge: 0.1m^3 at 2000 psi
Flow Velocity (m/s)	52.6	14.6 - 309	309	-	309 - 14.6	
Time/Volume Change	30 sec	114.3 sec - 1.05 hr	1.05 hr	-	1.05 hr - 114.3 sec	
Contaminant Infiltration Time Constants	46d / 218d	45.6d - 144 hr / 214d - 6.78 hr	1.44 hr / 6.78 hr	0 / 0	1.44 hr - 45.6d / 6.78 hr - 214d	H_2O M = 400 kg/mole
Fraction of Contamination Pressure in Volume at S. State	7.5×10^{-6} / 1.3×10^{-6}	$2.9 \times 10^{-5} - 0.42$ / $6.1 \times 10^{-6} - 0.13$	0.42 / 0.13	- / -	$0.42 - 2.9 \times 10^{-5}$ / $0.13 - 6.1 \times 10^{-6}$	H_2O M = 400 kg/mole
Pressure in Bay (torr)	-	$761 - 3 \times 10^{-2}$	3×10^{-2}	-	$3 \times 10^{-2} - 761$	Pressure produced by purge w/bay door closed
Purge Gas Mean Free Path (cm)	6×10^{-6}	$6 \times 10^{-6} - 4 \times 10^{-3}$	4×10^{-3}	-	$4 \times 10^{-3} - 6 \times 10^{-6}$	In cavity
Internal Outgass	6×10^{-5}	$6 \times 10^{-5} - 4 \times 10^{-2}$	4×10^{-2}	ϕ_{vac}	$4 \times 10^{-2} - 6 \times 10^{-5}$	Fraction of rates in vacuum
Gas Contamination Deposit	-	-	3 - 10Å	-	-	On 20°-degradation: 8% absorption at U.V.
Dia. of Particles Blown Away (cm)	1.2	$9.5 \times 10^{-2} - 1.4 \times 10^{-1}$	1.4×10^{-1}	-	$1.4 \times 10^{-1} - 9 \times 10^{-2}$	Maximum dia. for particles at 26 m/s (ground) & 155 m/s (orbit)

[a]Purge Cavity: Volume - 16 m^3, Length - 6 m, Cavity Door - 1.3 m^2, Accept. $\Delta P = 0.25$ psi
Shuttle Flight: Max $\Delta P/\Delta t = 0.457$ psi/s at 14 psi
Vent Requirements w/Cavity Door Closed: V/A = 10^3, depth d = 8 cm

[b]Ground purge as shown provides large protection at expense of large flow. Purge pressure can be lowered.

internal to the telescope, during the most critical periods of contaminant hazards. These periods are ground operations, launch, early hours in orbit, and return from orbit.

The results of the calculations which are indicated below have been shown in summary form in Table 1.

Telescope Tube Vent Area Requirements

The vent area required to prevent a hoop pressure differential across the tube can be estimated from Eq. (3). With the assumption that the tube structure can withstand a maximum $\Delta P = 0.25$ psi (13 Torr) while flying through the transonic regime of the flight at an ambient pressure of about 14 psi and the rate of pressure drop is about 0.457 psi/s (Ref. 13), the required volume vent area ratio with a coefficient of about 0.6 is $V/A = 1.67 \times 10^3 \; C_D \sim 10^3$. The required vent area for $V = 16 \; m^3$ is then $A \sim 159 \; cm^2$ (24.7 in^2). This total area will govern the entrance of contaminants and the purging flow through all of the operational phases of the telescope unless provisions are made to limit the area during other operational phases of the instrument.

Ground Purging Protection

If one decided to maintain a purge pressure of $P = 773$ Torr (0.25 psig) in the telescope and the entrance of contaminant is through the previously calculated 159.6 cm^2 with an assumed pipe depth of $d = 8$ cm, the time constant for a contaminant to infiltrate the passage and diffuse to a location $\ell = 6$ m away, is given by Eq. (12). The time constant for water vapor which has a diffusion coefficient $D_0 = 0.24 \; cm^2/s$ at 760 Torr and normal temperature has been calculated to be about 46.38 days where 39.2 days were taken by the contaminant to move through the passageway. The time constant for a contaminant having an $M = 400$ g/mole will be 218 days. These are the lengths of time required for the contaminants to build up to 64% of their external concentration if no purging flow existed. However, the required flow of purge gas to maintain 773 Torr in the telescope with the external pressure at 760 Torr (14.7 psi) will be $Q = 0.53 \; m^3/s$ at STP as calculated using Eq. (24). The exit velocity from Eq. (25) will be $v = 52.6$ m/s and the time for one change of volume will be $t_p = V/Q = 30$ s. The equilibrium maximum partial pressure of contaminant ($M = 400$ g/mole) which can exist in the telescope will be 7.5×10^{-6} of its external pressure and for water vapor 1.3×10^{-6}. These were calculated using Eq. (15). Therefore, there should be no contamination while the telescope is being purged.

The size of particles which will be rejected by the purge gas have been calculated using Eq. (22) with v_a = 52.6 m/s, ρ = 1.22 kg/m³, ρ_p = 2.6x10³ kg/m³ for sand, and $C_D \sim 0.5$ for the drag coefficient in the range of $10 < R < 10^5$ applicable to a spherical particle (Ref. 9). The calculation indicates that the particle size which could be blown away will be as large as 1.2 cm. The Reynolds number is about 21,730 for that size particle and for a dynamic viscosity for the air at normal temperature of $\eta = 18.1 \times 10^{-6}$ Pa·s.

It is apparent from these results that by using a 0.25-psig purging, one is overprotecting the volume at the expense of a large amount of purging gas. One may want to reduce the pressure and/or reduce temporarily the vent area during ground operation.

Launch Phase Protection

During the launch phase using a pressure controller, the purging pressure should be maintained at about 1-Torr higher than the decaying external pressure. This will provide a flow out of the volume which will provide an obstacle to external contaminants. The flow will be 0.14 m³/s with a velocity of 14.6 m/s at the start of the launch phase as calculated using Eqs. (24) and (25). When the external pressure $P_o < 0.53$ Torr (in orbit), the flow will be $Q = 4.2 \times 10^{-3}$ m³/s (Eq. 26) and the velocity will be 309 m/s (Eq. 27). The time constant for infiltration at 761 Torr (on the ground) will be τ 45.6 days for water and 214 days for M = 400, and with the purge period of t_p = 114.3 s the maximum pressure of these contaminants in the telescope will be $2.9 \times 10^{-5} P_o$ and $6.1 \times 10^{-6} P_o$, respectively. The particle sizes which will be rejected under these conditions will have diameters up to 9.5×10^{-2} cm with velocities up to 14 m/s. These results indicate that one could use the 1-Torr purging pressure during ground operations and the telescope would be sufficiently protected.

In Orbit Protection

At the end of the launch phase and in orbit, the purging pressure will be 1 Torr and the bay pressure will be less than 0.53 Torr. The in orbit purge pressure must provide a sufficiently high pressure to prevent incoming gaseous and particulate contaminants; arrest to a large extent the internal outgassing; and be as low as possible so that the required amount of gas is a minimum. This last requirement is important for the limitation on the amount of gas to be carried, and for the pressure which can be established in the Shuttle bay while the bay doors are closed. The 1-Torr pressure was selected with these considerations in mind.

The purge flow under the conditions, will be 4.2×10^{-3} m^3/s and its exit velocity 309 m/s. This flow rate is equivalent to $Q = 3.18 \times 10^3$ Torr l/s being emitted in the bay. The bay with closed doors has a venting area of about 0.5 m^2 which corresponds to a conductance in the viscous region of $C \approx 20 A = 10 \times 10^4$ l/s (Ref. 1). The equilibrium pressure in the bay is then $P = Q/C = 3.18 \times 10^{-2}$ Torr. This pressure may be cause for some concern. Voltage breakdowns under certain conditions could occur. However, other analyses have indicated that the bay pressure does not drop below this value until about 20 min in orbit. It drops to about 10^{-3} Torr and remains at that level for considerable time. Therefore, the additional pressure does not change substantially the bay pressure. After the bay doors open, the bay purging induced pressure will drop substantially. It appears that the flow is acceptable based on these considerations and the fact that the purging will last a limited amount of time as will be discussed later. If it is found unacceptable, one may provide an automatic partial closure of the vent area to limit the flow of gas into the bay. It is to be noted that the purging flow will also have some beneficial effects in arresting, mixing, and transporting the outgassing in the Shuttle bay.

The time constant τ as calculated for $P = 1$ Torr, for the infiltration of contaminants will be 1.44 h for water and 6.78 h for $M = 400$. These values and the volume change time of 1.05 h limit the maximum partial pressures of the water and that of the real contaminant ($M = 400$) to 0.42 and 0.13 respectively of their in bay pressures. These values were obtained by substitution in Eqs. (12) and (15) of $P = 1$ Torr, $V/A = 10^5$ cm, $d = 8$ cm, $\ell = 6 \times 10^2$ cm, $D_o = 0.24$ cm^2/s, $p_o = 760$ Torr, $T = T_o$, $M_o = 18$ g/mole, and $M = 400$ g/mole.

While this purge is on, at $P = 1$ Torr ($\lambda = 4 \times 10^{-3}$ cm), the outgassing of the internal surfaces will be about 4% of the values they would have if the total pressure on those surfaces was less than 4.6×10^{-2} Torr (Eq. 18). Also, some of the molecules outgassed will be transported out of the volume by the purge flow. One can reason that if the outgassing material inside the telescope is of the same nature of the many polymeric materials (such as RTV-566 adhesives, S-13G paints, and others) which have a saturated vapor pressure of about P_s = 3×10^{-8} Torr at normal temperature, than the actual vapor pressure in the purged volume will be less than $4\% \times 3\times10^{-8}$ = 1.2×10^{-9} Torr.

One must add to this pressure, the pressure created by the infiltration of the external contaminants. The in-bay partial pressure of a similar contaminant cannot exceed its saturated vapor pressure of 3×10^{-8} Torr. In fact, it must be less as dictated by its venting out of the bay, via the Shuttle side vents. It will be even less when the bay doors are open. If one assumes, however, conservatively that the outside contaminant partial pressure is 3×10^{-8}, then the infiltrated contaminant could have a maximum pressure of $3\times10^{-8} \times 0.13$ = 3.6×10^{-9} Torr in the volume. The total partial pressure of the contaminant in the volume will be summing the infiltrated and the self-produced, P = 3.6×10^{-9} + 1.2×10^{-9} = 4.8×10^{-9} Torr. This pressure determines the rate of impingement of contaminant on a surface in the volume. Since P is less than P_s, only an adsorption process can occur and the relationship to be used for the accumulation of deposits on a surface is given by Eq. (28). The polymeric material which is being considered as the contaminant (methyl-phenil-trisiloxane) has an activation energy of about 20 kcal/mole and M = 400 g/mole. Its sojourn time on a surface at T = 293 will be τ = $10^{-13} \, e^{20,000/1.98 \times 293}$ = 8.3×10^{1} s. The rate of impingement will be 5.83×10^{-2} $(400/293)^{\frac{1}{2}} \times 4.8\times10^{-9}$ = 3.269×10^{-10} g/cm^2/s and the deposit which can occur will be σ = $\gamma\phi\tau$ = $3.27\times10^{-10} \times 8.3\times10^{1}$ = 2.71×10^{-8} g/cm^2 with the sticking coefficient taken as one. The deposit will have a thickness of about 3 A assuming a density approximately equal to that of the water (ρ = g/cm^3). As the flux diminishes, some of this deposited contaminant may actually leave the surface.

With regards to the infiltration of particulates, those particles which are directed toward the vent at velocities of about 154 m/s and have diameters up to 1.4×10^{-1} cm will be rejected by the purge gas. The calculations have been carried out using Eq. (21) with η = 18.1×10^{-6} Pa·s, ρ_p = 2.65×10^{3} kg/m^3 and v = $v_a/2$ = 154.5 m/s. The Reynold's number was calculated to about 38, using a density ρ = $\rho_0/760$ = 1.6×10^{-3} kg/m^3 for the gas. These estimates indicate that the protection of the telescope surfaces against contaminants can be provided with 1 Torr. What remains to be determined is the length of time the purging should be carried out. This time is limited by the gas consumption and by the operational constraints. It is suggested that the purging be carried out for about an hour. The total amount of gas which will be used is about 0.1 m^3 at a pressure of 2000 psi. After 1 h in orbit, the total bay pressure with the bay doors open is estimated to be about 10^{-4} Torr (without purging) and the contaminant outgassing rates should drop by an order of magnitude (the outgassing rates often drop 1 order of magnitude after 1 h in vacuum and 2 orders in about 10 h). Further, with the bay open, the partial pressure of the characteristic contaminants discussed previously should be considerably less than 10^{-8} Torr. It is to be noted that many of the materials used will meet the criteria that their contaminant fractions be 0.1% of their mass loss when exposed at 125°C for 24 h. So, also based on this criteria, the partial pressure of contaminants should be 3 orders of magnitude lower than that of the total pressure. At the end of 1 h, while the purging is being terminated, the telescope should be raised from the bay, pointed away from the velocity vector, the cabin, the engines, and from the sun. The closure door should be fully open in this more rarefied environment which exists out of the bay. Full outgassing of the internal surfaces will result. However, the contaminant pressure in the telescope will be set by the conductance of the closure door (1.3-m diam.) and the outgassing surfaces, so $P \sim P_s(A_s/A_c)$, where P_s is the vapor pressure of the contaminant ($P_s \sim 10^{-8}$ Torr), A_s is the exposed surfaces of the outgassing sources, and A_c is the open door area. The pressure P cannot be greater than P_s and in fact, it will be much less due to the large vent area and the relatively small area of the organic materials. Under these conditions no additional deposits can occur on surfaces at 20°C or higher. Some deposits could occur only at a surface near and facing directly a source. This condition should be prevented whenever possible by blocking the field of view. This venting condition should be maintained as long as possible to ensure the depletion of most of the outgassing, before solar observation is carried out. Previous experiences would suggest that one should allow at least 24 h for the outgassing rate to become reasonably small.

As indicated, the maximum deposit of contaminant after having taken all the above precautions, should not exceed 3 A. Assuming, however, that the deposit is about 10 A, an estimate of the effect that this thickness would have on the absorption at 1215 A radiation indicates, using Eq.(30), that the loss in transmittance would be about 8.7%. The absorption coefficient was taken to be $\alpha = 9 \times 10^{-3}$ A^{-1} which was the average of the results on transmittance at that wavelength of many materials used in space application (Ref. 14).

De-Orbit and Landing Phase

A purging could be employed during the returning-phase of the flight. It would provide: 1) protection against particulates and other gases and vapors ingested by the Shuttle during the re-entry, 2) a heat conductive medium to transfer heat out of the telescope and 3) eliminate some of the expenses necessary for the refurbishment of the telescope. One would need to maintain a purging pressure slightly higher than the bay ambient pressure which would be increasing as the Shuttle descends. The purge supply system should have a response time as fast as the pressure increases in the bay and would require a pressure controller performing the reverse function of the one used for the launch phase. A method to restrict the telescope venting passage before the re-entry could facilitate this protective approach. This is possible because during re-entry, pressure changes occur more gradually than those during launch. The parameters for the two extremes of purging, in orbit and on ground, with 1-Torr pressure differentials are indicated in the summary table.

Conclusions and Recommendations

A method has been developed to evaluate the protection offered by gaseous purging of an instrument against external and internal gaseous and particulate contaminants. The protection is needed while the instrument is on the ground and in the Shuttle bay during launch, re-entry, and orbiting. Operational and other restrictions have been considered in the development.

The analysis of the results indicates that maintaining a pressure of about 1 Torr above the external environment in the instrument may afford sufficient protection against infiltration of gaseous and particulate contaminants and abate sufficiently the internal self-generated contaminant outgassing. The vent dimensions which are dictated by pressure differentials developed during launch, establish the purge gas flow and the gas quantity needed for the purging.

The prevention of contaminants infiltration has been evaluated in terms of the time constant for a gas to infiltrate a vent, diffuse in the volume, and the purge time for one volume change. The contaminant pressure which can be built in the volume is a function of the diffusion coefficient of a gas into another at different pressure, the size of the volume, the vent area and the length, and a characteristic dimension of the volume.

The self-contamination is prevented by the high pressure which reduces the outgassing rates of the materials in the volume. The total contaminant pressure in the volume is obtained by the superposition of the infiltrated and self-generated pressures. The total contaminant deposit on a surface can be estimated based on the pressure, chemical properties, and the temperature of the surface being contaminated. In turn, an estimate of the optical degradation of that surface is obtained using experimental data on the absorption of the contaminant at certain wavelengths.

The protection against particulate contaminants entering the volume is provided by the drag forces on the particles produced by the velocity, density, and viscosity of the purge gas at the vent exit.

The purge gas flow rates and velocities have been calculated in terms of the pressure differential, gas temperature, and molecular masses. Considerations on the pressures which the purge flow would produce in the Shuttle bay, on the amount of stored gas which can be carried along and on the length of time during which purging and free outgassing can be carried out without interfering with the instrument operational requirements, have been taken into account.

The findings of the analysis have been applied to a solar telescope which has a volume cavity of about 16 m^3 with an aperture door with an area of about 1.3 m to be opened on command. The summary Table 1 indicates the purging requirements, the suggested times for purging and free outgassing, and the protection which the purge will offer. The vent area which is needed because of structural integrity during Shuttle launch, has been calculated to be about 24 in^2. During ground operations and launch, the venting through the above area with a passageway about 8-cm long of the nitrogen gas at a few Torr higher than the ambient pressure, offer ample protection against gaseous and particulate contaminants. In orbit the purging at 1 Torr, will limit external and internal contaminant pressure to about 0.1 the pressures which would be otherwise expected. Particles of diameter up to 0.1 mm with speeds up to 150 m/s will be prevented from entering the volume. The purging is expected to induce the Shuttle bay with doors closed a pressure of about 2×10^{-2} Torr. The amount of purge gas for the recommended 1-h purging is equivalent to 0.1 m^3 at 2000 psi. The purging and the subsequent full venting are expected to allow contaminant deposits on volume surfaces of about 3-10 A. A 10 A deposit can produce about 8% absorption of uv radiations.

The following recommendations, based on the above analysis and other experiences, can be offered:

1) Internal and external surfaces should be maintained free of gaseous and particulate deposits during the various phases of manufacturing, assembly, tests, etc.

2) Materials employed with the instrument should be selected to have low outgassing rates with a minimal fraction of condensable products.

3) The exposed surfaces in the instrument of organic materials should be a minimum.

4) Critical surfaces should be shielded from known outgassing sources.

5) Critical surfaces should be kept (with heaters) at temperature slightly higher than the others. The heaters may also be needed to maintain the system at certain temperatures if it is exposed to the cold of space.

6) Ground purging should be carried out with pressures a few Torr higher than the total ambient pressure.

7) Launch and re-entry purging requires a variable pressure 1-Torr higher than the environment pressure.

8) The purge pressure while in orbit is about 1 Torr and the purge should continue for about 1 h. The amount of purge gas for the telescope will be about 0.1 m^3 at 2000 psi.

9) The volume, while the purge is being terminated, should be lifted from the bay and fully vented with the vent pointed away from the sun, velocity vector, engines, cabin.

10) The passive, full vent outgassing should be allowed to continue for about 24 h or longer before exposing the volume to the sun and initiate operations.

11) Methods to restrict the vent passage either manually or automatic should be explored. Partially closing the vent, on ground, in orbit, and re-entry could save large amounts of purge gas and offer additional protection. In orbit, the purge induced pressure in the bay would also be reduced.

12) The depth of the vent passage should be as long as permissible. This would extend the infiltration time constant.

13) The closure door of the volume should be tight during purging as to prevent additional loss of purging gas or pressure.

References

[1] Dushman, S., Scientific Foundations of Vacuum Techniques, 2nd Edition, edited by J. M. Lafferty, John Wiley & Sons, New York, pp. 21, 66, & 581.

[2] American Institute of Physics Handbook, 2nd Edition, McGraw-Hill Book Co., Inc., New York, pp. 2-235.

[3] Handbook of Chemistry and Physics, 46th Edition, The Chemical Rubber Co., Cleveland, Ohio, pp. F-43.

[4] Scialdone, J.J., "Water Vapor Pressure Control in a Volume," NASA Technical Paper 1172, March 1978.

[5] Malavasi, C., Vademecum per L'Ingegnere Construttore Meccanico, 14th Edition, edited by V. Hoepli, Milano, Italy, p. 1407.

[6] Tverskoi, P.N., Physics of the Atmosphere-A Course in Meteorology (Trans. from Russian), NASA TTF228, 1965, p. 286.

[7] Perry, R.H. and Chilton, C.H., Chemical Engineering Handbook, 5th Edition, McGraw-Hill Book Co., New York, 197-, p. 1358.

[8] Barenglotz, J. and Edgars, D., "The Relocation of Particulate Contamination During Space Flight," Jet Propulsion Laboratory, TM33-7317, Sept. 1975.

[9] Rouse, H., Elementary Fluid Mechanics, John Wiley & Sons, Inc., New York, p. 249.

[10] Leger, L.J., "Oxygen Atom Reaction with Shuttle Materials at Orbital Altitudes," AIAA Paper 83-0073, AIAA 21st Aerospace Science Meeting, Jan 10-13, 1983, Reno, Nevada.

[11] Scialdone, J.J., "Assessment of Shuttle Payloads Gaseous Environment," Proceedings ESA Symposium on Spacecraft Materials, the Netherlands, Oct. 1979, ESA SP-145, p. 101.

[12] Scialdone, J.J., "A Preliminary Assessment of the Self-Induced Environment and Contamination of the Space Telescope," Proceedings of International Symposium on Materials in Space, Toulouse, France, June 1982, ESA-SP 178, p. 107.

[13] Anon, "ICD-2-19001, Shuttle Orbiter/Cargo Standard Interface," NASA, JSC 07700, Vol. XIV, Revision G., 1980, Fig. 10.6.1.2-1.

[14] Muscari, J.A., "Non Metallic Materials Contamination Studies," Martin Marietta Corporation Technical Report, Contract 9554426, JPL, Denver, Co., Dec. 1980.

Chapter III. Properties and Effects of Contamination

Infrared Optical Properties of Thin CO, NO, CH_4, HCl, N_2O, O_2, N_2, and Ar Cryofilms

B. E. Wood*
Calspan Field Services, Inc., Arnold Air Force Station, Tennessee
and
J. A. Roux†
University of Mississippi, Oxford, Mississippi

Abstract

The real and imaginary portions (n,k) of refractive indices of thin condensed-gas films are essential for predicting the degradation of contaminated cryocooled optical surfaces. To identify and account for the effects from cryocontamination, the infrared spectral transmittance of cryofilms formed by CO, NO, CH_4, HCl, N_2O, O_2, N_2, and Ar were measured. These 0.25-15-μm-thick films were cryopumped onto a 20-K germanium substrate; the deposition pressure for the films was approximately 2×10^{-7} Torr. Transmittance spectra were obtained for the 500-3700-cm^{-1} range with a Fourier transform spectrometer. Values of n and k for these cryofilms were derived from the experimental data using a thin-film transmittance analytical model and the nonlinear least-squares method. Results from the least-squares method are compared with a Kramers-Kronig determination of the refractive index (n).

Introduction

Requirements to observe radiation sources at long ranges and infrared wavelengths have created severe design problems for infrared optical systems. These systems are often required to function at cryogenic temperatures that

---

Paper 83-0244 presented at AIAA 21st Aerospace Sciences Meeting, Reno, Nev., Jan. 10-13, 1983. This paper is declared a work of the U. S. Government and therefore is in the public domain.

*Research Physicist, AEDC Division.

†Associate Professor, Mechanical Engineering.

cause contamination of optical surfaces by atmospheric and rocket exhaust plume gases. These gases condense upon contact with cold optical surfaces and degrade system performance through thin-film interference and vibrational band absorption.

To identify and account for the effects from possible contamination by bipropellant [monomethyl hydrazine/nitrogen tetroxide (MMH/N_2O_4)] and monopropellant [hydrazine (N_2H_4)] engines in the infrared (i.r.), spectra of MMH, N_2O_4, N_2H_4, water (H_2O), carbon dioxide (CO_2), ammonia (NH_3), and mixtures of these constituents in a nitrogen (N_2) matrix were previously measured.[1,2] In the present work the i.r. spectra and optical properties of other contaminating species such as carbon monoxide (CO), nitric oxide (NO), methane (CH_4), hydrogen chloride (HCl), nitrous oxide (N_2O), oxygen (O_2), N_2, and argon (Ar) are reported. The normal transmittance spectra were measured using 20-K germanium as a substrate material. Germanium is a commonly employed optical component because of its high thermal conductivity and flat transmittance of 47% at 2-12 μm.

Complete experimental details are given in Ref. 3; thus, only a basic outline of the chamber and apparatus is presented here. The absolute transmittance of thin solid films ranging in thickness up to 15 μm is presented. Finally, a theoretical model of window plus film transmission is derived and is subsequently employed with the experimental results to determine the complex refractive index ($\bar{n} = n - ik$) of each of the above-mentioned species. The subtractive Kramers-Kronig treatment for calculation of the film refractive index has also been employed and results are compared to those of the nonlinear least-squares determination.

Instrumentation

Figure 1 shows the experimental apparatus and includes the i.r. interferometer (Digilab model FTS-14), the high-vacuum chamber containing the cryocooled substrate, and the i.r. source location. The chamber is a stainless steel cell equipped with a liner cooled with liquid nitrogen (LN_2) in which a vacuum of 10^{-8} Torr can be routinely obtained. The substrate holder can be actively cooled with either LN_2 (80 K) or gaseous helium (GHe) (20 K). The three platinum resistors located on the window holder yielded temperature measurements accurate to within 0.5 K.

INFRARED OPTICAL PROPERTIES

The spectral resolution of the interferometer system could be selected between 16.0 and 0.5 cm^{-1}, but 4-cm^{-1} resolution was sufficient for all of the work reported herein. Transmittance data were recorded in the 500-3700 wave number range. Transmittance measurements were made by rotating the germanium out of the beam and recording and storing a reference power spectrum. Generally, 16 interferograms were coadded to improve the signal-to-noise ratio before execution of the Fourier transform. Next, the window was rotated into the beam and the process repeated. The reference file was then divided into the sample file and plotted by a digital incremental plotter, producing the final data record on a linear ordinate scale of 0-100% transmittance.

Controlled contamination of the cryocooled germanium window was accomplished with a gas induction system that included a toroidal header with 36 orifices of 1/16-in.-diam spaced 10 deg apart to direct the gas toward the germanium window. The chamber pressure rose from 1×10^{-8} to 2×10^{-7} during deposition at 20 K. Gas was prevented from condensing on the back of the germanium window by a gas baffle positioned close to the back of the window holder. The thin-film thickness was uniform across the exposed

1. Pyroelectric detector and collection optics.
2. Stainless steel high vacuum chamber, 85 cm tall by 70 cm in diameter (33.5 in. by 27.5 in. in diameter).
3. Cryogenically cooled infrared window; germanium, 4 mm thick by 70 mm square (0.158 in. by 2.76 in.) and QCM.
4. Helium-neon laser (0.6327 µm) beam (one of two shown) employed to measure cryofilm thickness.
5. Infrared beam, 38 mm in diameter (1.5 in.).
6. 2-mw He-Ne laser.
7. Michelson interferometer.
8. Infrared source and collimator mirror.

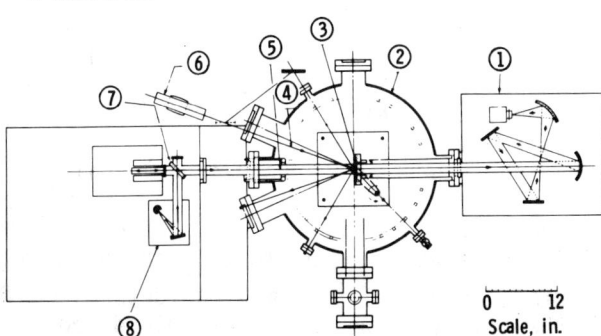

Fig. 1 Infrared optical transmission chamber with FTS-14 interferometer/spectrometer.

2-in.-diam window area; the color of the reflected light from the entire area of the cryofilm/germanium changed uniformly with increasing film thickness. Film uniformity and absolute thickness were two important parameters, since the objective of the experiment was to determine the complex refractive index of the thin film, a quantity derived by comparison of experimental transmittance data with theoretical model thickness data. Any error in absolute film thickness is directly introduced into the film complex refractive index results. A dual-angle laser interference technique[4,5] was used to monitor the film thickness and the film refractive index at $\lambda = 0.6328$ μm.

Two He-Ne laser beams are specularly reflected off the germanium window for two different, accurately measured incidence angles. As the gas is condensed, two interference patterns of different periods are monitored in the reflected laser light. If the ratio of pattern periods is termed[5] β, then the refractive index of the film, n, is given by

$$n = \left(\frac{\sin^2\theta_b - \beta^2 \sin^2\theta_a}{1 - \beta^2} \right)^{1/2} \tag{1}$$

where θ_a and θ_b (typically 18 and 68 deg, respectively) are the two laser beam incidence angles. Once n has been established, the thickness of the film d_1 is readily calculated from

$$m_a \lambda = 2nd_1 \left[1 - \left(\sin^2\theta_a / n^2 \right) \right]^{1/2}$$

where m_a is the order of the interference maxima for the 18-deg incidence angle. The dual-angle laser beam thickness monitor yielded thin-film refractive index values accurate to within 2%. A quartz crystal microbalance (QCM) was used in conjunction with the dual-angle laser beam interference technique to determine the density of each contaminant. The QCM operates on the principle that the crystal vibration frequency changes linearly with changes in the mass deposited on the crystal. The QCM was located adjacent to and just above the germanium window so that the mass deposition rate would equal that of the germanium window. The surface density (in g/cm^2) was determined from the QCM and the film thickness was determined from the interference patterns. From these two values the film density (g/cm^3) was determined.

Procedure

The chamber was initially pumped down to approximately 1.0×10^{-7} Torr using the diffusion pump and the LN_2 liner. Chamber pressures in the low 10^{-8}-Torr range were obtained when the germanium substrate, holder, and transfer lines were cooled to approximately 20 K with the cryostat.

Samples of research-grade test gas were obtained from lecture bottles for deposition. Condensation of the gas on the cold germanium window was monitored using two He-Ne laser beams with the two-angle interference technique. When film thickness corresponded to an interference maxima, the gas flow was shut off and transmittance measurements were made. Once a transmittance measurement was completed, the gas flow was again started and deposition occurred until a film thickness corresponding to the next interference maxima was reached, and so on. In some instances where very strong absorption bands were observed, transmittance measurements were alternately made for each interference minima and maxima. This was done since determination of n and k requires the use of as many thicknesses as possible to increase the accuracy of the values calculated.

After completion of a series of transmittance measurements for all the thicknesses, the cryogenic flow rate to the germanium substrate was shut off and the substrate and deposited film were allowed to warm up. Transmittance measurements were made at intervals during warmup. The time required for the interferometer to scan the sample (16 scans) was 2-3 min, with a similar time required for the reference beam. With the time required for obtaining the Fourier transform and the plotting of the data, this resulted in a time interval of about 10 min between measurements. The temperature given on each warmup data plot is the temperature at the end of the sample interferometer scan.

Transmittance measurements were made for as many thicknesses as possible during each gas experiment. The number of thicknesses obtained depended considerably on the particular gas. For "well-behaved" deposits, 16-25 thicknesses (interference maxima or minima) were obtained, whereas in some cases only a few thicknesses were possible because of scattering in the film (at $\lambda = 0.6328$ μm).

Results

Transmittance measurements were made of thin films of gases condensed on the cryogenically cooled germanium window and optical constants were derived for all of the gases except N_2O. The experimental data obtained for the N_2O showed that the absorption band locations shifted with film thickness and hence gave relatively large errors in the n and k determined in those wave number regions. This was the only gas exhibiting such behavior.

CO on 20-K Germanium

The CO transmittance spectra were taken on deposits 0.258-4.13-μm thick (Fig. 2). The calculated value for the

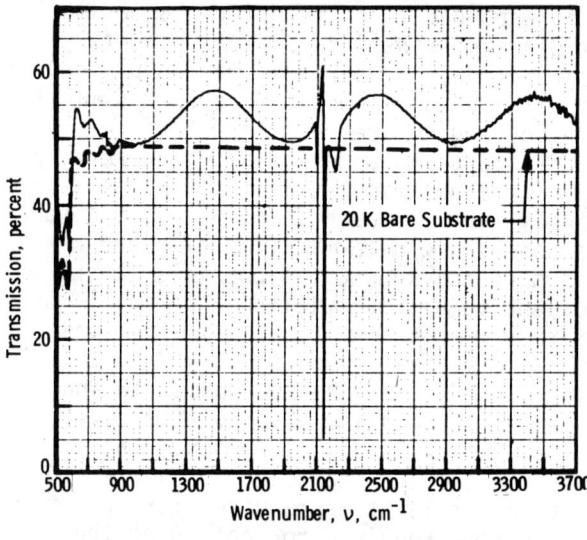

Fig. 2 Transmittance of 4.13-μm-thick CO film on 20-K germanium.

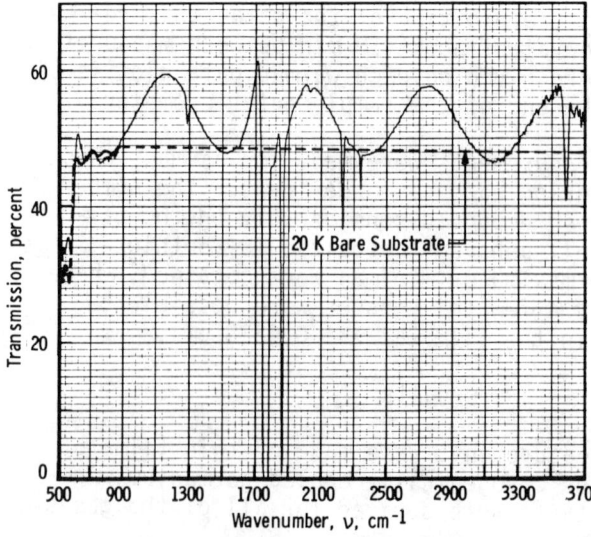

Fig. 3 Transmittance of 4.78-μm-thick NO film on 20-K germanium.

index of refraction at $\lambda = 0.6328$ μm is 1.27 and the density of the condensed gas is 0.80 g/cm^3. The corresponding Lorenz value of $(1/\rho)[(n^2 - 1)/(n^2 + 2)]$ is 0.21 cm^3/g. Absorptions are noted at 2095, 2140, and 2210 cm^{-1}. The fundamental stretching vibration of CO occurs at 2143 cm^{-1} in the gaseous state,[6] which is in very close agreement with the observed vibration at 2140 cm^{-1}. The broad absorption at 2210 cm^{-1} is felt to result from a combination of the fundamental and lattice vibration, similar to the vibration noted in CO_2 deposits. The absorption at 2095 cm^{-1} corresponds to the fundamental vibration of the isotope $C^{13}O^{16}$. The weak absorption at 2340 cm^{-1} is caused by traces of CO_2.

NO on 20-K Germanium

Transmittance spectra of NO deposited on 20-K germanium were obtained for thicknesses of 0.239-4.78 μm (20 interference maxima), an example of which is shown in Fig. 3. The refractive index at $\lambda = 0.6328$ μm was 1.37, and the density was determined to be 1.17 g/cm^3. The resulting Lorentz-Lorenz value is 0.191 cm^3/g at $\lambda = 0.6328$ μm. The fundamental vibration band of NO in the gaseous state[7] occurs at 1876 cm^{-1}. From the transmittance data, a narrow, fairly strong absorption is observed at 1860 cm^{-1},

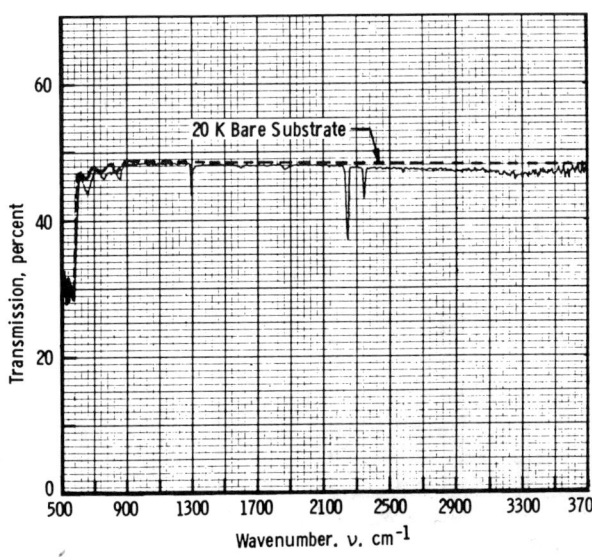

Fig. 4 Transmittance of NO film after warmup to 83 K.

corresponding to this fundamental vibration. At about twice this frequency (3580 cm^{-1}) a weaker absorption corresponding to $2\nu_1$ is noted.

Many other lines are observed in the spectra, the most obvious being the very strong absorption at 1760 cm^{-1}. Other absorptions are 1300, 1600, 2030, 2240, and 2340 cm^{-1}. Warming the sample to 83 K greatly decreased the magnitude of the bands at 1760, 1860, 2030, and 3580 cm^{-1}. Most of the NO had already sublimated. The spectrum taken at this temperature is shown in Fig. 4. Another feature of the spectrum at this temperature is the absence of the channel spectra, indicating that interference is no longer occurring in the sample. The absorptions at 1300, 1600, 2240, and 2340 cm^{-1} are still seen. At the chamber pressure of 2×10^{-6} Torr, NO sublimates at about 55 K. The probable

Table 1 Absorption lines of CH_4 on 20-K germanium

Frequency, cm^{-1}	Vibration	Frequency, cm^{-1}	Vibration
1300	ν_4	2820	ν_1
1350	ν_4 + lattice	3010	ν_3
1540	ν_2	3060	ν_3 + lattice
2590	$2\nu_4$	...	...

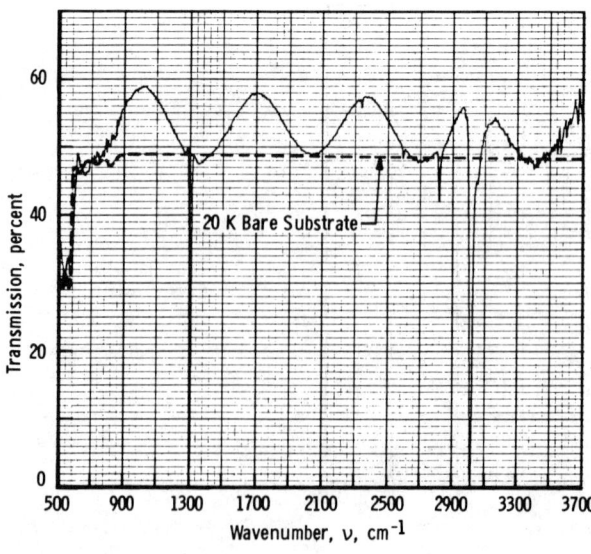

Fig. 5 Transmittance of 5.57-μm-thick CH_4 film on 20-K germanium.

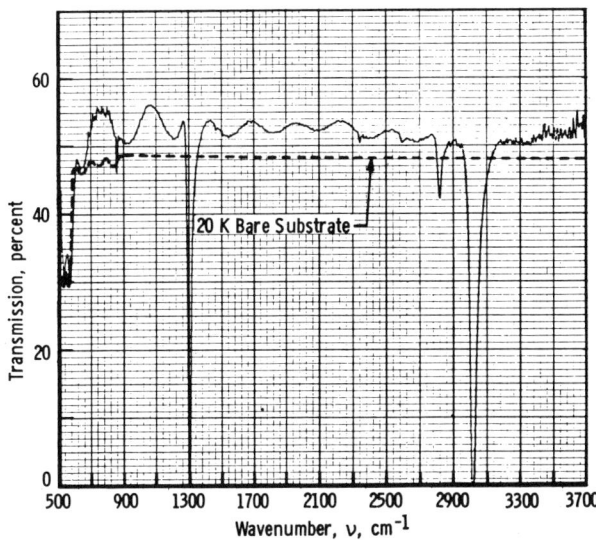

Fig. 6 Transmittance of 12.0-μm-thick CH_4 film after warmup to 54 K and recooling to 20 K.

contaminants causing the absorptions at 1600 and 2340 cm^{-1} were H_2O and CO_2, respectively. The probable contaminant causing the absorptions at 1300 and 2240 cm^{-1} was N_2O, which sublimates at about 84 K. The absorption at 1760 cm^{-1} was possibly N_2O_4, which has a very strong absorption at this frequency.

CH_4 on 20-K Germanium

Transmittance spectra were obtained for deposits of CH_4 on 20-K germanium. These deposits were 0.242-11.35-μm thick. The index of refraction was 1.35 at λ = 0.6328 μm, and the density was 0.426 g/cm^3. The resulting Lorentz-Lorenz value was 0.505 cm^3/g.

In Fig. 5 (d_1 = 5.57 μm) absorptions are observed at 1300, 1350, 1540, 2340, 2590, 2830, 3010, and 3060 cm^{-1}. These absorption lines have been identified as shown in Table 1.

The absorption at 2340 cm^{-1} is due to the presence of trace amounts of CO_2. After a thickness of 11.98 μm was deposited, the temperature was increased to 35 K. A spectrum taken at this temperature shows little change. Evidence that the bands at 1350 and 3060 cm^{-1} are combinations of fundamental and lattice vibrations is shown in Fig. 6. This spectrum shows the transmittance after the substrate has been raised to 54 K and recooled to 20 K. The bands at

1350 and 3060 cm^{-1} are no longer present, indicative of a change in the structure of the deposit.

HCl on 20-K Germanium

Transmittance spectra were obtained for 0.505-2.78-μm-thick HCl cryodeposited on germanium at 20 K, an example of which is shown in Fig. 7 for a 2.78-μm-thick film. The in-

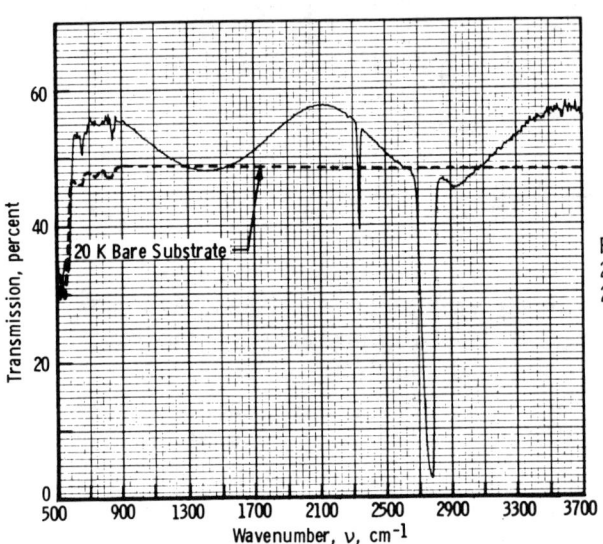

Fig. 7 Transmittance of 2.78-μm-thick HCl film on 20-K germanium.

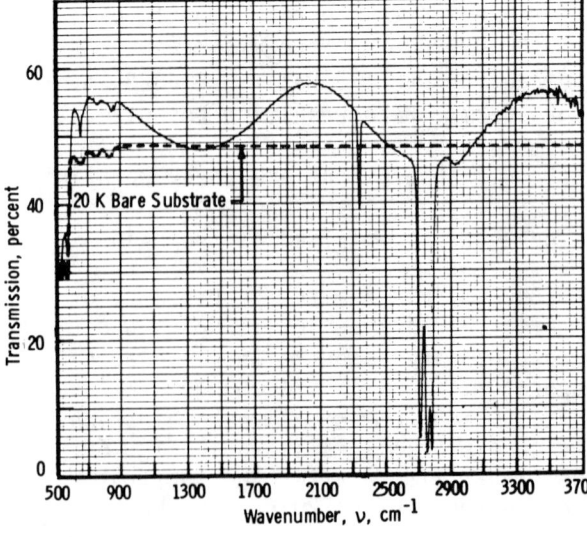

Fig. 8 Transmittance of 2.78-μm-thick HCl film after warmup to 55 K.

dex of refraction at $\lambda = 0.6328$ μm was determined to be 1.30 and the density of the deposit was 0.955 g/cm^3. The corresponding Lorentz-Lorenz value at $\lambda = 0.6328$ μm was 0.194 cm^3/g. In the spectra taken from this sample, absorption bands are observed at 2340, 2780, and 2920 cm^{-1}. The fundamental vibration of HCl occurs[6] in the gaseous state at 2886 cm^{-1}. Therefore, it seems likely that the absorption band at 2780 cm^{-1} is the fundamental vibration in the condensed state.

The sample was warmed to 55 K. A transmittance spectrum taken at this temperature (Fig. 8) shows that the absorption band at 2780 cm^{-1} has split into three distinct bands located at 2710, 2750, and 2780 cm^{-1}. As the sample was warmed further to 75 K, a new absorption was noted at 2620 cm^{-1} and the fundamental vibration had only two bands

Table 2 Absorption lines of N$_2$O on 20-K germanium

Frequency, cm^{-1}	Vibration	Frequency, cm^{-1}	Vibration
580	ν_2	2570	$2\nu_3$
1160	$2\nu_2$	2800	$\nu_1 + \nu_2$
1295	ν_3	3380	$\nu_1 + 2\nu_2$
2240	ν_1	3500	$\nu_1 + \nu_3$
2460	$\nu_3 + 2\nu_2$	...	...

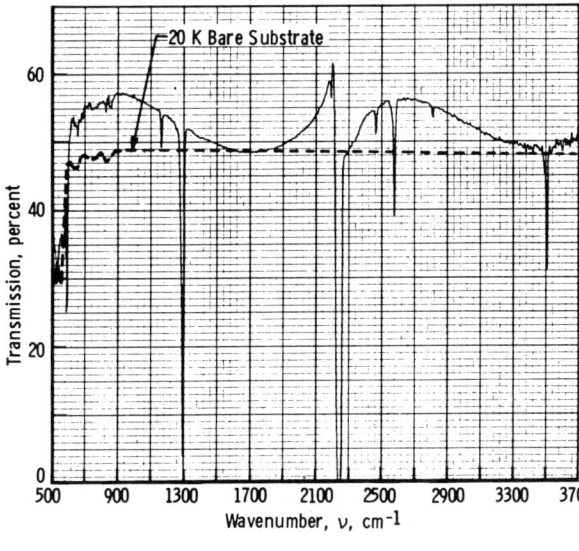

Fig. 9 Transmittance of 2.36-μm-thick N$_2$O film on 20-K germanium.

located at 2710 and 2750 cm^{-1}, whereas the absorption at 2910 cm^{-1} shifted up to 2990 cm^{-1}. The absorptions at the fundamental vibration are much stronger, approaching zero transmittance. Warming the sample to 92 K showed that the deposit was no longer present. The broad band occurring at 2920 cm^{-1} is probably a combination of the fundamental and lattice vibrations.

N_2O on 20-K Germanium

Transmittance spectra of N_2O deposited on 20-K germanium were obtained for thicknesses of 0.262-4.19 µm (16 interference maxima). The index of refraction at λ = 0.6328 µm and the density were determined to be 1.27 and 0.988 g/cm^3, respectively. The corresponding Lorentz-Lorenz value at λ = 0.6328 µm was 0.169 cm^3/g.

As shown in the spectra for the 2.36-µm-thick film (Fig. 9), N_2O has many absorptions in the infrared.[7] Absorptions are noted at 580, 1160, 1295, 2190, 2240, 2460, 2570, 2800, 3380, and 3500 cm^{-1}. All of these bands except the 2190-cm^{-1} band correspond within about 20 wave numbers to the absorption bands in the gaseous state. These assignments are shown in Table 2. The cause of the 2190 cm^{-1} absorption band is unknown, but possibly arises from an isotope of N_2O.

After a deposition of 4.19 µm, the temperature of the deposit was slowly increased. At about 29 K, the sample's structure changed, as evidenced by the scattering that began. Further evidence of this phase change was seen in a spectrum taken at 46 K in which the absorptions were significantly stronger than previously.

O_2 on 20-K Germanium

Transmittance spectra were obtained for thicknesses of O_2 deposited on 20-K germanium (Fig. 10a). These thicknesses were 0.263-14.45 µm. The index of refraction was 1.25 at λ = 0.6328 µm, the density was 1.22 g/cm^3, and the Lorentz-Lorenz value was 0.121 cm^3/g.

Condensed O_2 is slightly infrared active.[8] This absorption occurs at the fundamental frequency of 1550 cm^{-1}, is very weak, and is undetectable for the thicknesses examined in this experiment. In the spectra, two weak absorptions are seen at 2140 and 2340 cm^{-1}. These absorp-

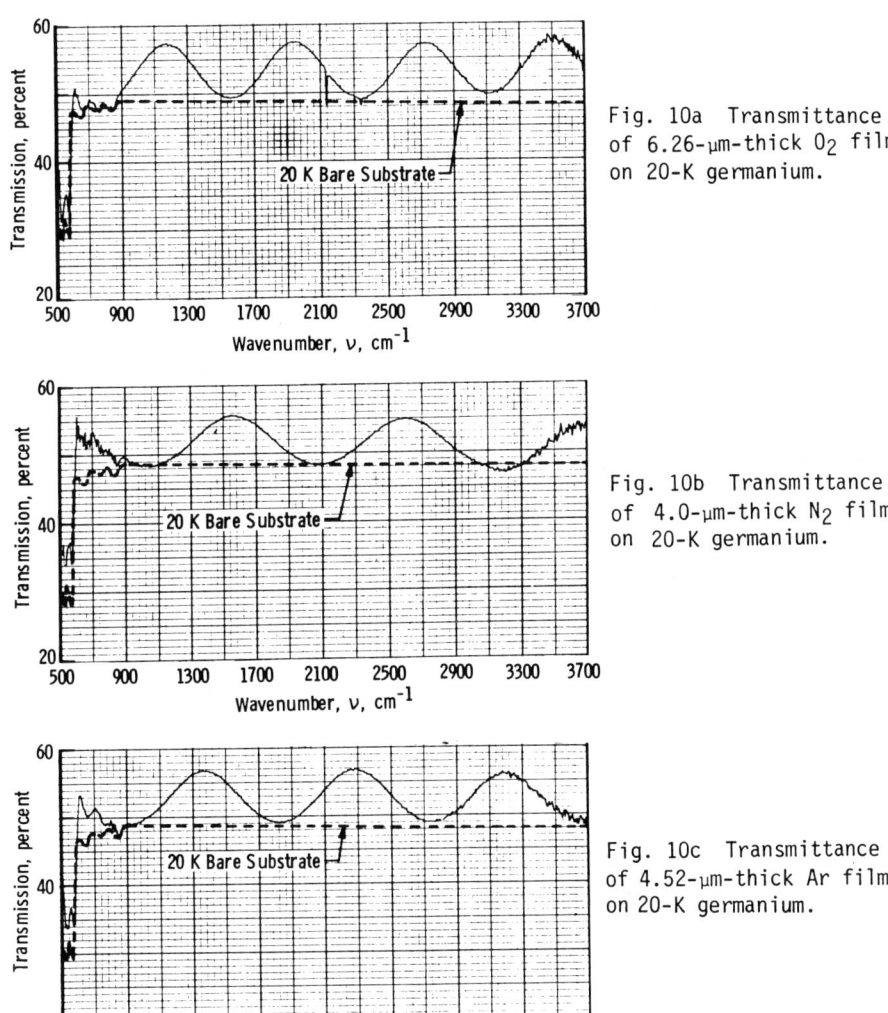

Fig. 10a Transmittance of 6.26-μm-thick O_2 film on 20-K germanium.

Fig. 10b Transmittance of 4.0-μm-thick N_2 film on 20-K germanium.

Fig. 10c Transmittance of 4.52-μm-thick Ar film on 20-K germanium.

tions are probably attributable to trace quantities of CO and CO_2, respectively. Warmup of the O_2 film to 30 K and subsequent recooling to 20 K showed no change in the spectral transmittance, indicating no change was observed for these thicknesses between the α and β phases.

N_2 on 20-K Germanium

Nitrogen is not active in the infrared (Fig. 10b). Spectra were taken for thicknesses of 0.261-5.22 μm. The index of refraction at $\lambda = 0.6328$ μm was 1.26, the density of the deposit was 0.83 g/cm^3, and the Lorentz-Lorenz value was 0.194 cm^3/g. The only noticeable N_2 effect on transmittance is the channel spectra.

Ar on 20-K Germanium

Argon, because it is not infrared active, should show no absorptions in the region of interest. Transmittance spectra were obtained for deposit thicknesses of 0.266-6.99 μm (Fig. 10c). The measured index of refraction at $\lambda = 0.6328$ μm was 1.23. Unfortunately, QCM data were not obtained; this prevented a determination of the density. As was expected, the deposits of Ar showed no absorptions, and the only noticeable effect was the presence of the channel spectra.

Optical Constants

To determine the complex refractive index ($\bar{n} = n - ik$) of the thin, solid film from the i.r. transmittance measurements for wave numbers between 700 and 3700 cm^{-1}, an analytical model of film plus window transmission was used. The optical constants of the cryocontaminants were determined by using this analytical transmission model[9,10] in conjunction with a nonlinear least-squares convergence routine. Also, the subtractive Kramers-Kronig relation between n and k was used in conjunction with the nonlinear least-squares determination of n. The subtractive Kramers-Kronig relation is given by

$$n(\nu) = n(\nu_m) + \frac{2}{\pi} P \int_{\nu_1}^{\nu_2} \left\{ \frac{k(\nu')\nu' - k(\nu)\nu}{(\nu')^2 - \nu^2} - \frac{k(\nu')\nu' - k(\nu_m)\nu_m}{(\nu')^2 - \nu_m^2} \right\} d\nu' \qquad (2)$$

where ν_m is a reference frequency (for CO $\nu_m = 2500$ cm^{-1}, n = 1.232; for NO $\nu_m = 2500$ cm^{-1}, n = 1.326; for CH$_4$ $\nu_m = 2200$ cm^{-1}, n = 1.330; for HCl $\nu_m = 1700$ cm^{-1}, n = 1.278; for O$_2$ $\nu_m = 2200$ cm^{-1}, n = 1.264; for N$_2$ $\nu_m = 2200$ cm^{-1}, n = 1.229; and for Ar $\nu_m = 2200$ cm^{-1}, n = 1.225). P indicates the Cauchy principal value of the integral. Integration was performed using the simple trapezoidal rule; the k values used in Eq. (2) were those determined by the nonlinear least-squares technique. Also in Eq. (2), $\nu_1 = 700$ and $\nu_2 = 3700$ cm^{-1}. The values of k outside 700 cm$^{-1} \leq \nu \leq 3700$ cm^{-1} were considered either as zero or as a constant at the end point values; both approaches showed a negligible effect upon the values of n that were obtained from Eq. (2).

Transmittance data recorded for all deposits discussed here were digitized every 2 cm^{-1}. The optical properties were computed every 10 cm^{-1} except in regions of strong absorption in which computations were performed every 2 cm^{-1}. The optical properties were initially computed by the nonlinear least-squares determination. However, in some instances, the program did not appear to converge upon a unique value of n. This usually occurred in regions of very strong absorption or low wave number or for cases in which it was possible to form only a few thin deposit thicknesses. The n value appears to be primarily defined by the period of the transmittance vs thickness curve at each wave number. At small thicknesses, high absorption, or low wave numbers, the transmittance vs thickness (for each wave number) curve is not well defined, i.e., the period of the interference as a function of thickness is not well defined. The k value, which is primarily defined by the magnitude of the transmittance, did not have this difficulty and was well defined over the whole spectral region (700-3700 cm^{-1}). Thus, to determine n, the k values were used with the subtractive Kramers-Kronig relationship; these new n values were then used in the analytical model (along with the k values) to see whether good agreement occurred with the transmittance data. For all wave numbers, the Kramers-Kronig n, along with the least-squares k, yielded good agreement when the analytical model and transmittance data were compared.

Errors in the refractive index, thickness, and density are all estimated to be within ±2%. The error in the absorption index varies from about ±2% at high k values (∼0.01 or greater) to a value that is inaccurate for values of k below the slanted dashed line at the bottom of the absorption index graphs. The value of k can vary over several orders of magnitude; therefore, the smaller the k value the larger the error. This is because of the error limits on the film-substrate transmittance measurements.

Generally the nonlinear least-squares results (thin-film model) did converge, and the Kramers-Kronig and thin-film results are compared for each species investigated. As stated, the optical properties were obtained through use of an analytical model and the nonlinear least-squares method. It was assumed that the germanium substrate (window) acted as a thick film and, hence, that no phase coherence occurred between multiple internal reflected rays (inside the germanium). The refractive index of the germanium was a modification of Herzberger and Salzberg[11] to account for the refractive index change due to the cryogenic temperatures. The thin film (cryocontamination layer) was considered to be an optical "thin film" with properties n and k. The overall transmittance through the contamination layer on the germanium substrate is given by

$$\tau = \tau(n,k,\phi,n_g,\nu,d_1) \tag{3}$$

where n and k are the optical properties of the contamination layer, ϕ the angle of incidence (here $\phi = 0$ which means normal incidence), n_g the germanium refractive index, ν the wave number, and d_1 the contamination-layer thickness. The exact expressions[1] for Eq. (3) are algebraically long and tedious but straightforward; these expressions involve Fresnel's equations for the interface reflectances and transmittances. The values of n and k were obtained by knowing ν, ϕ, n_g, and measuring the value of τ at known values of d (for CH_4, 24 values of τ were measured at 24 known values of d_1). Mathematically, it was required to have at least two values of τ at two known values of d to have two equations for the two unknowns, n and k. By using the nonlinear least-squares convergence method, the values of n and k that produce the least error (in the least-squares sense) between all the data (24 values for CH_4) and the analytical model [Eq. (3)] were determined. Using an initial guess for the solution for n and k, the nonlinear least-squares generated improved values for n and k; these were then resubstituted to find an even better solution. This was continued until convergence was obtained. Usually convergence was obtained at each wave number after about three iterations. Details of this standard method are available in Ref. 1. For all the results shown below, tabulated data are available in Ref. 12.

CO Optical Constants

The optical properties n and k, which are the refractive and absorption indices of carbon monoxide films, were determined from 15 deposit thicknesses. These curves are shown in Fig. 11. The refractive index n is essentially constant between 1.23 and 1.24, except for the region of the fundamental stretching vibration band centered at 2140 cm^{-1}. The values obtained using the nonlinear least-squares technique in this region are both higher and lower than obtained using the Kramers-Kronig technique. The k curve shows an absorption index maximum of 0.34 at 2140 cm^{-1}.

NO Optical Constants

The n and k of 20-K nitric oxide films were determined from 18 thicknesses (see Fig. 12). The refractive indices

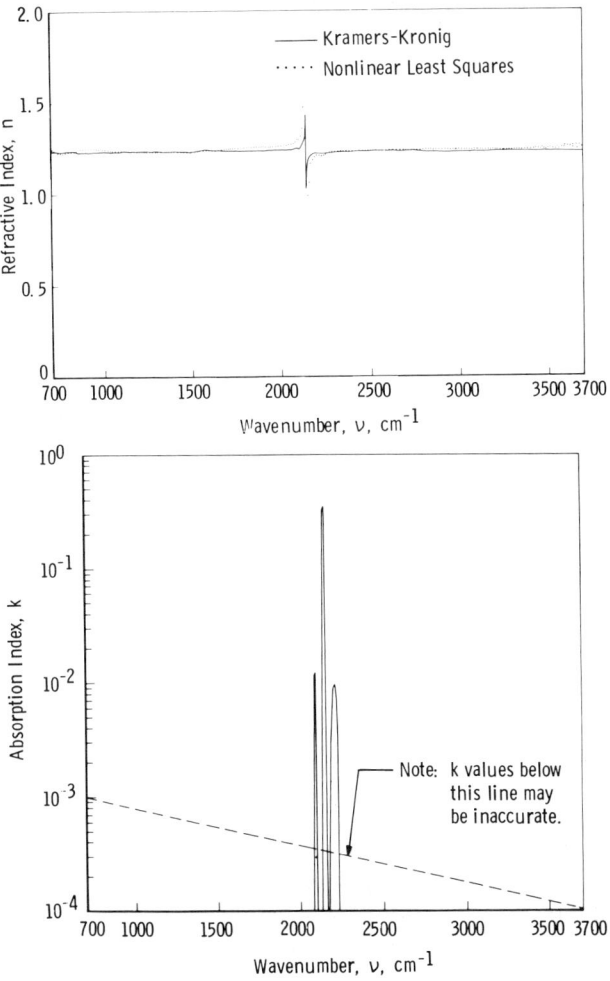

Fig. 11 Optical properties of CO condensed on 20-K germanium.

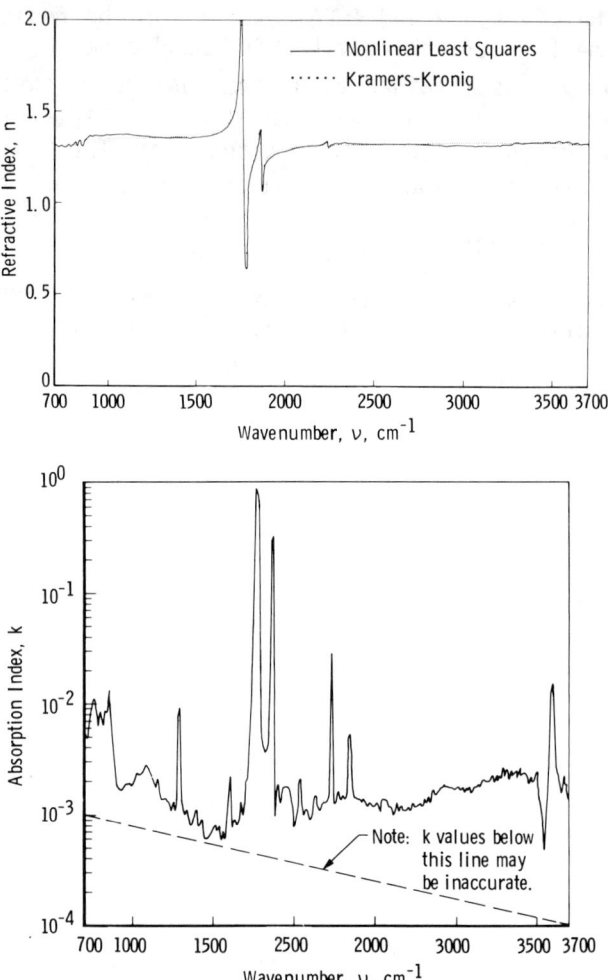

Fig. 12 Optical properties of NO condensed on 20-K germanium.

over much of the wave number range varied between 1.30 and 1.35. Between 1500 and 2000 cm^{-1} there was considerable variation, as shown in Fig. 12. Most of the major variation in k occurs between 1700 and 1900 cm^{-1}, with the strongest peak occurring at a k value of 0.9 at 1760 cm^{-1} and another peak at 1860 cm^{-1}. These bands occur in the region of the fundamental absorption band for NO. The agreement between the nonlinear least-squares and Kramers-Kronig techniques for determining n is quite good.

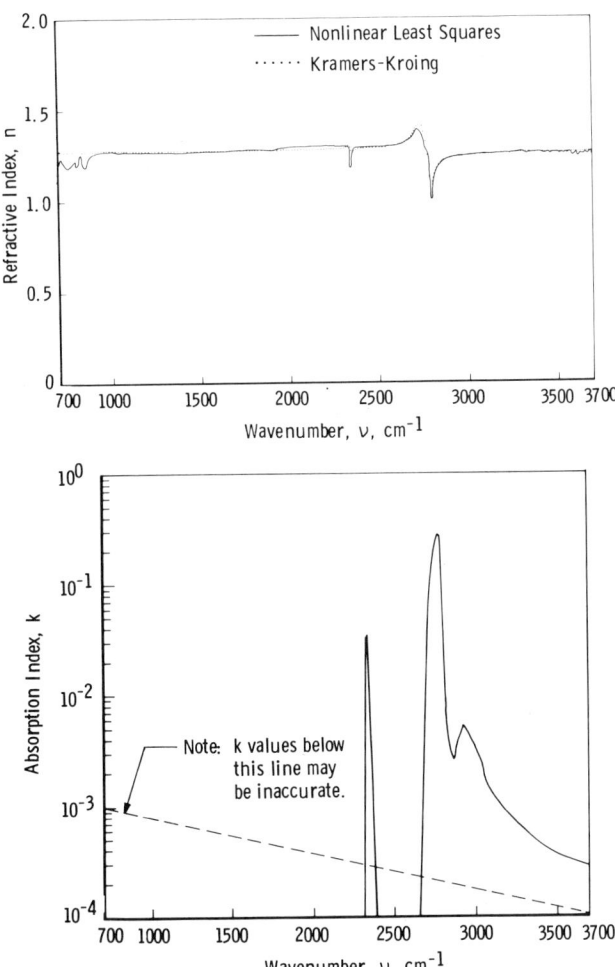

Fig. 13 Optical properties of HCl condensed on 20-K germanium.

HCl Optical Constants

The refractive and absorption indices shown in Fig. 13 were obtained for 10 thicknesses of HCl. The vibration band located at 2780 cm^{-1} is the only region of appreciable absorption, and this band shows an absorption index of only 0.28, which is considerably less than observed for the NO band. The refractive index varied between 1.25 and 1.29 over most of the wave number range, with an exception in the vicinity of the 2780-cm^{-1} absorption bands. The agree-

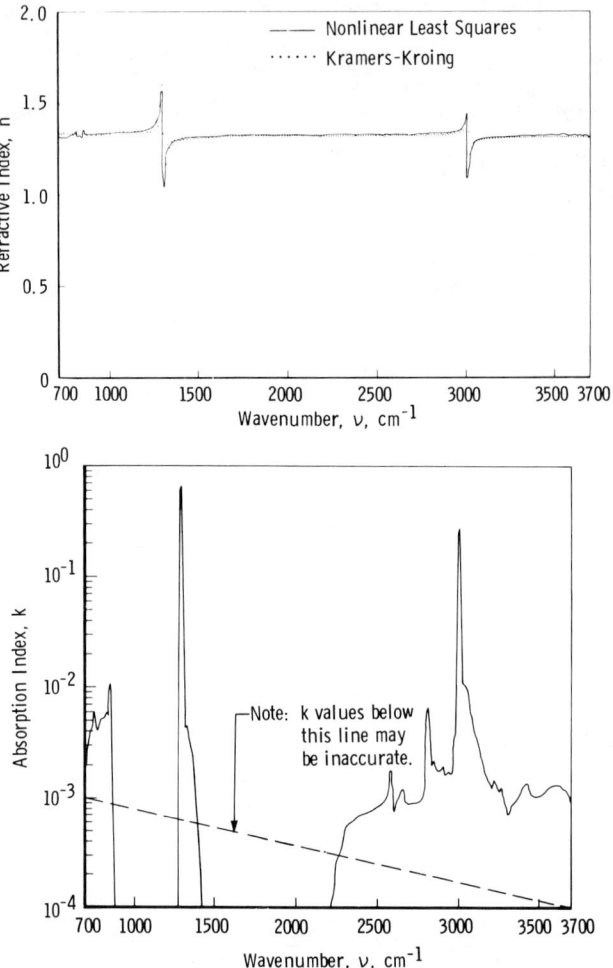

Fig. 14 Optical properties of CH_4 condensed on 20-K germanium.

ment between the two techniques of refractive index determination is again quite good.

CH_4 Optical Constants

The 20-K methane optical constants were determined using 24 film thicknesses and the results are shown in Fig. 14. The refraction index values were 1.32-1.35 over most of the 700-3700-cm^{-1} range. The two regions of highest absorption were at 1310 cm^{-1} (k = 0.65) and at 3020 cm^{-1}

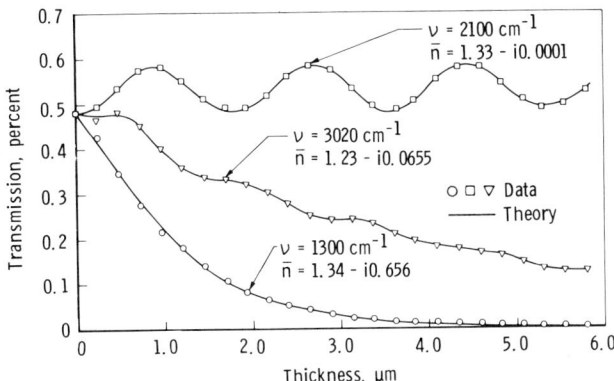

Fig. 15 Comparison of theory and data for 20-K solid CH_4 for three different wave numbers.

(k = 0.27), which correspond to the ν_4 and ν_3 vibration bands, respectively. The comparison of transmittance for theory and data for CH_4 films on 20-K germanium is shown in Fig. 15 for the 1300-, 2100-, and 3030-cm^{-1} regions. Good agreement between theory and data is again observed.

O_2, N_2, and Ar Optical Constants

Since the homonuclear gases O_2 and N_2 and the monoatomic gas Ar showed no absorption in the 700-3700-cm^{-1} wave number range, only the refractive index curves are presented. Figure 16 shows the refractive indices obtained for 20 thicknesses of O_2 condensed on 20-K germanium. The refractive index is essentially constant (between 1.26 and 1.27) over the entire wave number range.

Similar measurements were made for 15 N_2 film thicknesses condensed on 20-K germanium (Fig. 16). Refractive index values of 1.22 and 1.23 were observed for all wave numbers. The refractive indices for argon are also shown in Fig. 16 and were determined using 16 thicknesses of argon films condensed in 20-K germanium. As expected, the refractive indices are essentially constant with a value of 1.22 observed for most of the wave number range.

Summary

Experimentally determined transmittance measurements of condensed gases on a cryogenically cooled germanium window were made. Thin films of CO, NO, CH_4, HCl, N_2O, O_2, N_2, and Ar were deposited at 20 K. The infrared spectral

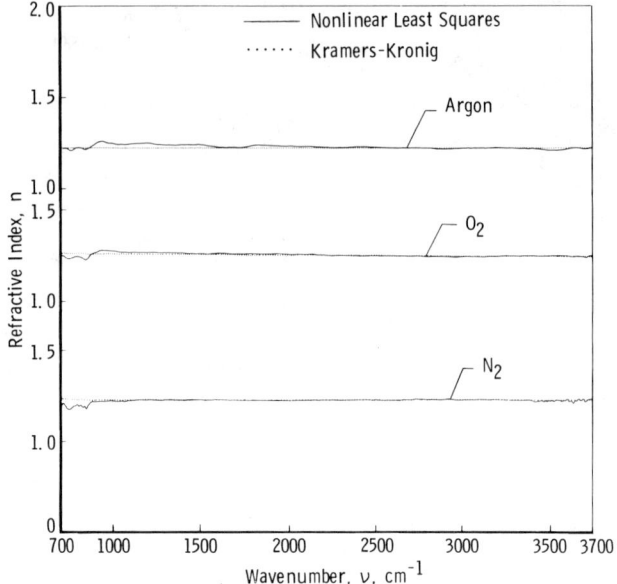

Fig. 16. Optical properties of O_2, N_2, and Ar condensed on 20-K germanium.

transmittance was studied over the 500-3700-cm^{-1} range (the 2.70-20.0-μm wavelength range). The film thickness was accurately determined using a two-angle laser-interference technique. The densities of the films were determined by using quartz crystal microbalances and by calculating the refractive index at 0.6328 μm to allow accurate thickness measurement. From the transmittance data, the optical properties n and k were determined for the 700-3700-cm^{-1} range. These results are available in graphical and tabular[12] form for a wave number interval of 2 cm^{-1}. These data are important for evaluating the influence of contamination on the performance of spacecraft optical surfaces and sensors.

Acknowledgment

The research reported herein was conducted by the Arnold Engineering Development Center (AEDC), U. S. Air Force Systems Command (AFSC). Work and analysis for this research were done by employees of ARO, Inc., a Sverdrup Corporation company, an operating contractor for the AEDC. Further reproduction is authorized to satisfy needs of the U. S. Government.

References

[1] Roux, J. A., Wood, B. E., and Smith, A. M., "Infrared Optical Properties of Solid Monomethyl Hydrazine, N_2O_4, and N_2H_4 at Cryogenic Temperatures," Journal of the Optical Society of America, Vol. 73, Sept. 1983, pp. 1181-1188.

[2] Roux, J. A., Wood, B. E., and Smith, A. M., "Infrared Optical Properties of Thin H_2O, NH_3, and CO_2 Cryofilms," Journal of the Optical Society of America, Vol. 72, June 1982, pp. 720-728.

[3] Pipes, J. G., Roux, J. A., Smith, A. M., and Scott, H. E., "Infrared Transmission of Contaminated Cryocooled Optical Windows," AIAA Journal, Vol. 16, Sept. 1978, pp. 984-990.

[4] Tempelmeyer, K. E. and Mills, D. W. Jr., "Refractive Index of Carbon Dioxide Cryodeposit," Journal of Applied Physics, Vol. 39, May 1968, pp. 2968-2969.

[5] Smith, A. M., Tempelmeyer, K. E., Müller, P. R., and Wood, B. E., "Angular Distribution of Visible and Near IR Radiation Reflected from CO_2 Cryodeposits," AIAA Journal, Vol. 7, Dec. 1969, pp. 2274-2280.

[6] Herzberg, G., Molecular Spectra and Molecular Structure, I. Spectra of Diatomic Molecules. D. Van Nostrand Company, Princeton, N.J., 1950.

[7] Herzberg, G., Infrared and Raman Spectra of Polyatomic Molecules. D. van Nostrand Company, Princeton, N.J., 1945.

[8] Cairns, B. R. and Pimentel, G. C., "Infrared Spectra of Solid α- and β-Oxygen," Journal of Chemical Physics, Vol. 43, Oct. 1965, pp. 3432-3438.

[9] Vasicek, A., Optics of Thin Films, North-Holland Publishing Company, Interscience Publishers, New York, 1960.

[10] Heavens, O. S., Optical Properties of Thin Films, Dover Publications, New York, 1965.

[11] Herzberger, M. and Salzberg, C. D., "Refractive Indices of Infrared Optical Materials and Color Correction of Infrared Lenses," Journal of the Optical Society of America, Vol. 52, April 1962, pp. 420-426.

[12] Roux, J. A., Wood, B. E., Smith, A. M., and Plyler, R. R., "Infrared Optical Properties of Thin CO, NO, CH_4, HCl, N_2O, O_2, N_2, Ar and Air Cryofilms," AEDC-TR-79-81, Aug. 1980.

Infrared Optical Properties of Solid Mixtures of Molecular Species at 20 K

K. F. Palmer*
Westminster College, Fulton, Missouri
J. A. Roux†
University of Mississippi, Oxford, Mississippi
and
B. E. Wood‡
Calspan Field Services, Inc., Arnold Air Force Station, Tennessee

Abstract

The real and imaginary portions (n,k) of refractive indices of thin condensed gas films are essential for predicting the degradation of contaminated cryocooled optical surfaces. To identify and account for the effects from cryocontamination, the infrared spectral transmittance of cryofilms formed from gaseous mixtures of CO, H_2O, NH_3, CO_2, N_2 and Ar were measured. These 0.25-15-μm-thick films were cryopumped onto a 20-K germanium substrate; the deposition pressure for the films was approximately 2×10^{-7} Torr. Transmittance spectra were obtained for the 500-3700-cm^{-1} range with a Fourier transform spectrometer. Values of n and k for these cryofilms were derived from the experimental data using a thin-film transmittance analytical model and the nonlinear least-squares method. Results from the least-squares method are compared with a Kramers-Kronig determination of the refractive index (n).

Presented as Paper 83-1452 at the AIAA 18th Thermophysics Conference, Montreal, Canada, June 1-3, 1983. This paper is declared a work of the U. S. Government and therefore is in the public domain.

*Chairman, Dept. of Physics.
†Associate Professor, Mechanical Engineering.
‡Research Physicist, AEDC Division.

Introduction

Cryogenic surfaces within infrared (i.r.) optical devices on spacecraft and on high-altitude aircraft can be contaminated by the condensation of engine exhaust and atmospheric gases. The degradation in performance of these optical devices by absorption and interference effects depends on the optical characteristics of the contaminating cryofilm. The optical properties are conveniently summarized by the complex index of refraction: $\hat{n} \equiv n + ik$, where n is the (real) refraction index and k is the absorption index. The plus symbol in this definition is appropriate for plane wave electromagnetic radiation with a time dependence proportional to $\exp(-i\omega t)$.

Roux et al. (Refs. 1-3) have identified common molecular contaminants from exhausts and in the atmosphere and obtained their optical constants. These include the i.r.-active molecular species CO_2, H_2O, NH_3, CO, NO, CH_4, HCl, and N_2O; unburned fuels such as MMH and N_2H_4, oxidizer N_2O_4, and the i.r.-inactive species N_2, Ar, and O_2. The i.r. transmittance spectra and optical constants of pure CO_2, H_2O, and NH_3 films on a 20-K germanium (Ge) substrate are given in Ref. 2, along with the transmittance spectra of several mixtures (also on a 20-K Ge substrate): N_2/CO_2, N_2/H_2O, N_2/NH_3, Ar/H_2O, and H_2O/CO_2. The transmittance spectrum of a simulated plume mixture ($N_2/H_2O/CO_2/CO$) appears in Ref. 1. In this report the optical properties of these mixtures are tabulated and discussed along with the spectra and optical properties of N_2/CO, CO_2/CO, $N_2/CO/CO_2$, and CO/H_2O cryofilms on 20-K Ge. Of particular interest are the changes in the optical properties of cryofilms as the proportions of their molecular constituents are varied. Changes in intermolecular interactions are shown below to be dependent on the type of molecular association (e.g., Van der Waals, or hydrogen bonding) and the concentrations of molecular species in the cryofilm.

Complete experimental details are given in Ref. 4; thus, only a basic outline of the chamber and apparatus is presented here. The absolute transmittance of thin solid films ranging in thickness up to 15 μm is presented. Finally, a theoretical model of window plus film transmission is derived and is subsequently employed with the experimental results to determine the complex refraction index ($\hat{n} = n + ik$) of each of the above-mentioned species. The subtractive Kramers-Kronig treatment for calculation of the film refractive index has also been employed, and results are compared to those of the nonlinear least-squares determination.

Instrumentation

Figure 1 shows the central portion of the path followed by the i.r. beam from the i.r. interferometer (Digilab Model FTS-14), through the 4-mm-thick Ge substrate (or window) of the stainless steel infrared optical transmission chamber (IROTC), to the pyroelectric detector. The interferometer is capable of 0.5-cm^{-1} resolution; its wave number accuracy is 0.02 cm^{-1}. A liquid-nitrogen (LN$_2$) cooled liner allowed a vacuum of 10^{-8} Torr to be achieved in the IROTC.

The window holder cools the Ge substrate with the use of either gaseous He (20 K) or LN$_2$ (80 K). For all of the measurements described in this report, the temperature of the Ge substrate was 20 K. Germanium is a commonly employed optical component because of its high thermal conductivity and flat transmittance of 47% between 2 and 12 μm. The three platinum resistors located on the window holder yielded temperature measurements accurate to within 0.5 K.

Controlled contamination of the cryocooled germanium window was accomplished with a gas induction system which

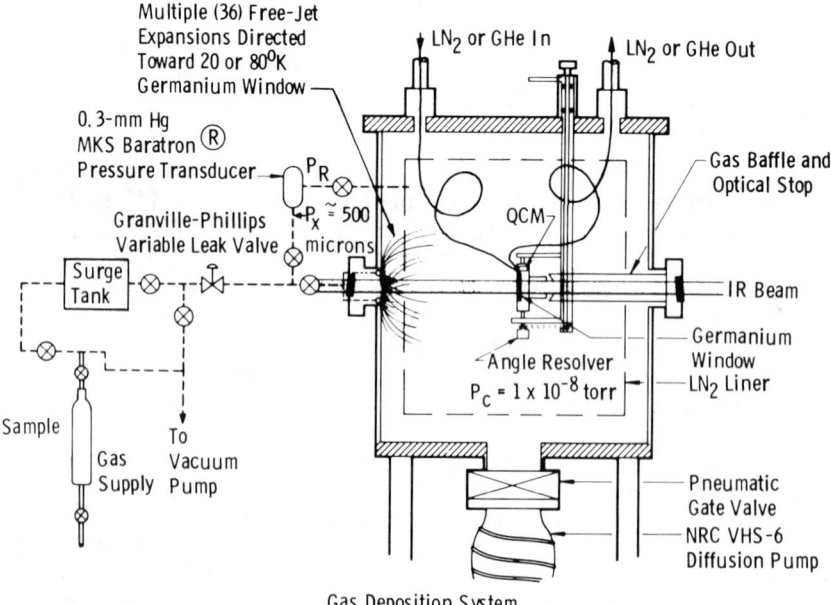

Fig. 1 Gas deposition system and i.r. Optical Transmission Chamber.

included a toroidal header with 36 1/16-in.-diam orifices spaced 10 deg apart to direct the gas toward the germanium window. The chamber pressure rose from 1×10^{-8} to 2×10^{-7} Torr during deposition at 20 K. Gas was prevented from condensing on the back of the germanium window by a gas baffle positioned close to the back of the window holder. The thin-film thickness was uniform across the exposed 2-in.-diam window area; the color of the reflected light from the entire area of the cryofilm/germanium changed uniformly with increasing film thickness. Film uniformity and absolute thickness were two important parameters since the objective of the experiment was to determine the complex refractive index of the thin film, a quantity derived by comparison of experimental transmittance data with theoretical model thickness data. Any error in absolute film thickness is directly introduced into the film complex refractive index results.

Procedure

When the LN_2 liner was filled and the cryostat was used to cool the Ge window, the holder, and the transfer line to nearly 20 K; the chamber pressure was in the low 10^{-8}-Torr range. After pressure and thermal equilibriums were reached, a reference i.r. power spectrum was recorded and stored by the interferometer while the Ge window was out of the i.r. beam. Before a Fourier transform was done, 16 interferograms at 4 cm^{-1} were usually co-added to ensure obtaining a large signal-to-noise ratio. The process was repeated with the Ge window in the i.r. beam path for zero and nonzero film thicknesses. The ratio of the reference power spectrum to the sample power spectrum (the absolute transmittance) was computed every 2 cm^{-1} and then plotted.

Research grade CO_2, CO, NH_3, or Ar gas in lecture bottles was introduced at the gas supply (Fig. 1). Before entering the chamber, water vapor, boiled off from distilled water in a vacuum, was purified further by a mechanical pump that removed foreign gases. For contaminant films of more than one constituent, the nominal mole fractions of a molecular species were found from the gas partial pressures. The chemical laboratory at AEDC revised these estimates of the mole fractions using chromatography methods on samples of the original gas mixture.

As the gases condensed onto the Ge substrate, the intensity changes (due to thin-film interference) of two helium-neon laser beams reflected from the film were monitored. The rays had incident angles of θ_a and θ_b (typically 18 and 68 deg). The refractive index at $\lambda = 0.6328$ μm was calculated from

$$n = \left[(\sin^2 \theta_b - \beta^2 \sin^2 \theta_a)/(1 - \beta^2)\right]^{\frac{1}{2}} \quad (1)$$

where β is the ratio of the period of the interference pattern at θ_b to that at θ_a. The film thickness d_1 was found for each experiment from the relation

$$m\lambda = 2d_1 (n^2 - \sin^2 \theta)^{\frac{1}{2}} \quad (2)$$

where m is the order of the interference maximum at incident angle θ. In practice, θ_a was the incident angle used in Eq. (2) because of its smaller periods. The orders m were usually integer, except for the occasional half-integral values used for highly absorbing films.

A quartz crystal microbalance (QCM) adjacent to the Ge window monitored the surface density (mass per unit area) of the condensed gases. The film surface density and the interference patterns from the two laser beams were used to calculate the film index of refraction at 0.6328 μm, the film thickness, and the film density (mass per unit volume). At a given wave number, the mathematical determination of the optical constants n and k requires transmittance measurements of at least two film thicknesses. However, in this work, measurements from more than two thicknesses were used to overdetermine n and k with a nonlinear least-squares computational technique so that more accurate constants could be obtained. Since some films fractured or shattered more easily than others at large thicknesses, the number of film thicknesses at which transmittance measurements were made varied from 8-26 for the different mixtures.

Optical Constants

The model used to interpret the normal absolute transmittance of cryofilms deposited on a substrate is pictured in Fig. 2, and has been discussed in detail in Ref. 2. The film is considered to be homogeneous, uniform in thickness d_1, and deposited on a homogeneous substrate having a uniform but much greater thickness d_2 than the film. The degree of coherence of the normally incident i.r. beam is assumed to allow thin-film interference in the cryofilm but not in the substrate. The infinite number of rays reflected from the 2-3 interface in Fig. 2 undergoes thin-film interference in the contaminant. The intensities of the infinite number of rays transmitted through the interface add to a finite sum.

The absolute transmittance at normal incidence, $T(\nu)$, of the film and substrate can be written as (Fig. 2)

$$T = \hat{t}\hat{t}* \qquad (3)$$

where

$$\hat{t} = \hat{t}(\nu, d_1, d_2, \hat{n}_0, \hat{n}_1, \hat{n}_2, \hat{n}_3) \qquad (4)$$

is the complex normal transmission coefficient of the film and substrate.

The exact expressions[4] for Eqs. (3) and (4) are algebraically long and tedious but straightforward; these expressions involve Fresnel's equations for the interface reflectances and transmittances. In the present work, media 0 and 3 were considered as vacuums (i.e., $\hat{n}_0 = \hat{n}_3 = 1$), the Ge substrate thickness was 4 mm, and its real refractive index was a modification of the Herzberger and Salzberg model which accounts for the refractive index change due to the cryogenic temperatures.

The reliable data domain lay between 700-3700 cm^{-1}. For wave numbers less than 700 cm^{-1}, the absorptions attributable to the substrate are large. Actually, k of the substrate is not precisely zero near 760 or 850 cm^{-1}, either. These weak Ge absorptions cause fictitous contributions of approximately 5×10^{-3} to the absorption index of the films, and should be regarded as spurious. At wave numbers greater than 3700 cm^{-1}, the low signal-to-noise ratio of the interferometer makes the transmittance data inaccurate. Measurements of the transmittance were obtained by interferometer every 2 cm^{-1} in the wave number domain from 500-4000 cm^{-1}. Least-squares computations of the optical constants of the cryofilms were normally made every 10 cm^{-1} between 700-3700 cm^{-1}, although the wave number domain was often extended to include absorption features slightly beyond the endpoints of the 700-3700-cm^{-1} domain. However, only the values between 700-3700 cm^{-1} are reported. The computations were normally performed every 2 cm^{-1} near noticeable absorptions.

In the nonlinear least-squares algorithm, estimated values of the optical constants were required at every wave number; the values used were usually the converged values of the preceding calculation. As noted by Roux et al. (Ref. 2), the least-squares values of k were more reliable than the n values, especially near strong absorptions, such as the narrow $^{12}CO_2$ ν_3 band, at which the optical constants changed rapidly with wave number. This result can be explained by the fact that portions of the i.r. beam pass through the film many times so that changes in the beam's attenuation are more strongly influenced by molecular absorptions, which determine k, than by reflection losses, which determine n.

As a consequence, there are many pairs of n and k values that, if introduced into our theoretical model, can yield transmittance values within the experimental uncertainties of the observed values for a given film thickness. The calculated n values ranged from near 0-10 in strong narrow absorptions, but k was much better determined, typically having a range of ±10% of the mean value. Occasionally, especially near the peak of a strong narrow absorption, the converged value of n given by the nonlinear least-squares algorithm was far out of line with the surrounding n values.

The uncertainties in the least-squares n values mentioned above have led Roux et al. (Refs. 1-4) and others to use Kramers-Kronig (KK) techniques to compute n from the least-squares values of k. In this work, the subtractive

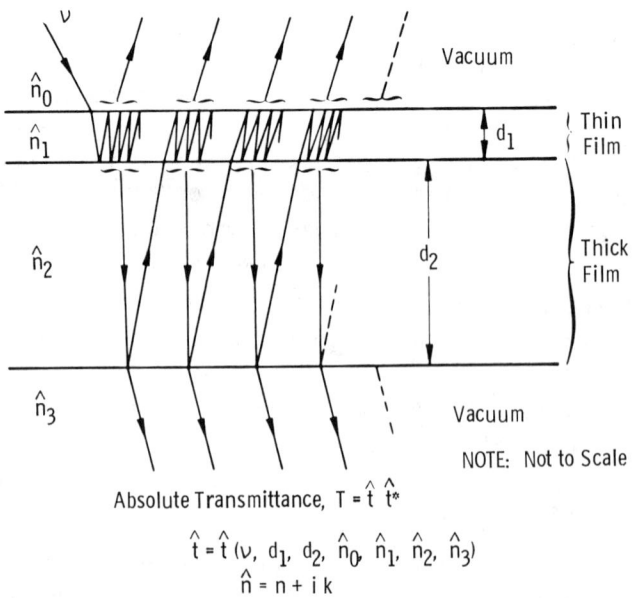

Fig. 2 Geometry depicting analytical model for a thin film formed on a thick film.

Kramers-Kronig (SKK) relation

$$n(\nu) = n(\nu_0) + \frac{2(\nu^2 - \nu_0^2)}{\pi} P\int_0^\infty \frac{\nu' k(\nu') d\nu'}{(\nu^2 - \nu'^2)(\nu_0^2 - \nu'^2)} \quad (5)$$

is also used to find $n(\nu)$. Here, the real index $n(\nu_0)$ is that found at wave number ν_0 by a separate measurement, and P denotes that the Cauchy principal value of the integral is to be taken. The reference wave number ν_0 was always chosen from a region where there were no strong absorptions and where n was very slowly varying. The least-squares determined value of $n(\nu_0)$ was used in the SKK computation. [One of the advantages of the SKK treatment of data is that an error of Δ in $n(\nu_0)$ merely shifts all of the computed n values by the same amount Δ, preserving the shape of the n curve.] Usually, the data domain in the SKK calculation ranged from 700-3700 cm^{-1}; however, if prominent features lay just beyond this domain, then these data were also included. The additional data, even if slightly inaccurate, help make the SKK determinations more precise near 700 and 3700 cm^{-1}. In every case, k was assumed always to be zero outside the data domain.

The least-squares values and the SKK values of n agreed except, as noted above, near the peaks of strong narrow absorptions. In all cases, the plotted SKK n values gave smoother curves than did the least-squares values, and the ranges of the SKK n values in regions of anomalous dispersion were not as large as the ranges of the least-squares values.

Errors in the refractive index, thickness, and density are all estimated to be within $\pm 2\%$. The error in the absorption index varies from about $\pm 2\%$ at high k values (≈ 0.01) (or greater) to a value which is dependent on k at the bottom of the absorption index graphs. The value of k can vary over several orders of magnitude; therefore, the smaller the k value the larger the error. This is because of the error limits on the film substrate transmittance measurements. For all the results shown below, tabulated data are available in Ref. 4.

Results

The intermolecular interactions encountered in the mixtures are basically the two types shown in Fig. 3. The attractive Van der Waals interaction is the relatively weak

1. Van der Waals:

2. Hydrogen Bonding:

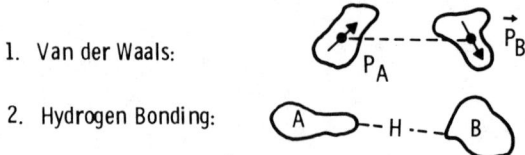

Fig. 3 Intermolecular interactions.

dipole-dipole force that is proportional to the separation distance raised to the -7th power. In the hydrogen bonding interaction, a proton acts as a bridge linking two molecules, one of which acts as a donor of the H atom. In many instances this association is strong enough to consider the two bonded molecules as a quasimolecule, which has noticeable changes in its spectra from the spectra of the separated molecular species.

CO_2 and 75% N_2/25% CO_2 Films

At the top of Fig. 4 appears the transmittance spectrum of a solid CO_2 film on 20-K germanium. Superimposed on the regularly spaced thin-film interference features, or channel spectra, are absorption bands of $^{12}C^{16}O_2$ and $^{13}C^{16}O_2$ molecules. Strong lattice combination bands give asymmetries to the ν_3 band of both $^{12}CO_2$ (2346 cm^{-1}) and $^{13}CO_2$ (2284 cm^{-1}). The ν_2 band (667 cm^{-1}) and the $2\nu_2 + \nu_3$ band (3602 cm^{-1}) of $^{12}CO_2$ also appear. There are prominent Ge substrate absorptions for wave numbers less than 650 cm^{-1}.

The bottom transmittance spectrum is that of a film containing three N_2 molecules for every CO_2 molecule. The intensities of the CO_2 absorption bands have decreased as expected, and the $^{12}CO_2$ lattice combination bands have almost disappeared, indicating a drastic reduction in the number of pure CO_2 microcrystals. Upon comparison of the two spectra, one can see their great similarity, except for the noted intensity differences, and can conclude the molecular interactions are of the Van der Waals type.

The mathematical model was utilized to obtain the optical constants k and n for N_2/CO_2 films and the results are presented in Fig. 5. The analysis essentially subtracts the effects of multiple reflections and the spectral characteristics of Ge substrate, as the k spectrum at the top shows. Two sets of n values are plotted on the bottom.

SOLID MIXTURES

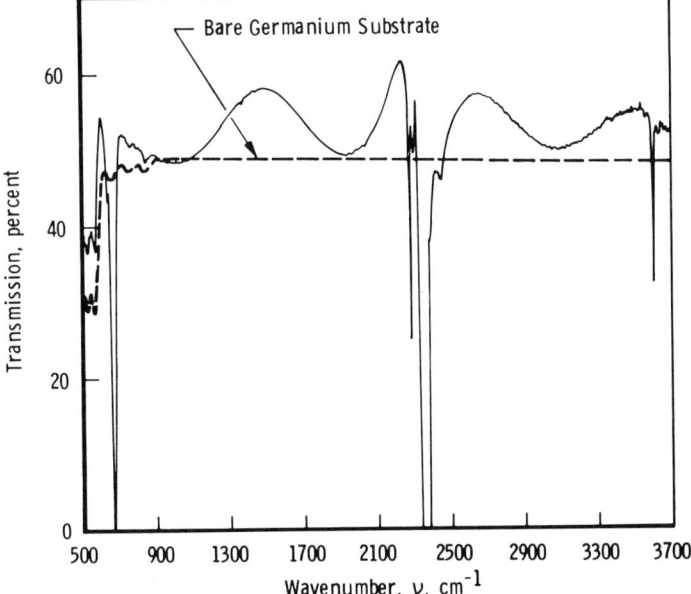

Transmittance of 3.88-μm-Thick Solid CO_2 on 20°K Germanium.

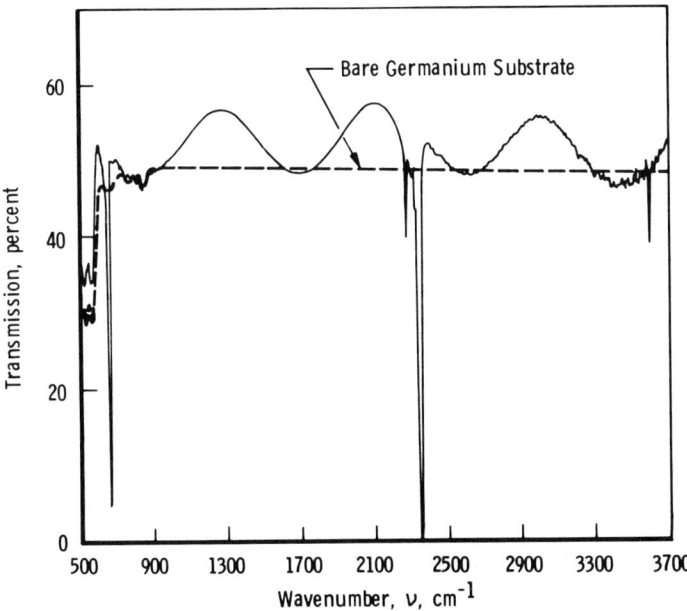

Transmittance of 4.74-μm-Thick Solid CO_2/N_2 Mixture (25%/75%) on 20°K Germanium.

Fig. 4 Transmittance of CO_2 and N_2/CO_2 films.

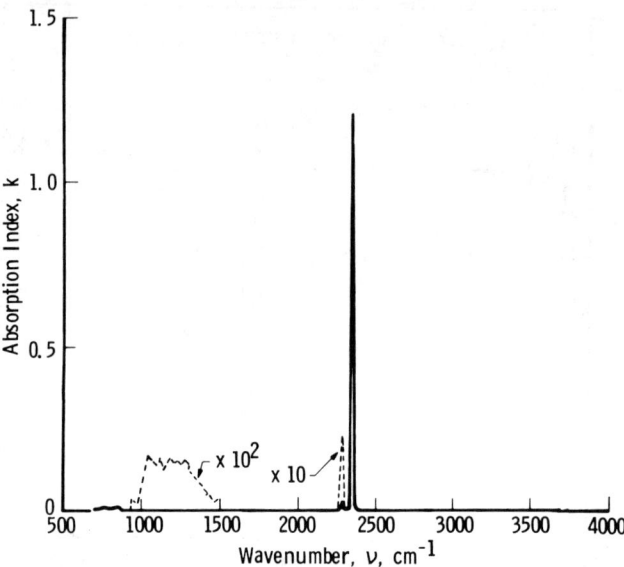

Absorption Index of Solid N_2/CO_2 Mixture (75%/25%) on 20°K Germanium

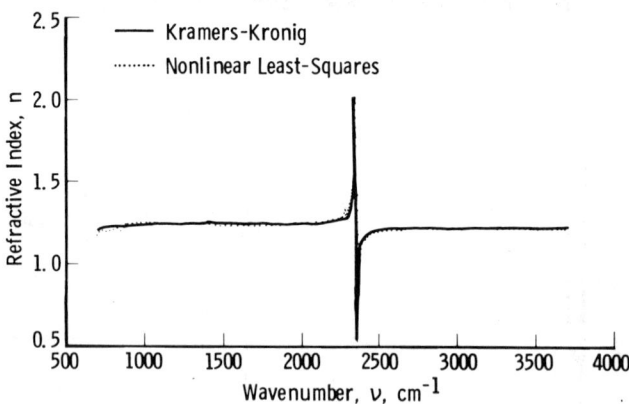

Refractive Index of Solid N_2/CO_2 Mixture (75%/25%) on 20°K Germanium.

Fig. 5 Optical constants of N_2/CO_2 films.

The dotted line is from the original least-squares analysis of the transmittance data. The solid line spectrum is the result of a Kramers-Kronig analysis of the least-squares k values seen above.

NH_3 and 85% N_2/15% NH_3 Films

Unlike CO_2, NH_3 molecules can participate in hydrogen bonding with adjacent molecules causing some unique spectral features. At the top of Fig. 6 there is a transmittance spectrum of a pure NH_3 film on 20-K Ge, and in the middle is the spectrum of a 20-K film mixture containing about 6 N_2 molecules for every NH_3 molecule. The overall reduction of NH_3 absorption in the N_2/NH_3 film is not surprising. However, in the pure NH_3 film, the ν_3 band (3380 cm^{-1}) is

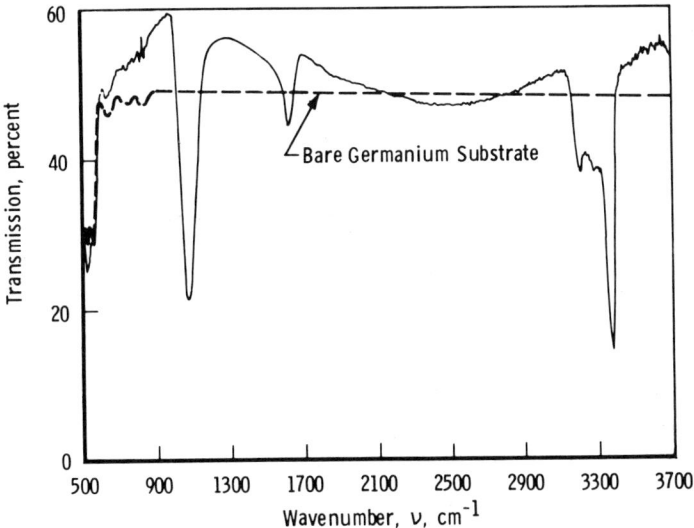

Transmittance of 1.43-μm-Thick Solid NH_3 on 20°K Germanium.

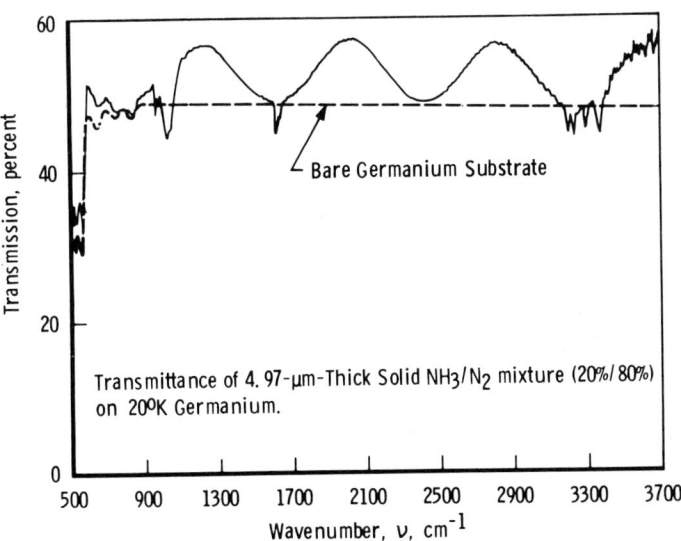

Transmittance of 4.97-μm-Thick Solid NH_3/N_2 mixture (20%/80%) on 20°K Germanium.

Fig. 6 Transmittance of NH_3 and N_2/NH_3 films.

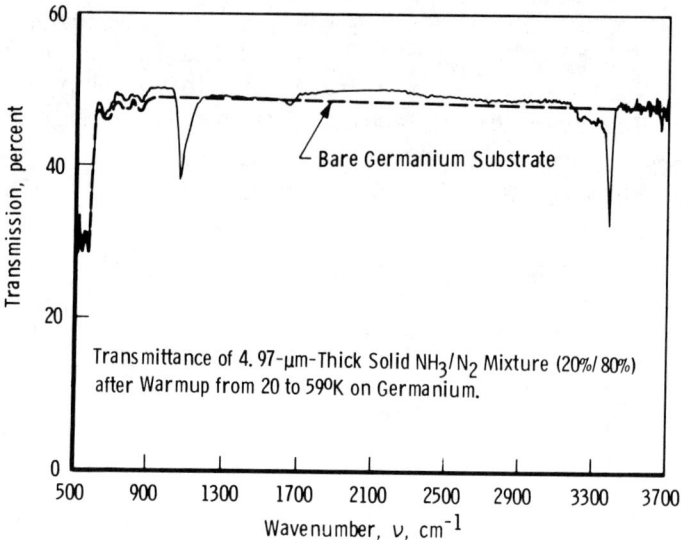

Fig. 6 (cont.) Transmittance of NH_3 and N_2/NH_3 films.

broader and has a much greater intensity relative to the neighboring $2\nu_4$ (3210 cm^{-1}) and ν_1 (3300 cm^{-1}) bands, than does the ν_3 band in the N_2/NH_3 spectrum. We attribute the greater intensity to a greater degree of hydrogen bonding among NH_3 molecules in the pure NH_3 film.

The bottom transmittance spectrum is that of the N_2/NH_3 film after it warmed up to 59 K. The channel spectra have disappeared indicating the film is no longer uniform due to evaporation and migration. More N_2 molecules than NH_3 should evaporate at this temperature and ambient pressure (10^{-6}-10^{-7} Torr). The higher concentration of NH_3 is expected to increase the degree of hydrogen bonding in the film. In fact the relative intensity and shape of the ν_3 band is now very similar to what it is in the pure NH_3 film at the top.

In general, the absorptions of H-stretching vibrations like ν_3 broaden and increase in intensity as more individual molecules form hydrogen bonds. In some instances sizeable shifts in frequency occur as well.

91% CO/9% H_2O Films

In Fig. 7 the transmittances of a thin film of H_2O molecules trapped in a matrix of CO are compared at two

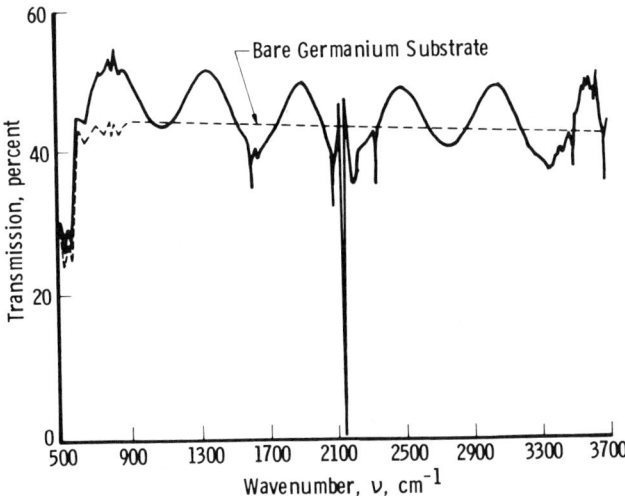

Transmittance of 6.54-μm-Thick Solid CO/H$_2$O Mixture (91%/9%) on 20°K Germanium.

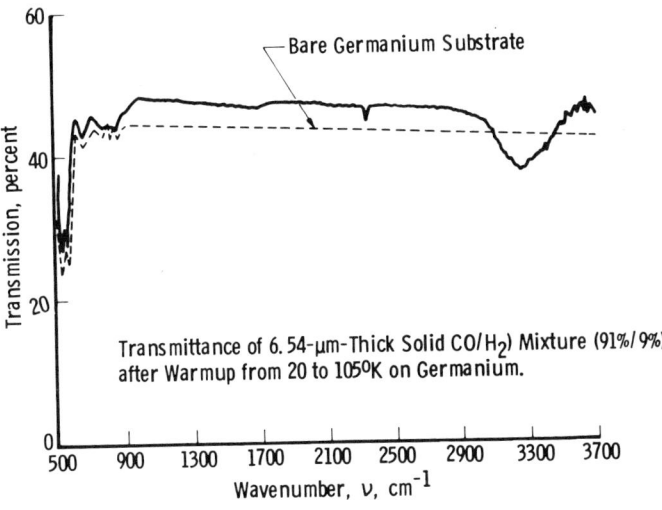

Transmittance of 6.54-μm-Thick Solid CO/H$_2$O Mixture (91%/9%) after Warmup from 20 to 105°K on Germanium.

Fig. 7 Transmittance of CO/H$_2$O films.

different temperatures, 20 and 105 K. In the 20-K spectrum at the top there is the intense $1 \leftarrow 0$ band of CO at 2141 cm^{-1}. There are also three weak bands near 3500 cm^{-1} due to H$_2$O; $2\nu_2$ (3236 cm^{-1}) and the fundamental O-H stretching bands ν_1 and ν_3, which are superimposed upon the larger channel spectrum.

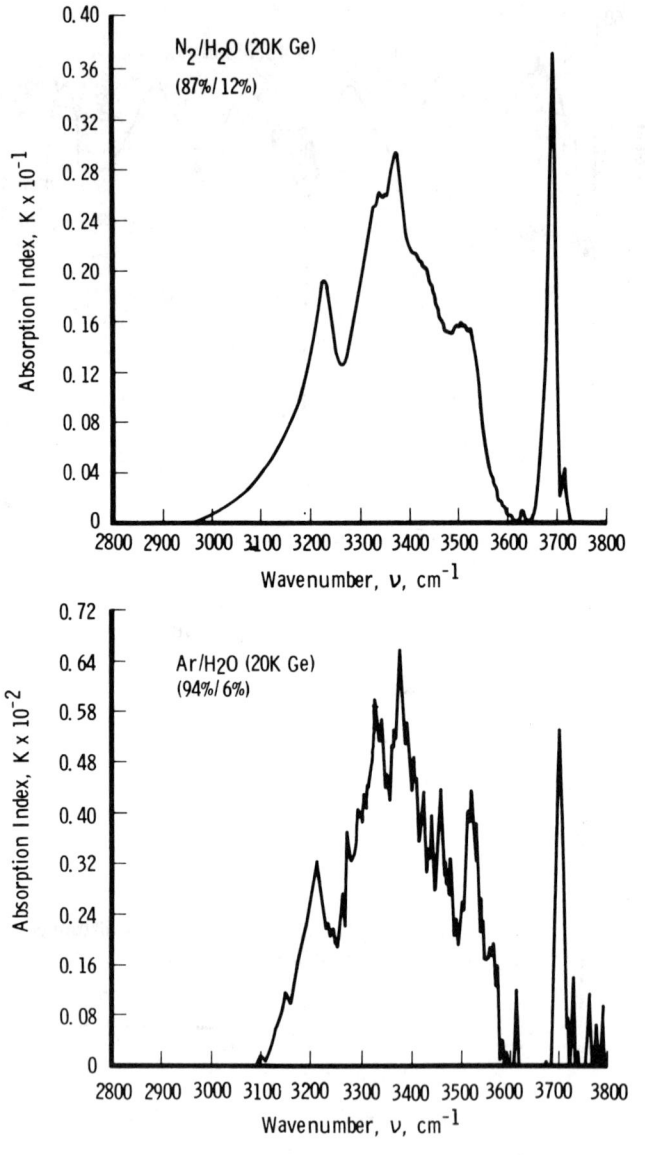

Fig. 8 k spectra (2800-3800 cm^{-1}) of Ar/H$_2$O and N$_2$/H$_2$O films.

In the 105-K spectrum at the bottom, the channel spectrum has disappeared because of evaporation, mostly by CO molecules. The broad absorption near 3300 cm^{-1}, which may look like a remnant of the channel spectrum, is actually due to hydrogen-bonded H$_2$O molecules. In addition to the

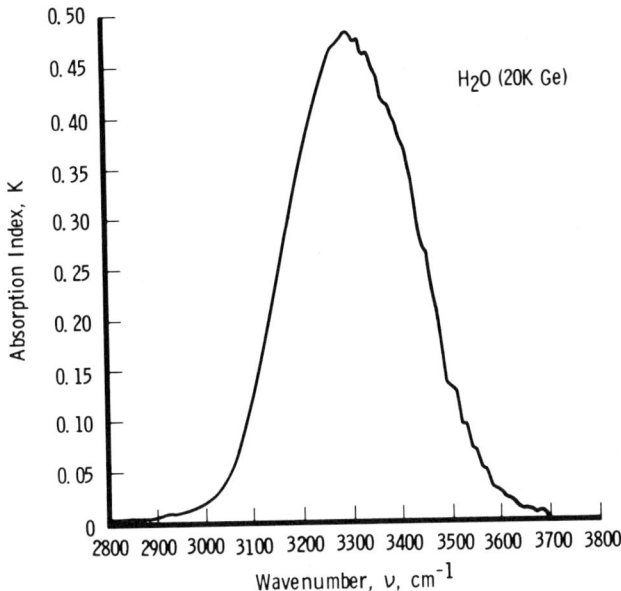

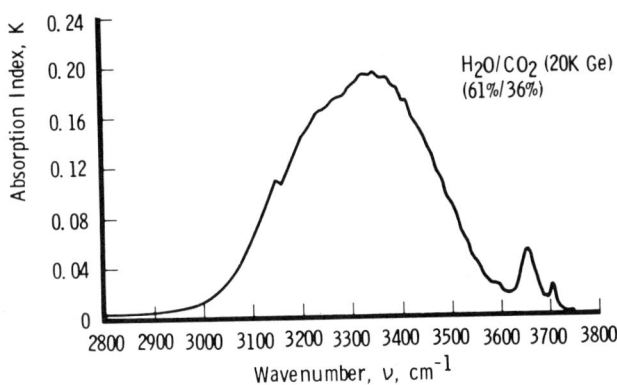

Fig. 9 k spectra (2800-3800 cm^{-1}) of H_2O/CO_2 and H_2O films.

increase in absorption intensity that was seen previously in hydrogen-bonded NH_3 spectra, the ν_1 and ν_3 bands have shifted downward in wave number by 200-300 cm^{-1}.

94% Ar/6% H_2O, 87% N_2/12% H_2O, 61% H_2O/36% CO_2 and Pure H_2O Films

Figures 8 and 9 indicate how the k-spectrum structure of the $2\nu_2$, ν_1, and ν_3 bands of H_2O changes with H_2O concentration in films. At the bottom of Fig. 8 is the k-spectrum of 94% Ar/6% H_2O films at 20 K. Some of the prominent peaks like ν_3 at 3730 cm^{-1}, are due to vibrational modes of unassociated, or monomer, H_2O molecules. Most of the other peaks are identified as perturbations of the monomer bands by dimer, trimer, and higher multimer associations of H_2O molecules.

The k spectrum at the top of Fig. 8 is that of 87% N_2/12% H_2O films which have twice the H_2O concentration as the Ar/H_2O films. Except for the ν_3 band, the peaks have blended together and increased in intensity. (Compare the scales of the Ar/H_2O and the N_2/H_2O spectra.)

At the bottom of Fig. 9 is the spectrum of 61% H_2O/36% CO_2/(2% N_2) films. There is a five fold increase from the N_2/H_2O films in the concentration of H_2O molecules, however, the integrated intensity has increased almost 20 times. The band at 3703 cm^{-1} is the $\nu_1 + \nu_3$ band of CO_2. The ν_3 band of H_2O has shifted to lower wave numbers and has blended with the other bands.

The k spectrum at the top of Fig. 9 is that of pure H_2O films at 20 K. The number of H_2O molecules per unit volume in pure H_2O is about 1.7 times that of the H_2O/CO_2 films. However, the integrated intensity of the pure H_2O spectra is about 2.4 times that of the H_2O/CO_2 spectra. Again, the intensities are not proportional to the concentration of H_2O molecules.

Summary

Experimentally determined transmittance measurements of condensed gases on a cryogenically cooled germanium window were made. Thin films of mixtures of CO, H_2O, CO_2, NH_3, N_2, and Ar were deposited at 20 K. The infrared spectral transmittance was studied over the 500-3700 cm^{-1} range (the 2.70 to 20.0-μm wavelength range). The film thickness was accurately determined using a two-angle laser interference technique. The densities of the films were determined by using quartz crystal microbalances (QCMs) and by calculating the refractive index at 0.6328 μm to allow accurate thickness measurement. From the transmittance data, the optical properties n and k were determined for the 700-3700 cm^{-1} range. These results are available in graphical and tabular[4] form for a wave number interval of 2 cm^{-1}. The spectral changes caused by Van der Waals interactions among molecules are predictable and very minimal. In fact a nearly linear relationship between the integrated absorption intensity of a non-hydrogen-bonded species and its concentration in a solid film was seen. Hydrogen-bonded species have spectra that can change drastically with the degree of hydrogen bonding, and this seriously hampers the interpretation of the spectra. More work needs to be done on the problem of hydrogen bonding in solid mixtures. It would be especially pleasing to find a method for relating intensities to the concentrations of the hydrogen-bonded species. These data are important for evaluating the influence of contamination on the performance of spacecraft optical surfaces and sensors.

Acknowledgments

The research reported herein was conducted by the Arnold Engineering Development Center (AEDC), Air Force Systems Command (AFSC). Work and analysis for this research were done by personnel of Calspan Field Services, Inc./AEDC Division, operating contractor for the aerospace flight dynamics test facilities at AEDC. Further reproduction is authorized to satisfy needs of the U. S. Government.

References

[1] Roux, J. A. and Wood, B. E., "Infrared Optical Properties of Solid Monomethyl Hydrazine, N_2O_4, and N_2H_4 at Cryogenic Temperatures," *Journal of the Optical Society of America*, Vol. 73, Sept. 1983, pp. 1181-1188.

[2] Wood, B. E. and Roux, J. A., "Infrared Optical Properties of Thin H_2O, NH_3, and CO_2 Cryofilms," *Journal of the Optical Society of America*, Vol. 72, June 1982, pp. 720-728.

[3] Roux, J. A., Wood, B. E., Smith, A. M., and Plyler, R. R., "Infrared Optical Properties of Thin CO, NO, CH_4, HCl, N_2O, O_2, N_2, Ar and Air Cryofilms," AEDC-TR-79-81 (AD-A088269), Aug. 1980.

[4] Palmer, K. F., Wood, B. E., and Roux, J. A., "Infrared Optical Properties of Solid Mixtures of Molecular Species at 20°K," AEDC-TR-80-30 (AD-A094214), Jan. 1981.

Measurements of Infrared Optical Properties of Al_2O_3 Rocket Particles

W. L. Konopka,* R. A. Reed,* and V. S. Calia†
Grumman Aerospace Corporation, Bethpage, New York

Abstract

High temperature (1700-3000 K) optical properties of Al_2O_3 rocket particles have been determined from mid-i.r. (1.3 to 4.5 μm) measurements of Al_2O_3/argon aerosols generated in a shock tube. The particles were collected from the exhausts of two different aluminized solid propellant rockets and subjected to comprehensive chemical and physical assays to determine contamination levels and particle size distributions. Observed rocket particle emissivities are low, but are nevertheless significantly greater than for pure Al_2O_3 particles. Comparison of the rocket particle absorption indexes against pure Al_2O_3 particles indicates that below the melt point, the mid-i.r. emission is determined by impurities, and above the melt point, it is determined by the bulk liquid properties of the particle material. For pure particles, this creates a jump in effective emissivities of over four orders of magnitude, whereas for contaminated particles the jump is smaller, depending upon contamination level.

Presented as Paper 83-1568 at the AIAA 18th Thermophysics Conference, Montreal, Canada, June 1-3, 1983. Copyright © American Institute of Aeronautics and Astronautics, Inc, 1983. All rights reserved.

*Senior Research Scientist, R&D Center.
†Head of Experimental Fluid Dynamics, R&D Center.

Introduction

The high-temperature, mid-i.r. optical properties of Al_2O_3 rocket exhaust particles have been determined from infrared measurements in controlled laboratory experiments with Al_2O_3/Ar mixtures (aerosol clouds). Specifically, the single particle spectral emissivities and spectral absorption indexes have been determined in the wavelength region from 1.3-4.5 μm for temperatures above and below the melt point of Al_2O_3, 2330 K. The particles investigated were captured in the exhausts of two full scale solid rocket motor firings at Arnold Engineeering Development Center, using a specially designed water ingestion probe.[1] These measurements were made because of the concern for the significant effect of contaminants on the thermal continuum radiation observed from rocket particles and because of the need for high-temperature optical data for liquid-phase Al_2O_3 (T > 2330 K).

It is widely agreed[2-8] that predictions based upon bulk aluminum oxide optical data are inadequate due to the presence of chemical impurities, and Pluchino and Masturso[9] have shown the importance of carbon and metallic aluminum impurities in a theoretical study of spherical core-mantle two-zone particles. Reliable data on realistically contaminated particles for comparison against theory have not been available. In addition, data for liquid phase Al_2O_3 has been determined exclusively using either flames or rocket motor exhausts, where control over particle properties is difficult.

Gryvnak and Burch[2] measured the high temperature optical properties of pure crystalline Al_2O_3 (sapphire) and also of 99% pure ceramic Al_2O_3 material (alumina) in the 0.6-4.0 μm spectral interval. As expected, the intrinsic absorption coefficient of the pure crystal was extremely small (sapphire is an excellant transmitting material for i.r. lenses and windows) and increased with increasing temperature. Upon reaching the melting point (2330 K), however, the absorption coefficient increased discontinuously by a factor of 30 to 40. Precise measurements of the liquid-phase properties were not possible in their experiment. The emissivity of the ceramic material was greater than that of the pure crystal at the same temperature, but showed the same general trend as it was heated up through its melting point. The effect of the impurities upon bulk ceramic Al_2O_3 emissivities was also studied in Ref. 3 and emissivities on the order of 0.2-0.4 were observed for relatively modest levels of contamination (1%).

Adams[4] measured the optical properties of micron sized Al_2O_3 particles injected into an H_2/O_2 flame and Mularz and Yuen[5] performed similar measurements in a propane jet. Carlson[6] performed another study using an Al_2O_3/water slurry injected into an H_2/O_2 fueled rocket motor operated as an atmospheric pressure burner (i.e., with nozzle removed). The data from these flame and rocket motor experiments show a considerable spread in values and, in some cases, they indicate disparate trends with temperature. Adams and Carlson observed a discontinuous increase in particle emissivity consistent with the observations of Gryvnak and Burch. Mularz and Yuen, on the other hand, observed no discontinuity in their investigation.

Worster and Kadomiya[7] collected Al_2O_3 particles in a high-altitude simulation chamber firing and measured room temperature optical properties of the powder material. They observed absorption coefficients corresponding to particle emissivities (1 μm radius) ranging from ~ 0.3 in the visible to ~ 0.15 at 5 μm, but recommended that their data be considered interim until laboratory studies on particles from real engines were available. Morozumi and Carpenter[8] determined empirical Al_2O_3 particle emissivities of approximately 0.2-0.3 from mid-i.r. measurements in the exhaust of two large-scale rocket engines.

These previous attempts to characterize rocket particles based upon analysis of rocket exhaust radiation or upon low-temperature (< 400 K) measurements of captured particles have all been of limited value because of uncertainties in particle size distribution, temperature, and contamination type and concentration.

The present investigation provides new data obtained under accurately controlled conditions, in the reflected region of a conventional shock tube, for two well-characterized particle samples. To enable the data to serve as a useful guide to future work in the area of contaminated particles, the infrared measurements for each sample are supplemented with chemical and thermophysical analyses.

Experimental Approach

Aerosol Generation

The experiments were carried out in the reflected zone of a conventional shock tube. Particles were introduced into an inert carrier gas by entrainment in the gas flow behind the incident shock wave. To illustrate the introduction method, a schematic wave-particle diagram is shown in Fig. 1. Prior to the test, the particles are distributed on a thin mid-plane plate located near the test section end wall (roughly 5 shock tube diameters away). When the incident shock wave passes the particle pile the particles are continuously entrained in the post-shock flow and rapidly spread across the shock tube. Subsequent interaction with the turbulent stagnant gas behind the reflected shock further distributes the particles to form the test medium. The resultant equilibrium aerosol cloud is uniform in space and time as has been demonstrated[10] by laser Doppler velocimetry (LDV) and extinction measurements of particle concentration at a variety of axial and radial locations within the aerosol cloud. In addition, the aerosol cloud is relatively free of particle agglomeration, which is essential in particle optical properties determination from cloud measurements, because the particle distribution must be accurately determined. This feature has been demonstrated[10] by comparison of LDV measurements of particle concentration

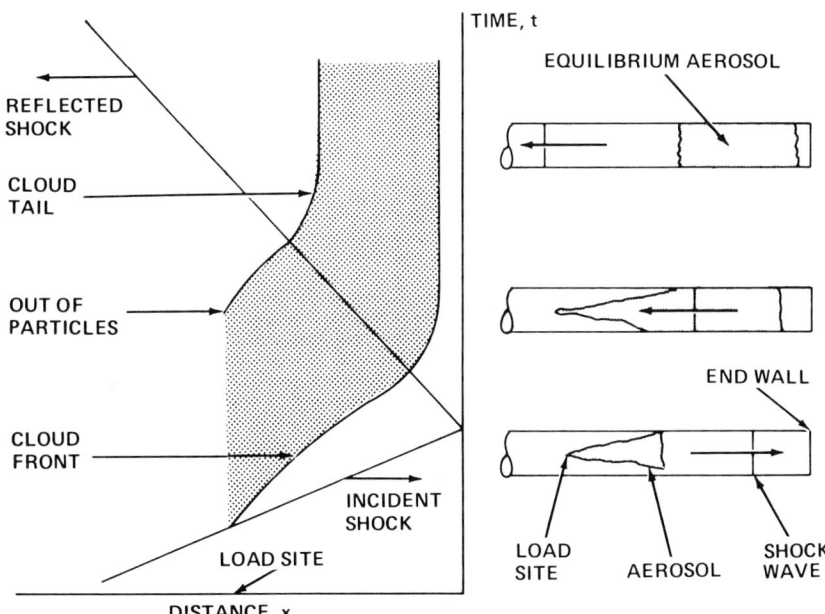

Fig. 1 Spatial & temporal development of aerosol cloud in a shock tube. Behind the incident shock, the particles spread with roughly 20° total included angle (measured from glow photographs). Mean velocity turbulence levels behind the incident shock (10 to 15 m/s, LDV) and in the stagnation zone (20 to 30 m/s, LDV) help to produce the uniform test medium.

with extinction measurements at 0.49, 0.63, 3.39, and 10.6 µm (i.e., no agglomeration was detected by enhanced laser extinction at long wavelengths where agglomerates have substantially larger cross sections than single particles).

The gas/particle mixture properties are predicted quite accurately from the usual shock-tube equations using measurements of the incident shock-wave velocity, the particle concentration, and the mean particle size. Work reported in Ref. 10 established that:

1) Particle temperature is controllable to better than 3% for lightly loaded gas/particle mixtures over the full temperature range of this investigation (i.e., 1500-3000 K).

2) Particle agglomeration is negligible, largely due to the high dynamic pressure employed to generate the aerosols and the removal of moisture by pre-evacuation prior to the test (i.e., size distribution is not biased by agglomeration).

3) The cool shock-tube walls permit measurement of low radiances from weakly emitting aerosols with a minimum of background interference.

Shock wave speed is measured by the time of arrival of the incident shock wave at two or more pressure gages a known distance apart. Particle concentration and size are determined by in situ multiwavelength extinction measurements and a posteriori by automated analysis of SEM

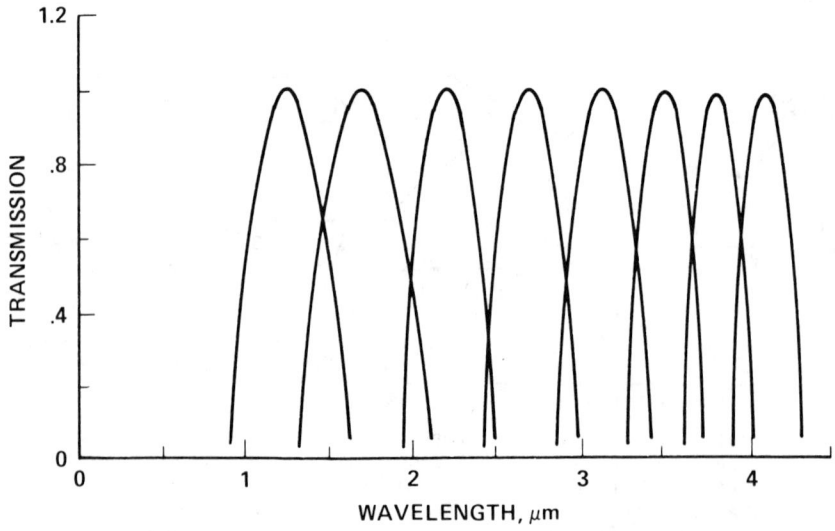

Fig. 2 Spectral response of 8 element spectrophotometer.

photographs of in situ collected samples. The samples were collected on small copper witness stubs located on the end wall of the shock tube. The aerosol cloud edge diffuses onto the stubs during and after the test.

Instrumentation

The shock tube was fully instrumented with 20 channels of pressure, emission, and extinction transducers. Spectral emissivity was measured using a low resolution 8-channel InSb spectrophotometer with the spectral response shown in Fig. 2. In addition, three radiometers operating at 1.3, 3.3, and 4.5 μm and three colinear extinction probes operating at 0.488, 0.632, and 3.39 μm were used for each test. All data with bandwidth requirements under 1 MH_z are recorded with a digital oscilloscope and time division multiplexer combination which is directly interfaced to our HP2108 RTE computer system. The oscilloscope contains two high-speed converters, each capable of a 12-bit conversion every 0.5 μs with sufficient memory to hold 2K word/channel. The multiplexer can divide each channel into 20 channels. Data with higher band width requirements are either recorded on 100 MHz analog scopes or an 8-bit A/D converter with a 2 K memory and a minimum sample time of 10 ns.

Radiometric calibrations for the fixed radio meters were made in a straightforward manner with a blackbody reference source via the Planck function $B(\lambda,T)$ and the center wavelength λ_0 of the radiometer narrow band (0.1 μm) filters. Array calibrations were somewhat more complicated due to the wide spectral response of each element. Average intensities were calculated by integrating the Planck function over the transmission envelope of each element $F(\lambda)$:

$$B(T) = \frac{\int B(\lambda,T) F(\lambda) d\lambda}{\int F(\lambda) d\lambda} \qquad (1)$$

In addition, the short wavelength element center wavelengths were corrected for temperature via:

$$\lambda_0 = \frac{\int \lambda B(\lambda,T) F(\lambda) d\lambda}{\int B(\lambda,T) F(\lambda) d\lambda} \qquad (2)$$

This latter correction was particularly important at the shortest wavelength $\lambda_0 = 1.26$ μm, where the apparent center shifted to λ_0 = 1.334 μm, changing the calibration by 25% (1260 K Blackbody). Radiometric measurements were typically accurate to 15% overall. Since the detectors were InSb, signal-to-noise ratio was always best at longer wavelengths.

Particle Optical Properties: Data Analysis

In each test, the shock-tube conditions were set up to produce a low particle opacity aerosol cloud (τ_p < 1) at the desired particle temperature. The aerosol cloud spectral emissivity and extinction behavior was measured in the wavelength region 1.3-4.5 μm. From, these data, the single particle spectral emissivity and absorption index was determined as follows. First the single particle emissivity $Q_a(\lambda)$ was determined from the aerosol cloud emission $I(\lambda)$ by:

$$\frac{I(\lambda)}{B_\lambda(T_a)} = \{1 - \exp(-Q_a \frac{\tau_e}{Q_e} \ell)\} f \qquad (3)$$

where $B_\lambda(T_a)$ is the Planck function at wavelength λ and temperature T_a, τ_e the extinction optical depth, Q_e the particle extinction efficiency, which approaches the asymptotic value of 2 in the visible, ℓ the path length, f is a correction factor, close to unity, based upon Monte Carlo analysis of multiple scattering.

The particle absorption index n_2 was then computed from $Q_a(\lambda)$ using the Dave' Mie code[14] and measured particle size distribution; n_2 is defined by n(complex) = $n_2 - in_2$. In the analysis, we have employed the recommendation of Gryvnak and Burch[2] for the real part of the complex index at elevated temperature. Derived values of n_2 are insensitive to the initial choice of n_1, ($dn_2/dn_1 \sim 0.01$).

The results are presented in the following section for two independently determined particle size distributions determined from in situ multiwavelength extinction measurements and from automated analysis of scanning electron microscope (SEM) photographs of in situ collected samples.

Particle Size Distribution Measurements

Uncertainty in the particle size distribution is the principal source of error in determining single particle optical properties from aerosol cloud measurements.[12] We have therefore taken care to characterize the particle size distribution by several methods and to estimate the uncertainty incurred in the final result (the absorption index n_2) by comparing derived values of n_2 for each size distribution.

Particle size distributions were determined from 1) in situ multiwavelength extinction measurements, 2) automated analyses of SEM photographs of in situ collected samples, and 3) the as-received particle material using a Coulter counter‡. These analyses show good agreement (Fig. 3). The mean particle radius in both rocket samples is small, approximately 0.1 µm, and the distribution decays roughly exponentially with increasing particle size. Within the accuracy of the measurements, the distributions can be approximated by (r in microns):

$$dn/dr = \exp\{-[\ln(10*r)]^2/(1.05)^2\}, \quad \text{rocket 1}$$

$$dn/dr = r^{0.46} \exp(-(6.0*r)), \quad \text{rocket 2}$$

Particle Physical & Chemical Properties

The Al_2O_3 particles used in our experiments were captured from the exhausts of two different full-scale solid-propellant rocket engines using a water ingestion probe. Particles from rocket 1 were captured from the core of the flow, and particles from rocket 2 in the periphery along a streamline tangent to the nozzle wall.

The particles were subjected to a variety of physical and chemical assays to determine their chemical form, and impurity content. The results are summarized in the following:

1) Visual inspection indicates that rocket 1 particles are grey and rocket 2 particles are brown (pure Al_2O_3 is white). SEM photographs show that the particles from both rockets are predominantly spherical in shape.

2) SEM/X-ray analysis and i.r. transmission spectra indicate that the particles are primarily Al and in the Al_2O_3 form. The presence of H_2O, CO_2, and possibly other hydrocarbons was observed in the i.r. transmission spectra.

3) Atomic emission spectroscopy, carried out at Cerac Inc.§ indicate a large number of metallic impurities for both rocket samples (Table 1). Unfortunately, the amount

---

‡Analyzed by Coulter Electronics, Inc, 590 West 20th Street, Hialeah, Florida.
§Cerac, Inc, 1316 W. St Paul Ave, Milwaukee, Wisconsin.

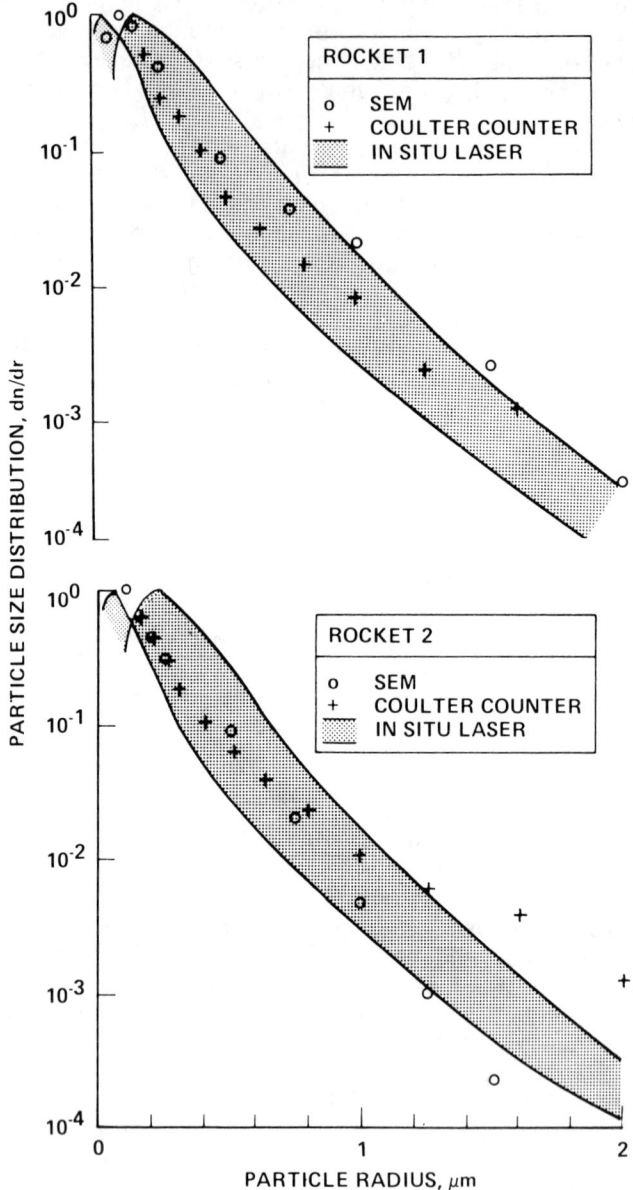

Fig. 3 Comparison of Al_2O_3 rocket particle size distribution determined from SEM, Coulter counter, and in situ multiwavelength extinction measurements. Range of distributions shown for the laser measurements correspond to log normal size distribution fits to the extinction data with distribution peak ranging from r = 0.05 to 0.20 μm.

Table 1 Spark spectrography analysis of metallic particle impurities in Al_2O_3

Metallic Impurity	Impurity weight classification,%		
	Rocket 1	Rocket 2	Commercial Al_2O_3
Ag	0.1-1.0	0.1-1.0	-
B	-	-	< 0.001
Ca	0.1-1.0	0.1	0.01
Cr	0.03	0.1-1.0	< 0.001
Cu	0.1	0.1-1.0	0.003
Fe	0.1	0.3-3.0	0.04
In	-	-	< 0.001
K	-	-	< 0.005
Mg	0.001	0.01	< 0.001
Mn	0.001	0.01	0.003
Na	-	-	< 0.01
Ni	0.01	0.1-1.0	0.006
Si	0.03	0.1-1.0	0.08
Sn	0.001	0.1	< 0.001
Ti	0.001	0.002	0.006

^a Analyzed at CERAC, Inc, 1316 W. St Paul Ave, Milwaukee, Wisconsin.

of free Al (not bound in the oxide form) is not specified in this test; i.e., unbound Al increases the absorption index of the material.

4) The carbon content of both rocket samples is below the threshold level (f 0.05% by weight) of a Leco carbon content analyzer (Cerac, Inc).

Results

The direct measurements of aerosol cloud i.r. radiance and laser beam extinction are related to single particle optical properties by

$$Q_a'(\lambda) = \frac{Q_a(\lambda)}{Q_e(vis)} = \frac{-\ln(1 - \varepsilon(\lambda))}{\tau_e(vis)} \quad (4)$$

where $\varepsilon(\lambda)$ is the aerosol cloud emissivity, λ is wavelength, τ_e is the extinction optical depth, vis is refers to visible wavelength (= 0.49 μm). $Q_a(\lambda)$ and $Q_e(vis)$ are the particle Mie efficiency factors for absorption and extinction, respectively, where Q_i is defined as the ratio of particle cross section for process i (i = absorption, extinction) to geometric cross section πr^2. For both rocket particle size distributions, $Q_e(vis)$

~ 2.50, so that Q_a' is roughly

$$Q_a' \simeq 0.40\, Q_a(\lambda) \qquad (5)$$

Figure 4 shows $Q_a'(\lambda)$ for the two rocket samples between 1700-2900 K. Both rocket particle samples have similar trends of Q_a' vs wavelength, but the temperature dependence is markedly different. Rocket 1 particles, the lesser contaminated of the two, have low emissivity at low temperature, and the emissivity rises at elevated temperatures to a plateau value above the melting point temperature (~ 2330 K). Rocket 2 particles, on the other hand, exhibit a nearly temperature independent emissivity which is almost as large below the melt point as above. Since the temperature accuracy of our data is 3%, we are unable to specify with certainty whether the measurements at the nominal temperature of 2375 K pertain to solid or

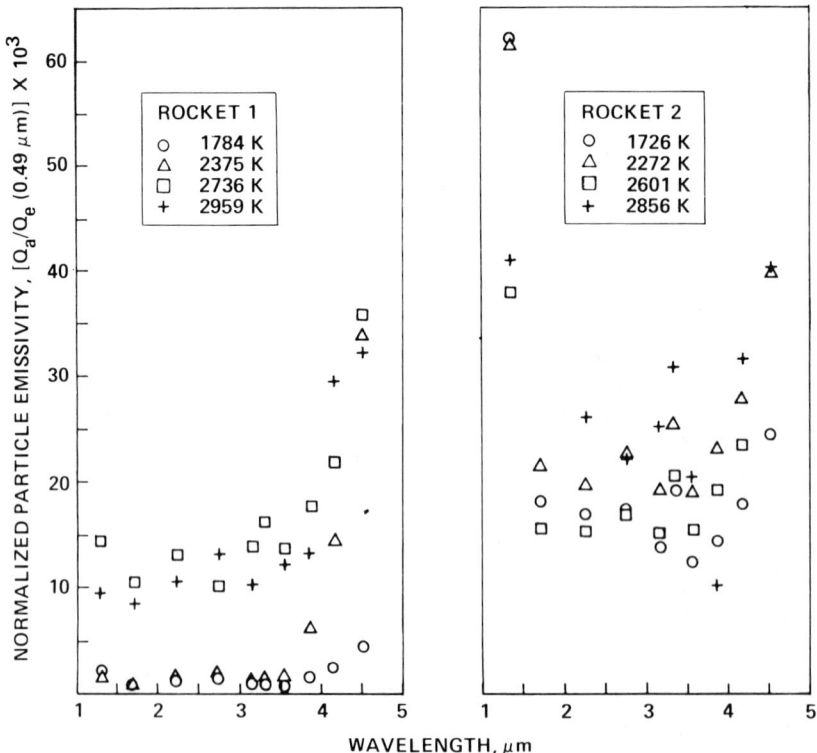

Fig. 4 Temperature and spectral dependence of Al_2O_3 rocket particle emissivity: normalization factor Q_e (0.49 μm) ≃ 2.5.

liquid particles. However, thermal analysis of particle melting lag times and comparison against the balance of our data at other temperatures, indicates that the aerosol particles were partially melted for this test.

We caution that since particle emissivity depends upon both the absorption index n_2 and particle size, the $Q_a(\lambda)$ values reported here are pertinent only to the specific particle sample we have investigated. Particle emissivities for different particle size distributions must be predicted using our absorption index data (to follow) and the appropriate size information.

Figure 5 shows the fully reduced absorption index for both types of rocket particles and pure Al_2O_3 particles. The rocket particle data points shown in Fig. 5 represent the average of n_2 values derived on the basis of the SEM and in situ laser size distributions, (Tables 2 and 3). Values of n_2 obtained using a Leitz TAS automated analysis of SEM photographs, in situ laser extinction, and Coulter counter size distributions all agree to within ± 30%. Additionally, since n_2 appears to be independent of

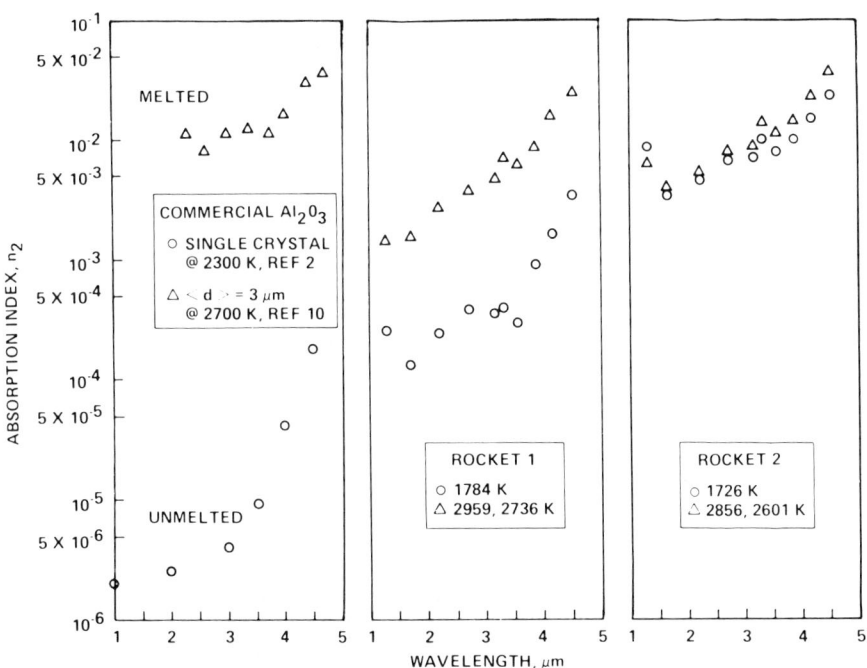

Fig. 5 Temperature and spectral dependence of Al_2O_3 absorption index for temperatures above and below the melt point (2330 K).

Table 2 Temperature and spectral dependence of Rocket 1 particle absorption index $n_2(\lambda,T)$ [a]

Wavelength, μm	Temperature, K							
	1784		2375		2736		2959	
	Particle Absorption Index, $n_2 \times 10^3$							
	Laser	SEM	Laser	SEM	Laser	SEM	Laser	SEM
4.50	4.00	2.50	30.50	19.00	32.00	20.00	29.00	18.00
4.15	2.00	1.20	11.40	7.10	17.00	10.60	23.00	14.40
3.86	1.10	0.70	4.20	2.60	12.10	7.70	9.00	5.70
3.54	0.43	0.27	0.81	0.52	8.00	5.20	7.10	4.50
3.30	0.47	0.31	0.65	0.43	8.30	5.40	8.50	5.50
3.16	0.43	0.28	0.57	0.38	6.40	4.20	4.90	3.20
2.17	0.44	0.31	0.68	0.47	3.70	2.60	4.70	3.30
2.20	0.27	0.20	0.39	0.28	3.40	2.50	2.70	2.00
1.70	0.14	0.11	0.14	0.10	1.90	1.40	1.50	1.10
1.30	0.27	0.22	0.22	0.17	1.90	1.50	1.20	0.97

[a] The absorption index, n_2 is shown for two independently determined particle size distributions: Distributions were determined from in situ extinction measurements and automated analysis of SEM photomicrographs of in situ collected particles.

temperature above the melt point, the results of the high-temperature runs have been averaged together. For pure Al_2O_3 below the melt point, the results of Gryvnak and Burch[2] are shown because the i.r. radiance of pure solid phase Al_2O_3 is below our system threshold and could not be measured in our experiments. Data for pure Al_2O_3 above the melt point is from our earlier work.[10] The precise value of n_2 for this case is somewhat uncertain because the particle size distribution in this prior experiment was not characterized as carefully as the rocket particle distributions of this paper. Additional measurements on this sample are planned.

The observed absorption indexes mimic the behavior already noted for the particle emissivity. Pure Al_2O_3 n_2 values are extremely low below the melt point, and even at 2100 K the particle cloud emission is below our system threshold. Above the melt point, n_2 for the pure material attains a significant temperature independent value in the range of 0.01-0.05. In the solid phase, rocket 1 n_2 values are larger than for the pure material, increase with temperature up to the melt point, and thereafter attain temperature-independent levels on the order of 0.003-0.03. Rocket 2 n_2 values for all temperatures are in the same 0.003-0.03 range, regardless of whether the material is solid or liquid.

Table 3 Temperature and spectral dependence of Rocket 2 particle absorption index n_2 (λ, T) [a]

Wave length, μm	Temperature, K							
	1726		2272		2601		2856	
	Particle Absorption index $n_2 \times 10^3$							
	Laser	SEM	Laser	SEM	Laser	SEM	Laser	SEM
4.50	19.0	27.0	31.0	45.0	-	-	31.0	45.0
4.15	12.0	17.0	19.0	27.0	16.0	22.0	21.0	30.0
3.86	8.5	12.0	13.0	19.0	11.0	16.0	5.2	7.6
3.54	6.4	9.2	9.6	14.0	7.8	11.0	10.0	15.0
3.30	8.5	12.0	11.0	16.0	9.1	13.0	14.0	19.0
3.16	5.8	8.4	7.7	11.0	6.0	8.7	8.1	12.0
2.71	5.4	7.7	7.1	10.0	5.3	7.6	7.8	11.0
2.20	3.9	5.5	4.5	6.3	3.6	5.0	5.1	7.2
1.70	2.9	3.9	3.5	4.7	2.5	3.4	4.2	5.7
1.30	7.5	9.5	7.4	9.4	4.6	5.8	4.9	6.3

[a] The absorption index n_2 is shown for two independently determined particle size distributions: Distributions were determined from in situ extinction measurements and automated analysis of SEM photomicrographs of in situ collected particles.

Discussion

The Al_2O_3 absorption index n_2, determined in our experiments, indicates large differences in the solid-phase. Absorption indexes for solid phase pure Al_2O_3, rocket 1, and rocket 2 are approximately in the ratio 1: $10^2:10^3$, respectively, with increasing emission corresponding to increased contamination level.

The occurrence of such large differences is not surprising when one considers that photon energies in the visible to mid-i.r. spectral region are too low to excite electronic interband transitions and too high to excite phonon lattice vibrations. Hence, the intrinsic absorption of the pure solid over a broad wavelength region (from uv to far i.r.) is near zero and, therefore, is determined almost exclusively by contamination. The general trend of the increased absorption at longer wavelengths is also typical of a wide variety of insulating materials and is due to free electron absorption introduced by the contaminants.

Obviously, there can be no one "good" set of Al_2O_3 rocket particle optical properties, at least in the solid phase, and the application of data obtained from one rocket engine to a different engine, or even to a different region in the nozzle exhaust of the same engine, should be viewed with extreme caution.

The discontinuous increase in the absorption index upon melting previously observed by Adams, by Gryvnak and Burch, and by Carlson is also observed in the present work for both the pure Al_2O_3 sample and rocket 1. The contamination level in rocket 2, however, appears to be sufficiently great to mask this transition, so that the absorption index for this particle sample is high regardless of whether the sample is melted or not. The sudden onset of strong absorption above the melt point is most likely due to the continuum of optically active modes that arises as soon as the crystal lattice structure is destroyed.

Mularz and Yuen are the only investigators who did not observe this discontinuity and attribute this to impurities (they used 99.9% pure Al_2O_3). Although this possibility cannot be ruled out, we believe that uncertainty in particle temperature due to thermal lag of molten droplets undergoing solidification is responsible. This view is supported by the appearance of a flat plateau region in their n_2 vs temperature data near 2300 K, which we associate with the temperature arrest of droplets releasing their heat of fusion. As previously noted, thermal lag also adversely impacts our own data in the 2375 K results, Fig. 4.

Above the melt point, the absorption indexes for all three samples are nearly temperature independent. Absorption indexes for rockets 1 and 2 agree to within a factor of 2 and both are in good qualitative agreement with the n_2 values for the pure material. Since the particle size distribution of the pure Al_2O_3 was not determined in detail in our previous investigation,[10] we attribute the observed differences in the liquid phase, at least between pure Al_2O_3 and rocket 1, to particle size uncertainty, and not to intrinsic differences in material properties.

Our absorption index determination for the liquid agrees well with the results of Adams, who derived a temperature independent value of n_2 = 0.0074 at a wavelength of 0.60 μm. The measurements of Mularz and Yuen in a propane jet at 1.2 μm and by Carlson in a rocket exhaust at 1.3 and 1.7 μm are approximately an order of magnitude below the present results. These investigators report a considerable dependence on temperature, corresponding to approximately an order of magnitude increase in n_2 between 2400 and 2800 K. We attribute this to inadequate control over particle temperature.

Our measurements have important implications for plume particulate emission models, such as that of Edwards and Bobco,[13] who assume that the emission of solid phase particles in the plume is negligible compared to that from the liquid particles. This assumption is evidently correct for rocket 1, but incorrect for rocket 2. Additionally, since the optical properties data of this model are based upon the early work of Carlson, the particulate emission from liquid particles is underpredicted by their analyses.

Conclusions

High-temperature optical properties of submicron sized Al_2O_3 particles captured from the exhausts of two full scale aluminized rocket motors were determined from i.r. measurements of accurately controlled Al_2O_3 aerosols produced in a shock tube. The measurements were made in the 1.3-4.5 μm wavelength range and 1700-3000 K temperature range. The rocket particle optical properties were compared to our earlier measurements of commercial grade pure Al_2O_3 particles. The results support the notion of two separate emission mechanisms for Al_2O_3. Below the melt point, the mid-i.r. emission is controlled by contaminants, and the ratio of absorption indexes from commerical grade Al_2O_3, rocket 1, and rocket 2 particles is approximately 1: 10^2: 10^3 respectively. Thus no one set of n_2 values can be recommended for solid phase Al_2O_3 particles. Above the melt point the absorption index appears to be controlled by the bulk properties of the liquid phase and the results for both rocket samples are within a factor of 2 of each other. The average value of these results is recommended for modeling liquid Al_2O_3. These observations are consistent with the general band structure of insulating materials.

Our measurements clearly demonstrate the need for systematic experimental characterization of contamination effects, along the lines of Pluchino and Al_2O_3 Masturso's[9] theoretical investigation. In particular, future measurements of captured rocket particles should be combined with additional experiments designed to identify and quantify the role of the principal types of impurities and structures on optical properties.

Acknowledgment

This work was partially supported by AEDC under Contract 04600-8-C-0007.

References

[1] Girata, P.T., McGregor, K., and Quinn, R., "Inertial Upper Stage (IUS) Solid Rocket Motor (SRM) Core Flow Sampling at High Altitudes," USAF AEDC-TR-83-1, (to be published).

[2] Gryvnak, D.A. and Burch, D.E., "Optical and Infrared Properties of Al_2O_3 at Elevated Temperatures," *Journal of the Optical Society of America*, Vol. 55, 1965, p. 625.

[3] Folweiler, R.C., "Thermal Radiation Characteristics of Transparent, Semi-Transparent and Translucent Materials under Non-Isothermal Conditions" USAF, ASD-TDR-62-719,1-115, 1964 [AD 600370].

[4] Adams, J.M., "A Determination of the Emissive Properties of a Cloud of Molten Alumina Particles," *Journal of Quantative Spectroscopy Radiative Transfer* Vol 1, 1967 p 273.

[5] Mularz, E.J. and Yuen, M.C., "An Experimental Investigation of Radiative Properties of Aluminum Oxide Particles," *Journal of Quantative Spectroscopy Radiative Transfer*, Vol 12, 1972, p. 1553.

[6] Carlson, D.J., "Emittance of Condensed Oxides in Solid Propellant Combustion Products," *Proc. 10th Int'l. Symp. Combustion*, 1965, p 1413.

[7] Worster, B.W., and Kadomiya, R.H., "Rocket Exhaust Aluminum Oxide Particle Properties," Aerodyne Research Report ARI-RR-30, Aug. 1973.

[8] Morizumi, S.J., and Carpenter,H.J., "Thermal Radiation from the Exhaust Plume of an Aluminized Composite Propellant Rocket," *Journal of Spacecraft and Rockets*, 501, Sept.- Oct. 1964, p 501.

[9] Pluchino, A.B. and Masturso, D.E., "Emissivity of Al_2O_3 *AIAA Journal.*, Vol. 19, Sept. 1981, p. 1234.

[10] Calia, V.S., Konopka, W., Reed R.A., and Oman, R.A., "Shock Tube Measurements of IR Radiation in Hot Gas/Particle Mixtures," *Proc. 13th Int'l Symp Shock Tubes and Waves*, July 1981, p. 676.

[11] Dave', J.V., "Subroutines for Computing Parameters of the Electromagnetic Radiation Scattered by a Sphere," Rept. 320-3237, IBM Scientific Center, Palo Alto, Calif, Rept. 320-3237, 1968.

[12] Egan, W.G., "Volumetric Scattering and Absorption by Aerosols: Parametric Sensitivity in Mie Modelling and Comparisons to Observations," *Applied Optics*, Vol. 21, 1982, p. 1445.

[13] Edwards, D.K. and Bobco, R.P., "Effect of Particle Size Distribution on the Radiosity of Solid Propellant Rocket Motor Plumes," AIAA paper 81-1052, 16th Thermophysics Conferences, Palo Alto, Calif., June 1981.

Improvements in Rocket Engine Nozzle and High Altitude Plume Computations

S. D. Smith*

Lockheed-Huntsville Research & Engineering Center, Huntsville, Alabama

Abstract

A knowledge of the structure of high altitude rocket exhaust plumes is necessary to solve on-orbit plume induced problems such as plume impingement heating, contamination, plume induced forces, and moments and base heating. Rocket exhaust flowfields are very complicated and are governed by many phenomena. Many simplifying assumptions are made to enable one to compute exhaust flows. However, many of these simplifying assumptions can compromise and invalidate the results, depending on the application for which the flowfield is intended. The purpose of this paper is to describe a computer code which allows the user to eliminate many of the simplifying assumptions and treat most of the phenomena which govern nozzle/plume solutions. Additionally, this code requires little or no interaction between the code and the user for the solution of a nozzle flowfield from the combustion chamber to the exhaust plume (including the backflow region).

Introduction

Rocket exhaust flowfields are very complicated and are governed by many phenomena. Many simplifying assumptions are made to enable one to compute exhaust flows. However, many of these simplifying assumptions can compromise and

---

Presented as Paper 83-1547 at the AIAA 18th Thermophysics Conference, Montreal, Canada, June 1-3, 1983. Copyright © American Institute of Aeronautics and Astronautics, Inc., 1984, All rights reserved.

*Staff Engineer, Computational Mechanics Section.

invalidate the results, depending on the application for which the flowfield is intended. Numerous codes are available that treat many of the governing phenomena, but no single code is available that treats reacting single- and multi-phase flows including boundary-layer effects as an integral part of the solution. Thus, previously it was necessary to use a multitude of codes to treat a nozzle/plume flow in detail. It is therefore desirable from both computational and economic standpoints to have a single code that can treat all the dominant phenomena in a rocket nozzle/plume flowfield. Additionally, it is possible to perform calculations which may range from the most simple (as for preliminary design studies) to the most complex as required for final design.

This paper will describe a nozzle plume flowfield code that has capabilities which do not presently exist in a single computer code. The RAMP code[1,2] which was developed by Lockheed under government funding was chosen as the basic code from which to work. The basic RAMP employs modular construction and has the following capabilities: 1) Two-phase with two-phase transonic solution, 2) Two-phase, reacting gas (chemical equilibrium, reaction kinetics), supersonic inviscid nozzle/plume solution, and is 3) Operational for inviscid solutions at both high and low altitudes.

During the course of the study the following capabilities have been added to the code:[3] 1) Direct interface with JANNAF SPF code,[4] 2) Shock capturing finite difference numerical operator, 3) Two-phase, equilibrium/frozen, boundary-layer analysis, 4) Variable oxidizer-to-fuel ratio transonic solution, 5) Improved two-phase transonic solution, 6) Two-phase real gas semiempirical nozzle boundary layer expansion, 7) Continuum limit criteria, and, 8) Sudden freeze free molecular calculation beyond the continuum limit.

Most of the above capabilities already exist in other computer codes. These codes were incorporated into the RAMP code to enhance its usefulness.

Future efforts to improve the code will be directed toward the following: 1) Simplified input, 2) More rigorous treatment of the boundary-layer/Prandtl-Meyer fan interaction, 3) Interface with plume impingement models, 4) Documentation, and 5) Verification of the applicability of the model in the backflow region.

This paper presents the present status of the code and presents some results of plume calculations for bipropellant and solid propellant motors.

ROCKET ENGINE NOZZLE AND PLUME COMPUTATIONS

RAMP Improvements

JANNAF Standard Plume Flowfield (SPF) Code Interface

To perform a plume calculation, the SPF code[4] requires nozzle exit properties as initial conditions. The RAMP code was modified to punch or put on tape exit plane data in the format that the SPF code uses. The code will punch data for both single- and two-phase cases.

Shock-Capturing Finite Difference Operator

It is desirable to have the capability to treat shocks that can occur in some nozzles. The original RAMP code has the logic for computing any number of right or left-running shocks using shock fitting techniques. This logic was partially verified under previous efforts. Shock capturing schemes require no special equations or program logic and are reliable. For most nozzle flows, existing shock capturing techniques are sufficiently powerful to treat the shocks. Additionally, in order to directly interface with the SPF code, it would be desirable to use a shock capturing numerical operator.

For the above reasons, a shock capturing algorithm was added to the code. The methodology, equations, and grid system which was incorporated into the code is identical to the SPF code.[4] To obtain more information on the scheme the reader should consult Ref. 4.

Equilibrium/Frozen Boundary-Layer Analysis

Nozzle boundary layers are known to influence certain regions of nozzle flow and high altitude exhaust plumes.[5,6,7] For nozzle designers, the nozzle boundary layer is important in determining the thermal loads to the nozzle, performance losses due to heat transfer and effects on the nozzle pressure distribution due to the displacement thickness effect on the inviscid flow structure. For spacecraft designers the nozzle boundary layer is important because of its effect on the exhaust plume. At high altitudes the nozzle boundary layer causes the plume to expand to large angles (approximately 180 deg). In these backflow regions, spacecraft and sensitive surfaces are subjected to unwanted contamination, forces, moments, and heating rates. For some applications the radiative properties of the expanded boundary-layer flow is important. Thus, it is easy to see that for many applications the nozzle wall boundary layer is an important factor.

The BLIMPJ boundary-layer code[8] was chosen for the solution of the nozzle wall boundary layer. BLIMPJ is a JANNAF standard boundary-layer code for determining boundary-layer effects on the performance of a rocket engine. The BLIMPJ code can treat nozzle flows for equilibrium or frozen chemistry, uses the JANNAF standard thermochemistry curve fits and has numerous ways to handle the nozzle wall boundary conditions that allow it to treat liquid engines as well as solid motors with and without ablative walls.

For most rocket motors the boundary layer is fairly thin (approximately 5-10% of exit diameter) and the resultant effect on inviscid flow properties is minimal. For these cases one pass through the inviscid nozzle solution and boundary-layer calculation is adequate. The boundary-layer results are then superimposed on the inviscid nozzle solution and an exit plane start line with boundary-layer effects is generated which can be used to perform a plume expansion.

Some low pressure or low thrust motors have boundary layers which contain a significant portion of the total mass flow of the system. In these cases the entire solution should be iterated by making two passes through the inviscid nozzle and boundary-layer calculations. After the initial nozzle and bounday-layer solution has been completed, the nozzle solutions will be calculated with the actual wall contour adjusted by the local displacement thickness which was determined from the boundary-layer calculation. The boundary-layer solution will then be rerun using the new edge conditions from the second nozzle calculation. The results will then be superimposed on the original nozzle contour and second flowfield solution, and an exit startline will be output or saved.

Particle Tracing Through Boundary Layer

Any particulate matter (Al_2O_3-solids or unburned propellant droplets in liquid motors) which might enter the nozzle wall boundary layer could be influenced by the boundary layer to expand into the backflow region. Thus, there may be certain motor/nozzle configurations in which it is necessary to track particles through the boundary layer in order to obtain the most accurate representation of the nozzle flow properties at the exit plane.

Lockheed-Huntsville added a particle streamline tracing module into the nozzle code. The code which was used as the basic building block of this module was used in predicting the IUS, SSUS particle distributions published in Ref. 9. This code uses initial particle properties (velocity flow angle temperature and density) and traces the particles through a known flowfield. This option is user-selectable.

Boundary Layer Expansion at Lip

Once a fully supersonic startline has been generated at the exit plane, the flow can be expanded around the lip using a Prandtl-Meyer expansion. The main problem with generating the startline is the treatment of the subsonic portions of the boundary layer and the boundary-layer lip interaction.

The present version of the code uses the replacement layer treatment of the subsonic portion of the boundary layer. This method conserves the mass of the subsonic flow using mass averaged properties at a slightly supersonic value, which results in a layer of constant properties near the wall. As experience is gained using the code, further effort will be spent on the treatment of the lip region so that more accurate predictions may be made in the lip dominated backflow region of the plume.

Variable Oxidizer-to-Fuel Ratio Transonic Solution

Solution of the subsonic transonic region of a liquid[10] rocket engine can vary in complexity from a simple one-dimensional variable O/F streamtube analysis to the most detailed model such as that of the distributed energy release (DER) model.[11] The streamtube analysis performs a multizone, one-dimensional calculation to the sonic point given a known O/F distribution just downstream of the injection face. The DER program is a complex model which is initiated upstream of the injector face and continues the solution up through the sonic line. The DER code was used to initiate nozzle solutions in Ref. 12 but is not particularly easy to use or input and requires a good bit of experience of the user to successfully execute. For these reasons it will not be utilized in the nozzle code. On the other hand, a one-dimensional streamtube analysis does not account for two-dimensional effects. A time-dependent scheme[13] is a compromise between these two schemes. The approach includes the radial momentum equation which results in a set of mixed partial differential equations. The solution procedure is an unsteady time-dependent finite-difference technique with equilibrium chemistry. This technique has both the equilibrium and variable O/F chemistry option and has been utilized[14,15] previously with excellent results. This code has been incorporated into the RAMP code as a module and has been executed for all combinations of ideal and equilibrium chemistry, constant, and variable O/F distributions.

Improved Two-Phase Transonic Solution

The original transonic module which was incorporated into the RAMP code could handle throat radii of curvature to throat radius ratios above 1.5. Many solid motors have radii of curvature ratios smaller than 1. To alleviate this limitation the improved approximate transonic module was taken from the new standard performance prediction program (SPP)[16] and put into the RAMP code.

Continuum Limit Criteria

There have been numerous studies over the past several years concerning methods of determining where continuum flow breaks down and free molecular flow begins. Examples of these criteria are the Knudsen number criteria, criteria based on Mach and Reynolds number, and the breakdown parameter as proposed by G. A. Bird.[17]

The present version of the code uses a Knudsen number criteria of 10 to check for free molecular flow transition. The code does keep track of the continuum breakdown criteria and locates the surface in the plume where the breakdown criterion of 0.05 is located. In the future the appropriate data at this surface will be stored so that this information may be passed along to a full Monte Carlo solution.

Free Molecular Flow Option

A sudden freeze free molecular flow option has been incorporated into the code. Once the Knudsen number criteria of 10 has been exceeded the flow is frozen. (Specific heat ratio, molecular weight, velocity and temperature along a gas streamline is held constant.) At this point the streamline is assumed to expand at a constant flow angle and the density varies according to the streamtube area. This option allows the calculations to be performed in the backflow region (>90 deg) of the plume. Numerous cases have been run for several hundred nozzle exit diameters.

Future Improvements

Input Simplification

Much of the input that is required by the RAMP code requires that the user have access to such things as particle size correlations, thermochemical curve fits, particle thermodynamics, reaction rate packages, and a step size control. Defaults for this information will be built into the code or stored on external files much as is done in the SPP code. The user will have the option to override these data. Additionally, the NASA-Lewis CEC program[18] will be put into the RAMP code as a module to facilitate generating the thermodynamic data necessary to perform a plume restart from an exit plane startline that includes a nozzle wall boundary layer. It is anticipated that input simplifications will make the RAMP code much easier and more reliable to use.

Boundary-Layer-Lip Interaction Model

Near the nozzle lip, the subsonic and a part of the supersonic portions of the boundary layer are influenced by the expansion process.[19,20,21] For highly underexpanded flows, the sonic line has been found to attach to the lip[19] so that the flow at the wall must rapidly accelerate when it gets near the lip and the static pressure rapidly decreases. For overexpanded flows the subsonic flow merely stays subsonic as it negotiates the lip so that downstream flow conditions can feed back up into the boundary layer.

These are exact solutions to the corner flow problem. Bird[19] uses a direct simulation Monte Carlo method. He set his model up to compute the entire nozzle starting at the region near the throat. He predicts the attachment of the sonic line for underexpanded nozzles. Baum[21] uses a finite-difference method along with boundary-layer equations to describe the subsonic portion of the boundary layer. His application was for base flow about blunt bodies and compares well with data. Finally, the GIM code[22] can be used to exactly solve the corner problem. All three of these methods are very complex and are outside the scope of this effort.

A more practical approach to adequately handling the lip problem is to develop a model of this region using integral relations, matched asymptotic or other semiempirical techniques. One of these methods will be selected and will be included in the nozzle/plume code.

There are existing published calculations of the lip region using Monte Carlo techniques. These results will be used to verify the model. Additional comparisons will be made using experimental data which exist for the lip dominated portion of the exhaust plume. The resultant model will provide a reliable, accurate representation of the boundary-layer-lip interaction zone.

Plume Contamination Model Surface Chemical Species Determination

Specification of the amount and type of plume exhaust products at a surface immersed in a rocket plume is important for determination of spacecraft surface degradation. After an adequate representation of the characteristics of the exhaust plume has been performed it is necessary to relate the spatial characteristics of the plume to a surface that is immersed in the plume.

Under previous studies, Lockheed-Huntsville developed the Lockheed plume impingement code.[23] This program will provide forces, moments, and heating rates to bodies immersed in the exhaust plume generated by the Lockheed method-of-characteristics program.[24]

The plume impingement program (PLIMP) will be modified so that it can use the results of the RAMP2 code to predict forces, moments, and heating rates due to plume impingement. Additionally, the PLIMP code will be modified to determine the types and amounts of plume species at any given location of a body immersed in the exhaust plume. Since the RAMP2 code has the capability of predicting free molecular flow properties, the PLIMP code will be modified so that free molecular flow data can be used to predict plume impingement surface characteristics (forces, moments, heating rates, and species distributions).

Present Program Capabilities

The present form of the RAMP code allows the user to perform a detailed nozzle and plume solution starting in the combustion chamber of the nozzle and proceeding several hundred nozzle diameters into the plume. The program will perform a first-order characterization of the backflow region of the plume. Any appropriate combination of the following capabilities may be used to perform the nozzle and plume solution:

(1) Single- or two-phase nozzles and plumes can be treated.

(2) The gas may be ideal or real. If the gas is real; frozen, equilibrium, or non-equilibrium chemistry assumptions can be made. The effects of oxidizer/fuel gradients may be considered.

(3) Two-dimensional or axisymmetric flow problem geometries can be used.

(4) Both upper and lower boundaries can be solid or free.

(5) Reacting gas solutions which are in chemical equilibrium have been facilitated by modifying the TRAN72[18] computer program as described in Sec. 2 of Ref. 2 to provide binary tape and punched output of its equilibrium or frozen real gas calculations at any desired O/F ratio(s) or total enthalpy(s).

(6) Hypersonic or quiescent approach flow options may be used.

(7) Exit to ambient pressure ratios from over expanded to highly under expanded are possible.

(8) A real gas nozzle boundary layer solution can be performed with no interface between the user and the RAMP code results. A frozen, equilibrium chemistry, turbulent or laminar solution is possible.

(9) The effect of a nozzle wall boundary layer on particles which enter the boundary layer is treated.

(10) An exit plane start line can be generated for the SPF code which includes two-phase flow and boundary-layer effects. Additionally exit plane data are available for use in other codes.

(11) Fuel striations (variable O/F) can be treated starting at the entrance to the nozzle throat region.

(12) The code will handle the two-phase transonic region for nozzles with throat radius of curvature to throat radius ratios of 0.5.

(13) Free molecular flow region is treated using a source flow approximation.

(14) Once the gas-particle flowfield solution has been obtained, the output tape may be used by the RAMP radial lookup program (described in Appendix A of Ref. 2) which determines the radial variations of flowfield properties across the nozzle and plume flowfields at constant axial stations. The plume impingement program (PLIMP)[23] may also be run to determine the effects of the rocket exhaust plume on objects immersed in the plume.

Backflow Model Applicability

The emphasis on the program development has been to provide an engineering tool that allows the designer to perform a detailed nozzle solution for most conceivable combustion chamber/nozzle/propellant combinations followed by a plume solution including the backflow region. The backflow region is calculated using as starting conditions the nozzle/boundary-layer solution at the exit plane and proceeding to perform the plume solution using either a continuum or continuum/sudden freeze free molecular approximation.

Previous studies[25,26] have shown that the continuum approximation (to within a few degrees of the limiting expansion angle) does fairly well for predicting mass flux distributions but Mach number and temperature distributions in the high Mach number (backflow) regions are not well characterized. For applications where contaminant mass fluxes are of interest the continuum approach is probably adequate. The "sudden freeze" source flow plume model that is presently in the program is a first order approximation of a collisionless plume. Temperature and Mach number predictions are more appropriate than continuum predictions for the backflow regions beyond the freeze line if the freeze line is adequately predicted. In order to establish how best to use the continuum or sudden freeze models which are presently in the code comparisons of program predictions with experimental data (other than QCM measurements) and flight data are necessary.

Examples of results of a continuum and continuum/sudden freeze free molecular calculation are shown in Figs. 1 and 2. Both plume solutions were initiated from an exit plane startline of a detailed solution of the Space Shuttle Vernier RCS nozzle.

Figure 1 presents contour plots of static pressure for the continuum assumption, while Fig. 2 presents static pressure plots for the continuum/sudden freeze assumption for the Vernier motor plume. A comparison of the two figures shows that pressure in the continuum region of both calculations are the same while pressure in the free molecular region is higher than that predicted by continuum theory.

Results

To date a multitude of problems has been solved using the RAMP code. However, three particular exhaust plumes have been selected to demonstrate the applicability of the new version of the code. These cases are: a 5-lbf bipropellant motor, the Space Shuttle reaction control system (RCS) motor and an early candidate first stage interim upper stage (IUS) solid motor.

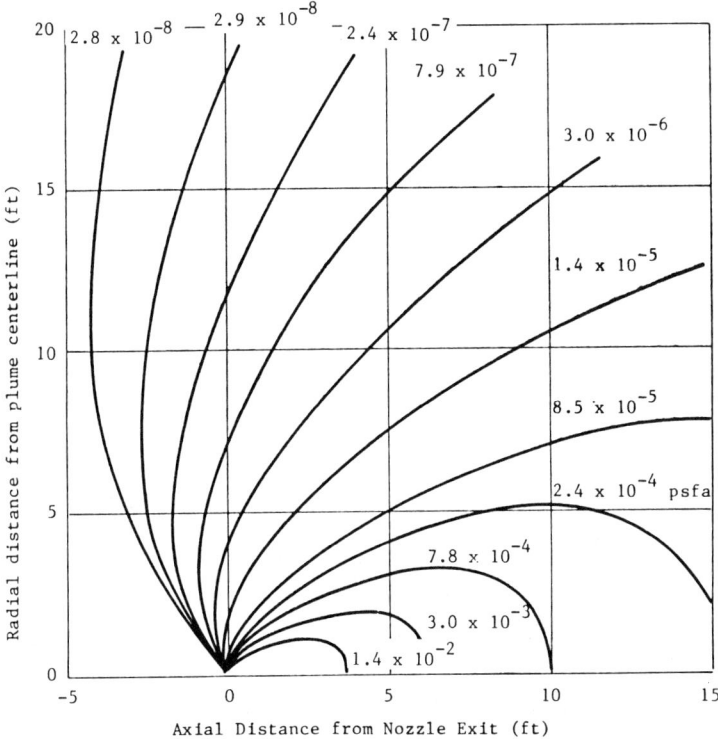

Fig. 1 Space Shuttle vernier motor continuum plume static pressure contours.

The 5-lbf bipropellant motor was used in a plume contamination test program[27] performed at AEDC. Part of the experimental data consisted of a survey of plume mass flux as a function of angle off the nozzle centerline. A comparison was made with the experimental data using results of a RAMP solution. Figure 3 presents the analytic and experimental results of normalized mass flux as a function of angle measured from the nozzle centerline. The analytic results compare very well with the measurements out to 150 deg. The flowfield solution was initiated at the nozzle throat assuming equilibrium chemistry at a constant oxidizer-to-fuel ratio. The thermochemistry was frozen chemically at a local/chamber pressure ratio of 0.2. Once the nozzle flowfield was completed, the boundary-layer module was executed and predicted laminar boundary-layer conditions at the exit plume. The boundary-layer results were merged with the inviscid exit plane conditions and the plume solution was performed. The plume solution considered variations in

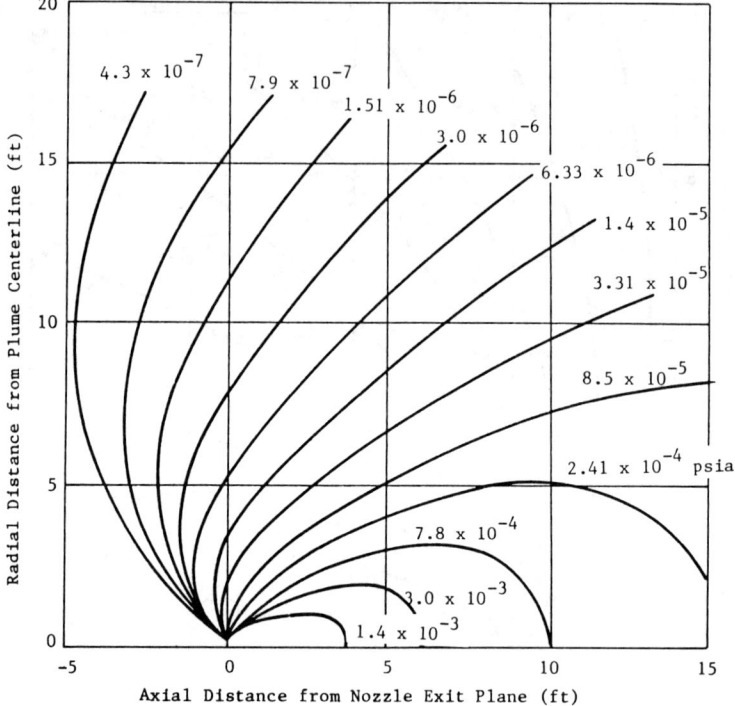

Fig. 2 Space Shuttle vernier motor free molecular plume static pressure contours.

total enthalpy due to boundary layer effects. A Knudsen number of 10 was used to determine where the plume went free molecular.

An accurate description of the Space Shuttle RCS motor is important due to the effect of the RCS exhaust on the large number of various payloads which will be placed in orbit by the Space Shuttle. The RCS motor is a bipropellant motor producing 870 lbf of thrust with a 20:1 area ratio contoured nozzle. The injector pattern of the combustion chamber was designed to produce an overall oxidizer-to-fuel ratio of 1.6, while providing film cooling of the combustion chamber walls. The injector pattern results in a distribution of O/F ratio across the inlet of the nozzle. Previous studies[14] have shown that O/F distributions can have a significant effect on the above characteristics. The RCS O/F distribution was inferred from the injector flow pattern and applied at the inlet to the throat region. The transonic module was utilized to give a supersonic start line downstream of the

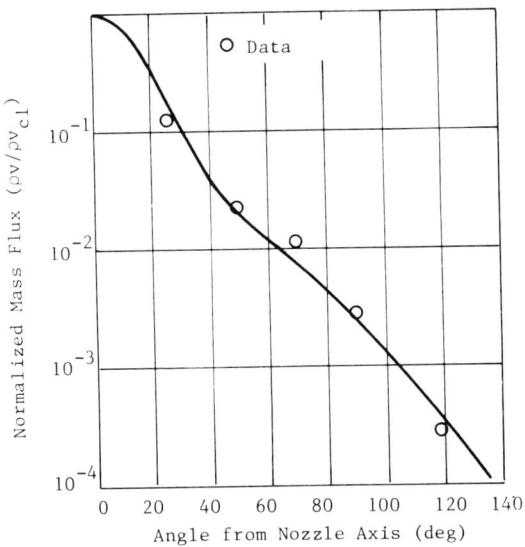

Fig. 3 Mass flux normalized by centerline mass flux versus angle from plume centerline for 5 lbf bipropellant motor.

throat. Equilibrium chemistry was used in the solution with chemical freeze points specified as a function of O/F ratio and local static pressure. The pressure to freeze the flow was determined from previous studies[14] which utilized finite rate chemistry results. The remainder of the solution was performed identically to the 5-lbf bipropellant motor except that a turbulent boundary layer was predicted and the plume was calculated via an ideal gas approximation (including total enthalpy variations) since the code presently cannot handle both variations in O/F ratio and total enthalpy.

Figure 4 presents density contour plots of the RCS exhaust plume plots of the RCS exhaust plume assuming the continuum flow approximation throughout the plume. Reference 28 presents a detailed definition of the RCS plume. Figure 5 shows a comparison of plume pitot pressure survey data with the results of the RCS plume calculations assuming both constant and variable O/F distributions. The constant O/F assumption results in an overprediction of the pitot pressure while the variable O/F results compare very well with the data. The effects of variable O/F ratio will be even more pronounced in the backflow region of the plume (> 90 deg) where the O/F ratio is considerably lower than the mean value of the motor.

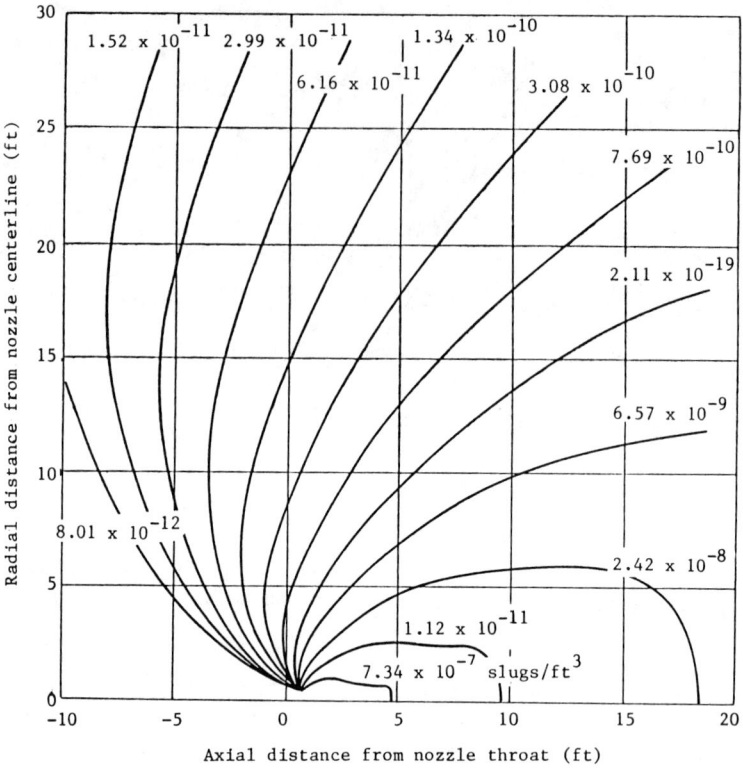

Fig. 4 Space Shuttle RCS motor exhaust plume density contours.

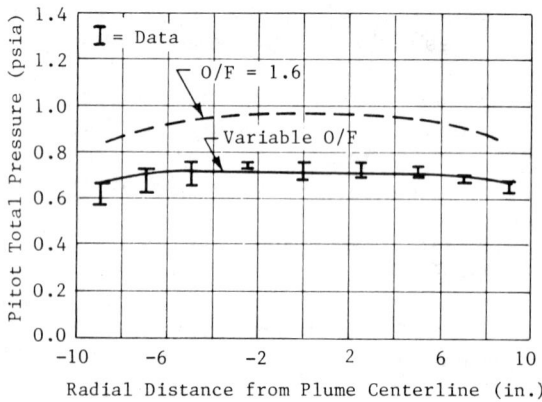

Fig. 5 Comparison of RAMP calculation for Space Shuttle RCS motor plume with pitot pressure survey taken at 45 in. from nozzle exit plane.

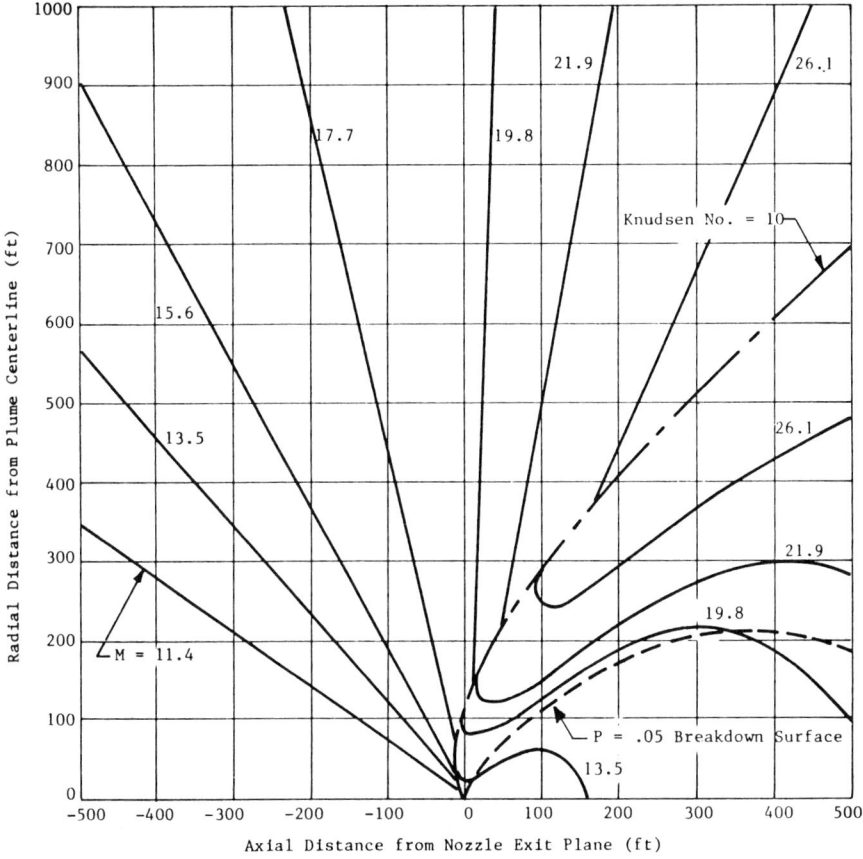

Fig. 6 Large Boeing motor exhaust plume Mach number contours.

The interim upper stage (IUS) is used to boost payloads from near Earth orbit to higher orbits. This stage uses solid rocket motors as propulsion. From a payload and stage design standpoint, it is necessary to define the exhaust plume. The RAMP code is especially suited to calculating the IUS exhaust plume including all the driving phenomena: equilibrium/frozen thermochemistry with total enthalpy variations, two-phase flow, nozzle wall boundary layer, and free molecular flow. Figure 6 presents contour plots for Mach number in the exhaust plume. Also Knudsen number of 10 included in this figure is the continuum breakdown surface corresponding to the Bird breakdown criteria of 0.05 and the boundary beyond which free molecular flow is treated. A more complete definition of the first-stage IUS plume will be published soon.

Conclusions

A computer code has been described which can be used to model the dominant phenomena which affect the prediction of a rocket nozzle and orbital plume flowfield. With a single computer run it is possible for the designer to predict a high altitude plume starting back in the combustion chamber. Further work is planned to verify the results of plume calculations to orbital and high altitude applications. The final version of the code will be a user-oriented design tool that will allow rocket nozzle/plume calculations that can range from the most simple preliminary design calcuations to final design predictions.

Acknowledgment

This work was performed under Contract NAS9-16256 for the National Aeronautics and Space Administration, Johnson Space Center, Houston, Texas.

References

[1] Penny, M. M. et al., "Supersonic Flow of Chemically Reacting Gas-Particle Mixtures-Volume I-A Theoretical Analysis and Development of the Numerical Solution," LMSC-HREC TR D496555-I, Lockheed Missiles & Space Company, Huntsville, Ala., Jan. 1976.

[2] Penny, M. M. et al., "Supersonic Flow of Chemically Reacting Gas-Particle Mixtures-Volume II-RAMP, a Computer for Analysis of Chemically Reacting Gas-Particle Flows," LMSC-HREC TR D496555-II, Lockheed Missiles & Space Company, Huntsville, Ala., Jan. 1976.

[3] Smith, S. D., "Improvement of Rocket Engine Plume Analysis Techniques-Final Report," LMSC-HREC TR D784753, Lockheed Missiles & Space Co., Huntsville, Ala., Jan. 1982.

[4] Dash, S. M., Pergament, H. S. and Thorpe, R. D.,"A Modular Approach for the Coupling of Viscous and Inviscid Processes in Exhaust Plume Flows," 17th Aerospace Sciences Meeting, New Orleans, La., Jan. 1979.

[5] Penny, M. M., "Analytical Data for the APS R-1E Engine Plume Impingement Forces and Moments on the S-IVB Protuberances and the OWS Panels (Small Version), LMSC-HREC D148628, Lockheed Missiles & Space Company, Huntsville, Ala., Jan. 1969.

[6] Boynton, F. P., "Exhaust Plumes from Nozzles with Wall Boundary Layers," Journal of Spacecraft and Rockets, Vol. 5, Oct. 1968, pp. 1143-1147.

[7] Simons, G. A., "Effect of Nozzle Boundary Layers on Rocket Exhaust Plumes," AIAA Journal, Vol. 10, Nov. 1972, pp. 1534-1535.

[8] Evans, R. M., "BLIMP-J User's Manual," UM-75-64, Aerotherm, Mountain View, Calif., July 1975.

[9] Smith, S. D., "Definition of Solid Particle Plume Flowfields of the Space Shuttle Interim Upper Stage (IUS, SSUS-A, SSUS-D) Solid Booster Motors," LMSC-HREC TN D568087, Lockheed Missiles & Space Company, Huntsville, Ala., Nov. 1977.

[10] Prozan, R. J., "Striated Combustion Solution," LMSC-HREC A791356, Lockheed Missiles & Space Company, Huntsville, Ala., May 1968.

[11] Combs, L. P., "Liquid Rocket Performance Computer Model with Distributed Energy Release," NASA CR-11462, June 1972.

[12] Ring, L. R., "Evaluation of Combustion/Nozzle Gasdynamic Models for Liquid Rocket Engine Applications," AFRPL-TR-77-46, Air Force Rocket Propulsion Laboratory, Edwards, Calif., July 1977.

[13] Stephens, J. T., and Ratliff, A. W., "Studies of Rocket Engine Combustion Chamber Geometry Using an Equilibrium Reacting Gas Transonic Flow Program," LMSC-HREC A784898, Lockheed Missiles & Space Company, Huntsville, Ala., Nov. 1967.

[14] Ratliff, A. W. et al., "Analysis of Exhaust Plumes from Skylab-Configuration R-4D Attitude Control Motors," LMSC-HREC D162171, Lockheed Missiles & Space Company, Huntsville, Ala., March 1970.

[15] Ring. L. R. et al., "Analysis and Correlation of High Altitude Rocket Exhaust Plume Experimental Data," JANNAF 7th Plume Technology Meeting, Huntsville, Ala., April 1973.

[16] Nickerson, G. R. et al., "Solid Rocket Motor Performance Predictions Using the Improved SPP Computer Model," 16th JANNAF Combustion Meet- ing, Monterey, Calif., Sept. 1979.

[17] Bird, G. A., "Breakdown of Translational and Rotational Equilibrium in Gaseous Expansions," AIAA Journal, Vol. 8, Nov. 1970, pp. 1998-2003.

[18] Gordon, S., and McBride, B. J., "Computer Program for Calculation of Complex Chemical Equilibrium Compositions, Rocket Performance, Incident and Reflected Shocks, and Chapman-Jouget Detonations," NASA SP-273, Lewis Research Center, Cleveland, Ohio, 1968.

[19] Bird, G. A., "The Nozzle Lip Problem," Rarefied Gas Dynamics: Proceedings of the 9th International Symposium, Vol. 1, 1974, pp. A.22-1-A.22-8.

[20] Wiernbaum, S., "Rapid Expansion of a Supersonic Boundary Layer and Its Application to the Near Wake," AIAA Journal, Vol. 4, 1966, pp. 217-226.

[21] Baum, E., "An Interaction Model of a Supersonic Laminar Boundary Layer on Sharp and Backward Facing Steps," AIAA Journal, Vol. 6, March 1966, pp. 440-447.

[22] Spradley, L. W., and Pearson, M. L., "GIM Code User's Manual for the STAR-100 Computer," NASA CR-3157, Nov. 1979.

[23] Penny, M. M., and Wojciechowski, C. J., "User's Manual and Description of a Computer Program for Calculating Heating Rates, Forces and Moments Acting on Bodies Immersed in Rocket Exhaust Plumes," LMSC-HREC D162867-II, Lockheed Missiles & Space Company, Huntville, Ala., March 1971.

[24] Smith, S. D., and Ratliff, A. W., "User's Manual-Variable O/F Ratio Method of Characteristics Program for Nozzle and Plume Analysis," LMSC-HREC D162220-IV, Lockheed Missiles & Space Company, Huntsville, Ala., June 1971.

[25] Guernsey, Carl S., "Effects of Translational Nonequilibrium on Vacuum Plume Expansions," JANNAF 13th Plume Technology Meeting, Houston, Texas, April 1982.

[26] Chirivella, J. E., and C. S. Guernsey, "Nozzle Lip Flow and Self-Scattering Molecular Collisilns as Contributors to Plume Backflow," JANNAF 13th Plume Technology Meeting, Houston, Texas, April 1982.

[27] Alt, R. E., "Bipropellant Engine Plume Contamination Program-Vol. I; Chamber Measurements - Phase I," AEDC-TR-79-28, Arnold Engineering Development Center, Arnold Air Force Station, Tenn., Dec. 1979.

[28] Alred, J. W., "Flowfield Description of the Reaction Control System Jets of the Space Shuttle Orbiter," AIAA Paper 83-1548, AIAA 18th Thermophysics Conference, Montreal, Canada, June 1983.

α_s/ϵ_H Measurements of Thermal Control Coatings over Four Years at Geosynchronous Altitude

D. F. Hall* and A. A. Fote†
The Aerospace Corporation, El Segundo, California

Abstract

P78-2 was inserted into a 27,600 × 43,300 km orbit on 2 February 1979 after 3 days in a transfer orbit. The ML12 experiment on board the P78-2 spacecraft includes 16 thermal control coating (TCC) samples, each mounted in a calorimeter. The solar absorptances α_s of these samples are deduced from on-orbit measurements of temperatures and the prelaunch-measured values of thermal emittances ϵ_H and residual heat leaks. The time-dependent α_s of each sample is reported for the first 4 yr on orbit. With one exception the early orbital values of α_s are in good agreement with prelaunch values indicating there was little contamination during prelaunch activities. The very small change in α_s for a fused silica mirror and other space-stable samples indicates that the SCATHA spacecraft was exceptionally clean. As a result, large α_s changes for other TCCs are attributed to the effects of radiation damage.

Introduction

This paper presents the analysis of data acquired during 4 yr on orbit of a thermal control coating (TCC) experiment. This experiment, designated ML12,[1] is on board the United States Air Force (USAF) space test program P78-2 spacecraft as part of the spacecraft

---

Presented as Paper 83-1450 at the AIAA 18th Thermophysics Conference, Montreal, Canada, June 1-3, 1983. Copyright © American Institute of Aeronautics and Astronautics, Inc., 1983. All rights reserved.

*Research Scientist, Chemistry and Physics Laboratory.

†Member of the Technical Staff, Chemistry and Physics Laboratory.

charging at high altitudes (SCATHA) investigation.[2] The P78-2 was launched on 30 January 1979 into a 176 × 43,278 km transfer orbit. On 2 February 1979, it was injected into a 27,600 × 43,300-km, 7.9-deg inclination final orbit. Thus, it passes through geosynchronous altitude (35,786 km) twice per day and is always within 23% of that altitude. The vehicle (Fig. 1) is a spin-stabilized right cylinder approximately 1.75 m in both length and diameter. The spin axis of P78-2 is held approximately normal to the sun line.

Previous publications[3,4] have given details regarding the instrument designs, sample selection and cleanliness, data reduction, preflight calibration, error analysis, and early flight results. This paper presents recent flight results.

Instrument Design

The instrument consists of three interconnected units designated ML12-3, ML12-4, and ML12-5. These units are a modification and augmentation of hardware originally designed and fabricated at TRW Systems.[5] The ML12-3 and ML12-4 are sample trays that each carry eight

Table 1 Flight Samples

ML12-3 Tray	MIL12-4 Tray
Grafoil (Union Carbide, GTA grade)	10-nm Au on 5 mil Kapton/Al (Sheldahl Lot 105788, coated by SRI)
Black paint (3M401C10)	Astroquartz silica fabric (J. P. Stevens 581)/FEP/Al
OSR (OCLI SI-100 mirror)	In_2O_3/OSR (OCLI SI-100 mirror)
Vacuum deposited Au (Optically opaque) on Al	22-nm 10% SnO_2, In_2O_3 on 2 mil Kapton/Al (prepared by General Electric)
Diamond polished 2024 T3 Al	Diamond polished 2024 T3 Al
FEP (2 mil)/Ag (Sheldahl)	Black paint (3M401C10)
FEP (5 mil)/Ag (Sheldahl)	Astroquartz silica fabric (J. P. Stevens 581)/tape (Sheldahl 405900)
Yellow paint (NASA-Goddard No. NS43G)	Kapton (5 mil)/Al (Sheldahl)

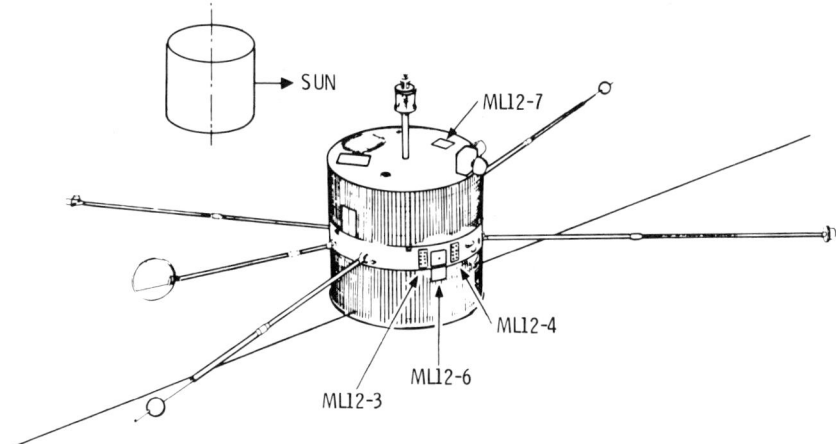

Fig. 1 P78-2 spacecraft showing locations of ML12-3 and -4 TCC trays.

calorimetrically mounted 1.25-in.-diam samples. The ML12-5 carries the electronic circuitry that monitors the TCC coupons. The location of the trays on the P78-2 vehicle is shown in Fig. 1.

Samples, Data Reduction, and Preflight Calibration

The 16 samples chosen for testing are listed in Table 1.

The ratio of solar absorptance to infrared hemispherical emittance α_s/ε_H is determined for each sample from on-orbit sample temperature measurements that are corrected for several factors. For instance, the solar radiant power falling on the sample is a function of the sun azimuth at the time of measurement and any shadowing that may be caused by the spacecraft booms. The temperatures are also significantly affected by the sample mass and heat capacity and by conductive and radiant heat exchange between the sample and its mount. The appropriate thermal coupling constants were carefully measured in the laboratory along with heat capacity and hemispherical infrared emittance, as described in Refs. 2 and 3. The α_s values reported herein are calculated under the assumption that these quantities have not changed with time. Since final orbit was achieved, Earth albedo and emission have been taken to be negligible. Following the second year on orbit, data collection has been restricted to local dawn and dusk periods. During such periods, the samples have no view of the earth so that this procedure introduces no error.

Error Analysis

Uncertainties affecting the accuracy of the calculated α_s values include those of the laboratory calibrations and the orbital measurements of sample and mount temperature. Other error sources are associated with the solar power incident on the sample: boom shadowing, sun azimuth, and solar irradiance.

Error analysis has demonstrated that for those spacecraft axis-sun line angles that produce no boom shadowing, the major contributor to uncertainty in α_s is quantization errors in sample mount temperature measurements. These errors are introduced by the 8-bit telemetry system used on the spacecraft. The random errors range from 0.008-0.016 α_s units depending on the sample.[2,3]

Flight Results

Plots of measured sample $\alpha_s(t)$ on orbit are presented in Figs. 2-17 as individual data points. In addition, a curve representing the least-squares fit of the equation

$$\alpha_s(t) = \alpha_s(o) + Kt + \Delta\alpha (1 - e^{-t/\tau}) \qquad (1)$$

to the data is plotted as a continuous curve. Whenever it was possible to produce a good fit to the data with either K or $\Delta\alpha$ set to zero, this was done. Of course α_s of all samples must eventually stabilize to values less than 1, and so the linear dependence is an approximation to an exponential curve with a very long time constant. Therefore, extrapolation of linear fits

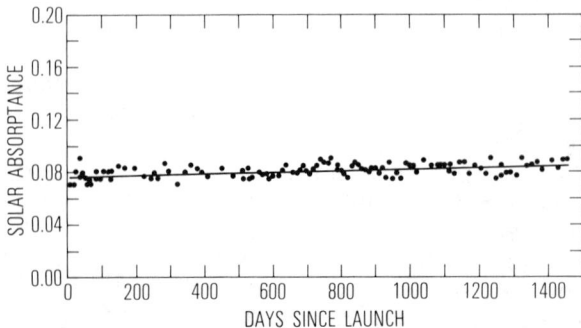

Fig. 2 Solar absorptance of fused-silica mirror sample vs time on orbit.

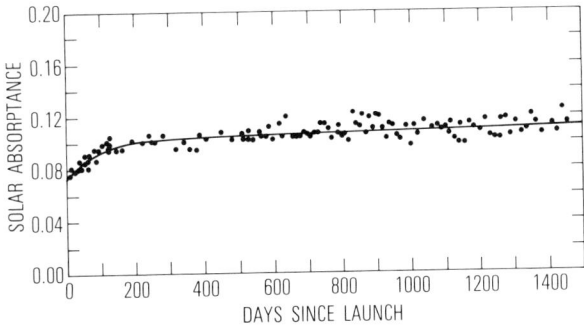

Fig. 3 Solar absorptance of indium oxide coated fused-silica mirror sample vs time on orbit.

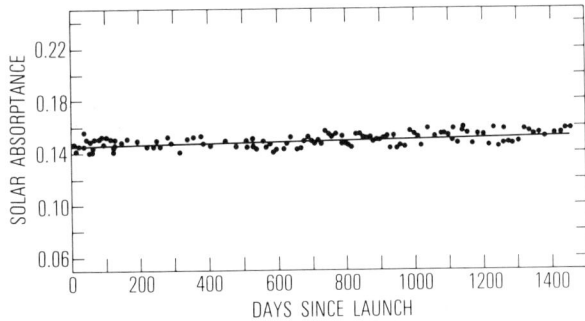

Fig. 4 Solar absorptance of polished Al sample on ML12-3 vs time on orbit.

to the data very far into the future will produce an overestimate of α_s.

Except for the two black paint samples, the scatter of the data points about the fitted curves is almost completely within the range of the calculated random errors.[2,3] The data points scatter evenly above and below the curves except for the early yellow paint data. Thus, except for the yellow paint sample, it is unlikely that any other equation would fit the data better. The parameters of the equation for each sample, along with the preflight measurements of the solar absorptance α_s' and the average sample temperatures on orbit, are given in Table 2. The quantities labeled S in Table 2 are values for the standard errors of the parameters denoted by the respective subscripts and are calculated from the statistics of the data. S_E is the standard error of $\alpha_s(t)$ on t and is a measure of the scatter about the curve.

Table 2 Parameters and their standard errors giving best bit to Eq. (1) along with prelaunch values of ε_H and α_s'. Suppressed digits are not fully significant

	Sample No.	Temp. on Orbit (°K)	Prelaunch measurement		$\alpha_s(0) \pm S_{\alpha(0)}$
			ε_H	α_s'	
Grafoil	3-1	331	0.34	0.66 ± 0.01	0.664 ± 0.001
Black paint	3-2	301	0.87	0.97 ± 0.01	0.911 ± 0.02
OSR	3-3	234	0.81	0.08 ± 0.01	0.076 ± 8 E-4
Gold	3-4	329	0.03	0.21 ± 0.01	0.192 ± 8 E-4
Polished Al	3-5	315	0.05	0.1_4 ± 0.05	0.14_5 ± 7 E-4
FEP (5 mil)	3-6	242	0.80	0.11 ± 0.01	0.11 ± 0.03
FEP (2 mil)	3-7	246	0.68	0.06 ± 0.01	0.07 ± 0.01
Goddard paint	3-8	260	0.78	0.31 ± 0.01	0.33_4 ± 0.005
Au film/Kapton	4-1	312	0.42	0.53 ± 0.01	0.52_7 ± 0.004
Fabric (FEP/Al)	4-2	257	0.68	0.20 ± 0.01	0.20_2 ± 0.004
Coated OSR	4-3	238	0.76	0.09 ± 0.01	0.07_3 ± 0.004
Coated Kapton	4-4	278	0.71	0.40 ± 0.01	0.38_5 ± 0.004
Polished Al	4-5	314	0.04	0.1_4 ± 0.05	0.138 ± 0.001
Black paint	4-6	300	0.92	0.97 ± 0.01	0.966 ± 0.002
Fabric (Tape)	4-7	267	0.60	0.19 ± 0.01	0.19 ± 0.006
Kapton	4-8	282	0.81	0.48 ± 0.01	0.50_3 ± 0.004

(Table continued on next page)

Table 2 (cont.) Parameters and their standard errors giving best fit to Eq. (1) along with prelaunch values of ε_H and α'_s. Suppressed digits are not fully significant

	Sample No.	Curve fit parameters to orbital data			
		$K(yr^{-1}) \pm S_K$	$\Delta\alpha \pm S_{\Delta\alpha}$	$\tau(day) \pm S_\tau$	S_E
Grafoil	3-1	$-1.04 \text{ E-2} \pm 4.6 \text{ E-4}$	0	0	0.005
Black paint	3-2	$-5.5 \text{ E-3} \pm 1.2 \text{ E-3}$	0	0	0.01
OSR	3-3	$2.2 \text{ E-3} \pm 3.5 \text{ E-4}$	0	0	0.004
Gold	3-4	$2.0 \text{ E-4} \pm 3.9 \text{ E-4}$	0	0	0.005
Polished Al	3-5	$1.8 \text{ E-3} \pm 3.2 \text{ E-4}$	0	0	0.004
FEP (5 mil)	3-6	0	0.28 ± 0.02	$2.0 \text{ E3} \pm 0.2 \text{ E3}$	0.005
FEP (2 mil)	3-7	0	0.24 ± 0.01	$1.4 \text{ E3} \pm 0.1 \text{ E3}$	0.005
Goddard paint	3-8	0	$6.9 \text{ E-2} \pm 3 \text{ E-3}$	$7.1 \text{ E2} \pm 0.9 \text{ E2}$	0.006
AU film/Kapton	4-1	$1.35 \text{ E-2} \pm 7 \text{ E-4}$	$1.3 \text{ E-2} \pm 4 \text{ E-3}$	$1.1 \text{ E2} \pm 0.9 \text{ E2}$	0.005
Fabric (FEP/Al)	4-2	$5.3 \text{ E-3} \pm 6 \text{ E-4}$	$5.0 \text{ E-2} \pm 3 \text{ E-3}$	$1.2 \text{ E2} \pm 0.1 \text{ E2}$	0.004
Coated OSR	4-3	$2.8 \text{ E-3} \pm 7 \text{ E-4}$	$2.8 \text{ E-2} \pm 4 \text{ E-3}$	$8 \text{ E1} \pm 2 \text{ E1}$	0.005
Coated Kapton	4-4	$3.7 \text{ E-2} \pm 4 \text{ E-3}$	$4.7 \text{ E-2} \pm 4 \text{ E-3}$	$9.8 \text{ E1} \pm 1.8 \text{ E1}$	0.007
Polished Al	4-5	$2.65 \text{ E-3} \pm 4.5 \text{ E-4}$	0	0	0.006
Black paint	4-6	$-2.2 \text{ E-3} \pm 1 \text{ E-3}$	0	0	0.009
Fabric (Tape)	4-7	$4.3 \text{ E-3} \pm 1.5 \text{ E-3}$	$1.2 \text{ E-1} \pm 4 \text{ E-3}$	$2.0 \text{ E2} \pm 0.2 \text{ E2}$	0.007
Kapton	4-8	$2.7 \text{ E-2} \pm 9 \text{ E-4}$	$3.43 \text{ E-2} \pm 4 \text{ E-3}$	$1.2 \text{ E2} \pm 0.3 \text{ E2}$	0.006

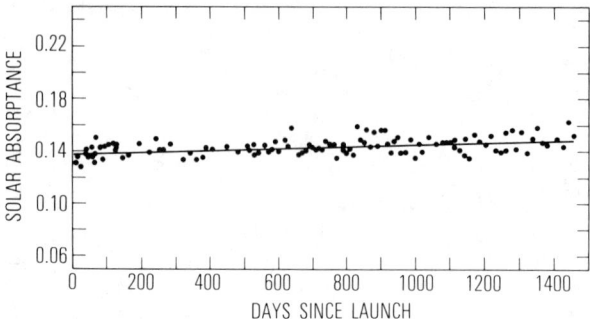

Fig. 5 Solar absorptance of polished Al sample on ML12-4 vs time on orbit.

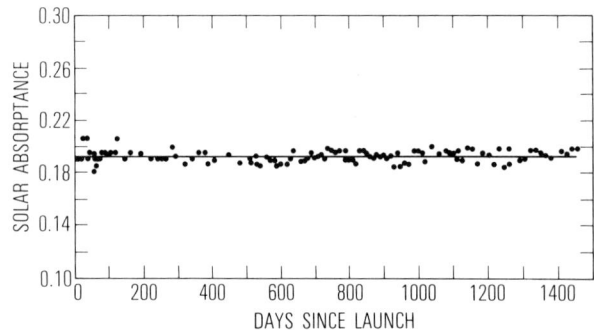

Fig. 6 Solar absorptance of vacuum deposited Au sample vs time on orbit.

Discussion

Comparison with Prelaunch Measurements of α_s

The values $\alpha_s(o)$ can be compared with the prelaunch values α_s' in Table 2. When measurement errors are considered, $\alpha_s(o)$ and α_s' agree for all but one of the black paints. Because all 16 samples were exposed to the same environments from the time the calorimeters were mounted on the trays, it is concluded that any contamination acquired in tray calibrations, testing, and launch was not sufficient to affect any of the $\alpha_s(o)$ values measurably. The apparent decrease in the black paint sample $\alpha_s(o)$ on ML12-3 is likely the result of a problem in the flight instrumentation, since the 0.92 value is lower than expected and not confirmed by the sample of the same paint on ML12-4.

MEASUREMENTS OF THERMAL CONTROL COATINGS 223

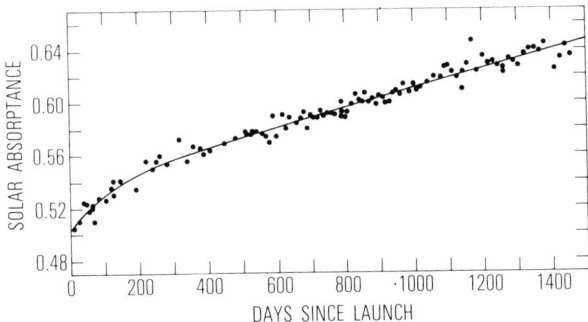

Fig. 7 Solar absorptance of 5-mil-thick aluminized Kapton sample vs time on orbit.

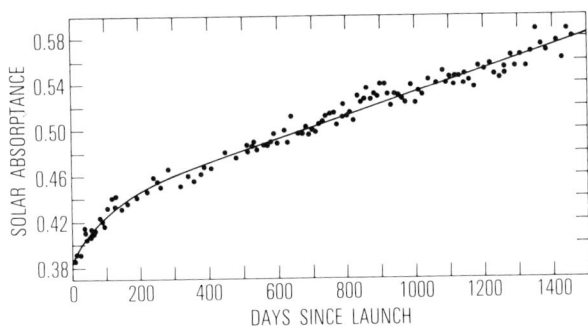

Fig. 8 Solar absorptance of indium tin oxide coated aluminized Kapton sample vs time on orbit.

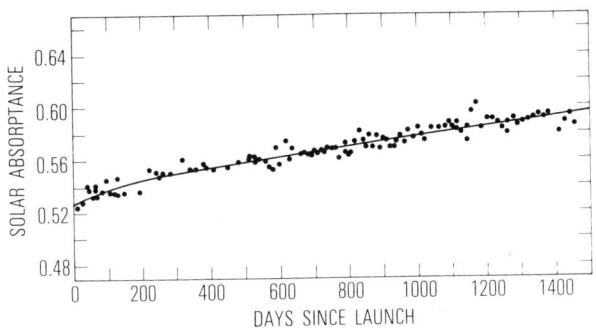

Fig. 9 Solar absorptance of 10-nm Au-coated aluminized Kapton sample vs time on orbit.

Absorption of Contaminants

The best monitor for the influence of condensable contaminants is the uncoated optical solar reflector (OSR) sample, which is the coldest sample of the group and one of the least likely to experience material degradation. The very small rise in α_s for this sample, amounting to 2.2×10^{-3} yr^{-1} and representing an upper limit for contaminant effects, indicates that contamination was minimal and that P78-2 was a fairly clean spacecraft. This is further discussed elsewhere.[6] Note that the changes in α_s of the two aluminum samples and the gold sample are also quite low. The large changes in α_s observed on other samples must then be considered to result from material degradation.

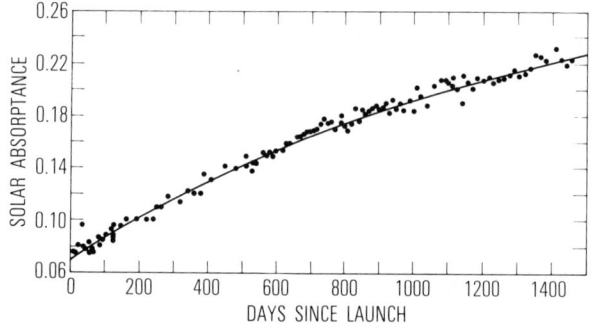

Fig. 10 Solar absorptance of 2-mil-thick silvered Teflon FEP sample vs time on orbit.

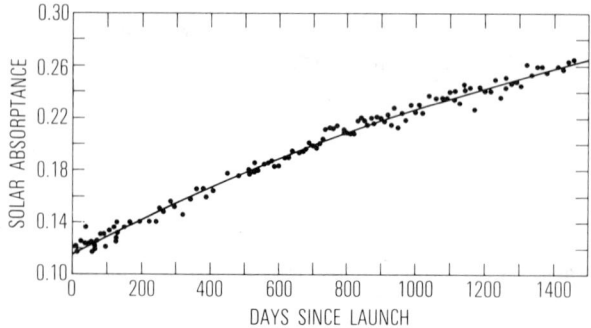

Fig. 11 Solar absorptance of 5-mil-thick silvered Teflon FEP sample vs time on orbit.

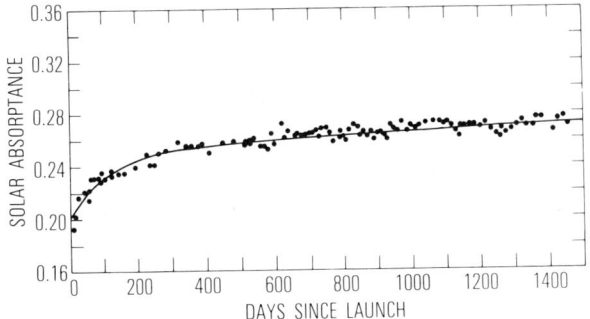

Fig. 12 Solar absorptance of Astroquartz fabric backed by aluminized Teflon FEP sample vs time on orbit.

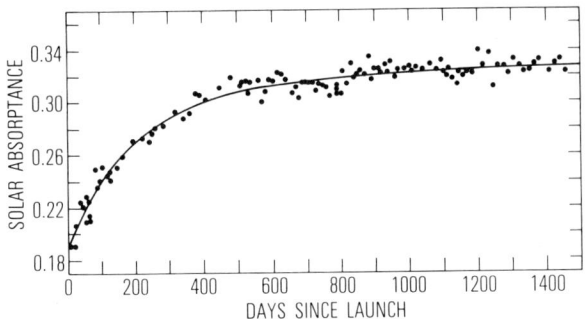

Fig. 13 Solar absorptance of Astroquartz fabric backed by double adhesive tape sample vs time on orbit.

Degradation of the In_2O_3 Conductive Coating

Nearly transparent indium oxide coatings were carried on two of the samples: a fused silica mirror and an aluminized Kapton sample. The indium oxide coating is electrically conductive, and when it is properly connected to the spacecraft frame it should prevent the dielectric samples from acquiring a large electrostatic charge during the magnetic substorms that occasionally occur at synchronous altitude. (No attempt was made to ground the coating on these samples in this experiment; therefore, the surfaces may charge and affect the kinetic energy of arriving charged particles just as with uncoated samples.) A comparison of the $\alpha_s(t)$ curves of the conductively coated samples with their uncoated counterparts indicates that during the first 9 months the coating is responsible for a higher degradation.

(When only 2-1/2 yr of orbital data were available, it appeared that the long-term $d\alpha_s/dt$ of the coated mirror, though small, was twice that of the uncoated mirror. The more extensive data now available indicate that after an initial transient period of about 9 months, the $d\alpha_s/dt$ values of the coated and uncoated mirrors are the same. The earlier result was a consequence of unbalanced scatter in the data during the period between 850 and 910 days after launch. This is a dramatic example of the importance of collecting material degradation data over a period approaching that of mission design life.)

The indium tin oxide (ITO) coated Kapton sample exhibits a transient degradation with a greater amplitude than that of the OSR with a similar indium oxide coating. However, the other Kapton samples also have a degradation time dependence of this form, so at least

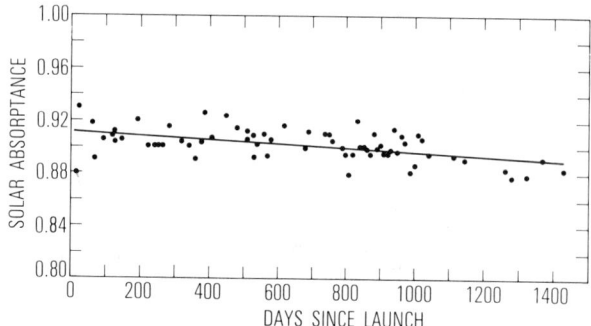

Fig. 14 Solar absorptance of 3M 401C10 black paint sample on ML12-3 sample vs time on orbit.

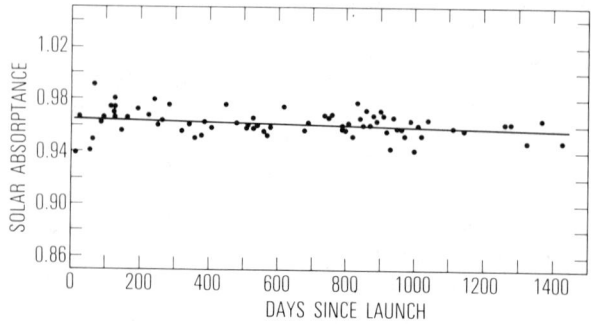

Fig. 15 Solar absorptance of 3M 401C10 black paint sample on ML12-4 sample vs time on orbit.

some of the coated sample degradation must be assigned to the Kapton itself.

Degradation of the Kapton Samples

Comparison of the data for the three Kapton samples also provides evidence that material degradation of the polymer is responsible for the large changes in solar absorptances. The gold coated sample exhibits the smallest change of the group. It should be the least affected both because its optical properties are partly due to the semitransparent gold, and because this film may partially shield the underlying Kapton from the damage producing environment.

If the contribution by the ITO coating to degradation is completed by the time $\alpha_s(t)$ becomes linear, as in the case of the IO coating on the OSR, then why is the subsequent degradation approximately 32% more rapid than that of the uncoated Kapton sample? The analysis in the Appendix implies that this is a result of the difference in thickness of the two samples. In particular, if the radiation damage is restricted to a layer of the Kapton which is thinner than the thinnest sample, a larger rate of degradation is expected for that sample than for the thicker sample. This is a consequence of the smaller α_s of the thinner material. The notion that the damaged layer is thin is supported by both theoretical[7] and experimental[8] investigations.

When the ratio in Eq. (A7) of the Appendix is computed with the values of α_s at 300 days, we obtain the relative rate of α increase in Kapton:

$$(\dot{\alpha}_s, 2 \text{ mil})/(\dot{\alpha}_s, 5 \text{ mil}) = 1.21$$

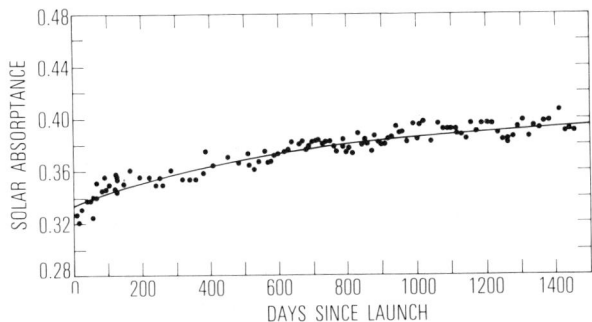

Fig. 16 Solar absorptance of NASA/GSFC NS43G yellow paint sample vs time on orbit.

Therefore, 2-mil-thick aluminized Kapton is expected to degrade significantly faster than 5-mil-thick aluminized Kapton, as is observed in Figs. 7 and 8 and in laboratory simulations of the synchronous environment.[8] The observed ratio ranges from 4-16% larger than that predicted by Eq. (A7). The reasons for this are not known. Of course, if the Kapton in the two samples differs in susceptibility to radiation damage, the ratio would be affected.

Degradation of the Silvered Teflon Samples

Figures 10 and 11 reveal that the 2- and 5-mil-thick FEP samples degrade at similar rates. This is probably the result of their relatively small α_s values. At first the thinner sample degraded 22% more rapidly than the thicker sample. However, the degradation rate of the 2-mil sample diminished more rapidly than that of the 5-mil sample, and by 1000 days had become the smaller of the two. Over the 1500 days of data, the average increase in α_s of the 2-mil sample was 6.8% greater than the 5-mil sample. This compares fairly well with the forecast of Eq. (A7) that the 2-mil sample would lead by about 4% over the whole flight data period. As in the case of Kapton, ground simulation of FEP damage[8] is consistent with the ML12 flight result.

Degradation of the Astroquartz Fabric Samples

The most severe degradation of these materials occurred over the first 2 yr, and the magnitude of the degradation is comparatively large. Three processes are

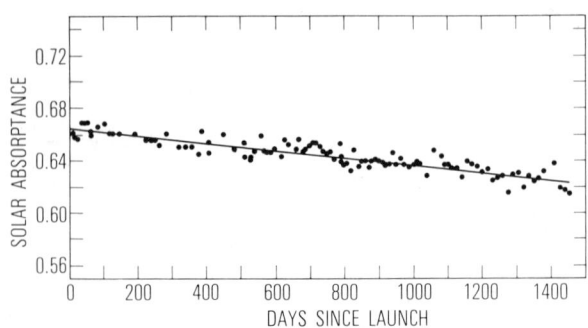

Fig. 17 Solar absorptance of Union Carbide Grafoil sample vs time on orbit.

believed to contribute. Unlike the smooth samples, the fabrics were not cleaned before launch. Because of their high surface areas, they are especially susceptible to collection of contamination. Therefore, uv darkening of prelaunch contamination is probably an important process. Secondly, some sunlight reaches the backing material on the fabrics. As the backing material darkens, the sample α_s increases. Finally, the samples acquire some contamination on-orbit.

Data from the other samples provide a basis for estimating the relative importance of these processes. The polished aluminum and OSR samples degrade very slowly, so it seems likely that the contribution of on orbit contamination to fabric degradation is small also. Since the time constant of the degradation from the silvered Teflon FEP samples is much longer than that of the transient portion of the aluminized FEP backed fabric degradation, this transient is probably primarily due to darkening of prelaunch contamination. It is, therefore, probable that the larger and longer degradation of the adhesive tape backed fabric is caused by degradation of the backing material.

Degradation of the Paints and Grafoil

There has been some bleaching of the two black paint and Grafoil samples. Although the k values (coloration rate) of the two black paint samples agree within their statistical uncertainties, the value of $\alpha_s(0)$ for sample 3-2 is low compared with the prelaunch value α_s' and with α_s' and $\alpha_s(0)$ of sample 4-6. A problem with the thermistor on sample 3-2 is suspected, and the absolute $\alpha_s(t)$ values for this sample are regarded with suspicion.

The Goddard paint displays both a transient and a long-term degradation.

Comparison with Other Flight Experiments

Our data indicate that P78-2 was a rather clean spacecraft. Furthermore, the field of view of our calorimeters, except for two booms with very small view factors, was devoid of line of sight contaminant sources. Most data published from other synchronous altitude spacecraft exhibit much higher changes in the solar absorptions of low α_s samples than that from ML12. These changes are believed to be caused by contamination rather than material degradation. For example, $d\alpha_s/dt$ for the OSRs on board three COMSTAR satellites[9] fell in the second year on orbit to values between 0.012 and 0.021 per year. These are 5-10 times greater than we observed on P78-2. By contrast, calorimeter data on the change in α_s for silvered Teflon FEP and OSRs on a reasonably clean satellite are comparable to those reported here.[10]

The high rates of degradation for the two quartz fabrics and the conductive paint coatings are frequently observed in both flight and terrestrial experiments. In particular, the fabric performance is quite similar to samples flown at much lower altitude on the P78-1 vehicle,[11] as well as other samples flown at synchronous altitude.[10]

Summary

With one exception, the early orbital values of α_s are in good agreement with prelaunch values obtained from optical reflectance measurements. Apparently, little contamination was acquired by the samples during launch and prelaunch activities.

Temporal data have been presented for the 16 TCC samples over 4 yr on orbit. Since at least some of the changes observed are caused by photon damage, when these data are compared with other test data or used as a basis for prediction, the exposure of the materials to sunlight should be taken into account. In particular, because P78-2 spins, π hours are required to accumulate one equivalent sun hour. Sample property changes resulting from vacuum and energetic particle bombardment effects are not "slowed" in this way. However, contamination related change may be slowed because of the important role of sunlight on contaminant adhesion[12] and optical properties.[13]

Measurements on the space-stable samples, and on the uncoated OSR sample in particular, reveal that contamination by condensable materials was minor on P78-2 and contributed an α_s change of less than 0.0022 yr^{-1}. Thus, the large changes observed on other samples have been attributed to intrinsic degradation of the material.

In the case of the silvered Teflon and aluminized Kapton, damage by radiation of low energy and short penetration seems to be responsible for the change in α_s. The additional degradation of the In_2O_3 coating on two of the samples may also be due to this mechanism.

The two Astroquartz fabric samples on board P78-2 both showed large changes in α_s. Our data indicate, however, that the problem is in the optical or thermal properties of backing material and contamination acquired before launch.

The two black paint samples and the Grafoil sample all exhibited bleaching. The Goddard Space Flight Center conductive paint has both transient and long-term degradation.

Appendix

This appendix develops the expected dependence of metallized polymer mirror degradation rate on polymer thickness. It is shown that to predict the dependence observed in the flight data, it is necessary to postulate that the polymer damage does not extend throughout its entire thickness.

Figure A-1 illustrates a beam of light of intensity I_0 incident on a metallized polymer mirror. The solar absorptance of such an opaque optical system α_s is given by

$$\alpha_s = (I_0 - I_R)I_0^{-1} \qquad (A1)$$

where I_R is the reflected light intensity. If first surface and interlayer reflections are negligible, and the light were to encounter damaged polymer along the entire path between the outer surface and the metal film,

$$I_R = I_0 R e^{-2(a_1 x_1 + a_2 x_2)} \qquad (A2)$$

Here a_1 and a_2 are the solar spectrum weighted average absorptances of the film coating and the damaged polymer, respectively; x_1 and x_2 are the thicknesses of these layers; and R is the reflectance of the metallization. (In the case of an uncoated mirror, $a_1 = x_1 = 0$.)

Substitution of Eq. (A2) into Eq. (A1) gives an expression relating the solar absorptance of the system

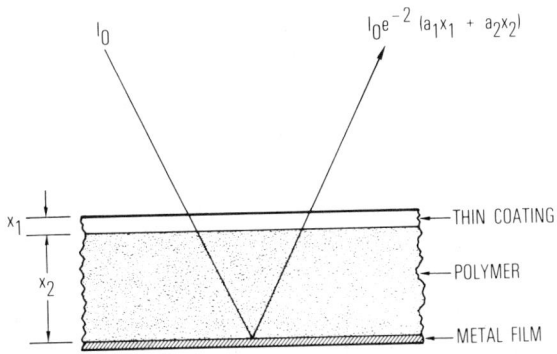

Fig. A-1 Model of light reflecting from a coated polymer mirror.

to the optical properties of its components.

$$\alpha_s = 1 - Re^{-2(a_1x_1 + a_2x_2)} \tag{A3}$$

The time derivative of this expression $\dot{\alpha}_s$ is

$$\dot{\alpha}_s = 2Re^{-2(a_1x_1 + a_2x_2)} \frac{d}{dt}(a_1x_1 + a_2x_2)$$

$$= 2(1 - \alpha_s) \frac{d}{dt}(a_1x_1 + a_2x_2) \tag{A4}$$

The case of particular interest here is where $\dot{a}_1 = 0$ and $a_2 = a_2(t)$, i.e., where an indium tin oxide coating is either not present or has stopped darkening but the Kapton or Teflon FEP is still changing. For this case, Eq. (A4) becomes

$$\dot{\alpha}_s = 2(1 - \alpha_s) \dot{a}_2(t) x_2 \tag{A5}$$

If two mirrors of different thicknesses x_2 and x_2' are made of the same material, then $\dot{a}_2(t) = \dot{a}_2'(t)$, and

$$\frac{\dot{\alpha}_s'}{\dot{\alpha}_s} = \frac{(1 - \alpha_s')}{(1 - \alpha_s)} \frac{x_2'}{x_2} \tag{A6}$$

Substitution of thickness and solar absorptance values of both the ML12 Kapton and Teflon FEP polymer mirror pairs into Eq. (A6) yields the prediction that the thicker mirror will increase in α_s more rapidly than the thinner mirror. This is the inverse of what is shown by the data.

However, if it is assumed that the darkening does not extend throughout the entire thickness of the polymer, but is limited to a thin layer regardless of the film thickness (as would be the case if the damaging radiation did not penetrate the sample fully), the outcome is quite different. In the case $x_2 = x_2'$, then

$$\frac{\dot{\alpha}_s'}{\dot{\alpha}_s} = \frac{(1 - \alpha_s')}{(1 - \alpha_s)} \tag{A7}$$

Because α_s increases with thickness, Eq. (A7) correctly predicts our observation that thinner polymer mirrors degrade more rapidly than thicker mirrors made of the same polymer.

Acknowledgments

This work was largely supported by the United States Air Force Wright Aeronautical Laboratory (AFWAL/MLBE) under Space Division Contract F04701-82-C-0083. D. Prince of AFWAL helped select, characterize, and supply most of the thermal control coating samples. Special thanks are due to J. Kordan and J. N. Wakimoto of The Aerospace Corporation for assistance with data analysis, and the MCC-F Mission Control Team of the Air Force Satellite Test Center for flight operations.

References

[1] Hall, D. F., Borson, E. N., Winn, R. A., and Lehn, W. L., "Experiment to Measure Enhancement of Spacecraft Contamination by Spacecraft Charging," National Technical Information Service, Springfield, Va., NASA SP-379, Nov. 1975, pp. 89-107.

[2] Stevens, J. R. and Vampola, A. L., eds., "Description of the Space Test Program P78-2 Spacecraft and Payloads," National Technical Information Service, Springfield, Va., SAMSO-TR-78-24, Oct. 1978, pp. 1-12.

[3] Hall, D. F. and A. A. Fote, "α_s/ϵ_H Measurements of Thermal Control Coatings on the P78-2 (SCATHA) Spacecraft," Heat Transfer and Thermal Control. Progress in Astronautics and Aeronautics, New York, 1981, pp. 467-486.

[4] Hall, D. F. and A. A. Fote, "α_s/ϵ_H Measurements of Thermal Control Coatings During Three Years in Geosynchronous Orbit," to be submitted to Journal of Spacecraft and Rockets.

[5] Luedke, E. E. and Kelley, L. R., "Development of Flight Units for Thermal Control Coatings Experiment," AFML/MBE, Wright-Patterson Air Force Base, Ohio, AFML-TR-72-233, Oct. 1972.

[6] Hall, D. F., "Current Flight Results from the P78-2 (SCATHA) Spacecraft Contamination and Coatings Degradation Experiment" (ESA SP-178), Proceedings of an International Symposium on Spacecraft Materials in Space Environment, Toulouse, France, June 1982.

[7] Bourrieau, J. and Paillous A., "Effect of Radiations on Polymers and Thermal Control Coatings," ESA SP-145, Proceedings of an ESA Symposium on Spacecraft Materials, Dec. 1979, pp. 81-90.

[8] Kurland, R. M., "Properties of Metallized Flexible Materials in the Space Environment, Final Report," Air Force Space and Missile Systems Organization, SAMSO TR 78-31, Jan. 1978.

[9] Hyman, N. L., "Solar Absorptance Degradation of OSR Radiators on the COMSTAR Satellites, AIAA Paper 81-1185, 16th Thermophysics Conference, Palo Alto, Calif., June 1981.

[10] Ahern, J. E. and Karperos, K., "Calorimetric Measure-ments of Thermal Control Surfaces on Operational Satellites," AIAA-83-0075, 21st Aerospace Sciences Meeting, Reno, Nev., Jan. 1983.

[11] Prince, D. E., "ML-101 Thermal Control Coating Spacecraft Experiment," AFML-TR-75-17, AFML/MBE, Wright-Patterson AFB, Ohio, Aug. 1975.

[12] Hall, D. F., "Flight Experiment to Measure Contamination Enhancement by Spacecraft Charging," Optics in Adverse Environments, SPIE, Bellingham, Wash., Vol. 216, 1980, pp. 131-137.

[13] Fleischauer, P. D., and Tolentino, L., "The Far Ultraviolet Photolysis of Polymethyl-phenylsiloxane Films on Quartz Substrates," NASA SP-336, U.S. Government Printing Office, Washington, D.C., 1973, pp. 645-650.

Calorimetric Measurements of Thermal Control Surfaces on Operational Satellites

John E. Ahern* and Kurt Karperos†
Aerojet ElectroSystems Company, Azusa, California

Abstract

The results of flight temperature measurements for a variety of thermal control surfaces on long-life operational satellites in geosynchronous orbit are presented. Solar absorptance values were developed from calorimetric measurements as a function of equivalent sun hours of exposure for second surface mirrors, silvered and aluminized FEP Teflon, white paint, silica cloth, and a silver-alumina-silica surface. Solar absorptance values are presented in the form of curves and exponential equations for up to 10,000 hr of equivalent sun exposure. The dependence of solar absorptance degradation upon time and thermal control surface material is demonstrated.

Nomenclature

a	= albedo
A	= area
C	= capacitance
F_a	= view factor, albedo
F_e	= view factor, Earthshine
H_e	= earthshine heat rate
n_m	= nanometer
S_o	= solar heat flux
t	= time
T	= temperature
α_s	= solar absorptance

Presented as Paper 83-0075 at the AIAA 21st Aerospace Science Meeting, Reno, Nev., Jan. 10-13, 1983. Copyright © American Institute of Aeronautics and Astronautics, Inc., 1983. All rights reserved.
*Senior Technical Specialist.
†Member Technical Staff, Space Surveillance Division.

β = angle between satellite axis and Earth sun plane
ϵ = emissivity
θ = angle between surface normal and sun vector
σ = Stefan-Boltzmann constant
τ = time constant

Introduction

It is apparent from many recent papers that the general warming trend with time of orbiting satellite systems is due to degradation of the thermal control surfaces.(1-3) In view of the economic benefit derived from reliable longlife satellite operation the design of certain thermal control systems should consider the impact of this time-dependent parameter. A primary aspect of the thermal design for long-life is the lack of reliable data on thermal control surfaces exposed to the real space environment. Data that are available consist of two types. The first is developed from temperature measurements made on operational thermal control surfaces and other components that contribute to the heat input and output of the measured radiator surface. The reliability of the performance evaluation of the thermal control surface is dependent to a significant degree on the complexity of the system and the sophistication of the thermal modeling. Because of the influence of the other components in the system and the usually limited temperature measurements available on spacecraft, this type of data must be considered to be approximate when considering it for use in designing other types of systems. The second type is that developed from calorimeters where the thermal control surface under study is thermally isolated from potential input and output heat flows except those involving solar heating and space heat rejection. In this latter approach the test sample temperature provides a direct indication of the $\alpha s/\epsilon$ value when exposed to solar radiation and of the ϵ value when the sample is shaded from solar input. This paper presents results of orbital flight measurements made using the latter calorimetric method. The data presented includes measurements of solar absorptance as a function of equivalent sun hours of exposure so as to be directly applicable to design purposes. Laboratory measurements of emissivity as a function of temperature are given for some thermal control surfaces. Flight data indicates that emissivity of these thermal control surfaces does not change with time.

The data presented in this report is a continuation of that in Ref. 1. The phase II calorimeter described in Ref. 1, and shown in Fig. 1, was used to obtain the present data. It was installed on several spacecraft in the same clean location (-T axis of Ref.1 satellite) in similar geosynchronous orbits. The cleanliness of the calorimeter location is attested by the almost negligible degradation of the second surface mirror reference samples. Each calorimeter had four surface samples one of which was a reference sample to provide direct correlation between the calorimeter results on the different spacecraft. Twenty samples were evaluated on five calorimeters. The samples are listed and described in Table 1.

The degraded performance of thermal control surfaces can be attributed to two factors. One is the change in the thermal control surface properties caused by the deposition of outgassing spacecraft contaminants. This subject is addressed elsewhere in this volume in the paper entitled "Analysis of Contamination Degradation of Thermal Control Surfaces on Operational Satellites". The other factor, and the one addressed in this paper, is the change in control surface characteristics produced by the exposure to the space environment consisting of electrons, protons and ultraviolet rays from the solar radiation flux. The data presented in this paper are for clean uncontaminated surfaces and can serve as baseline data for selection of optimum thermal control surfaces. Where contamination by the outgassing

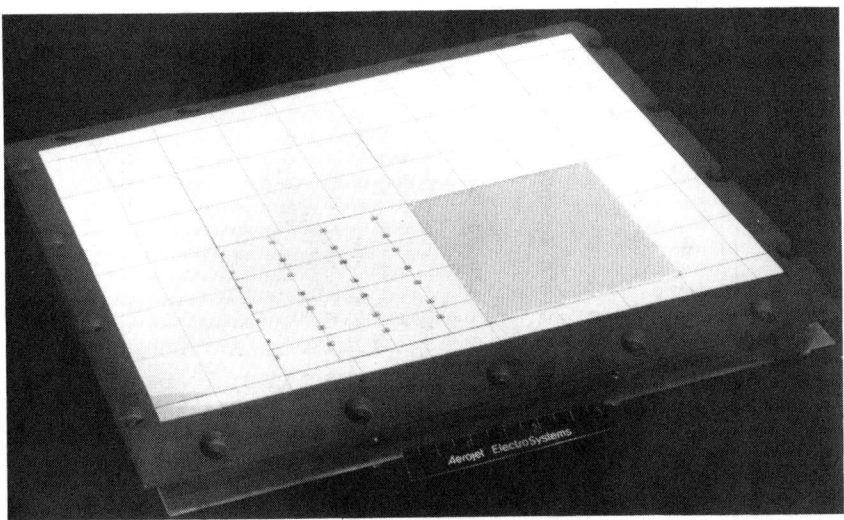

Fig. 1   Flight calorimeter.

Table 1 Calorimeter sample description

Flight sample	Method of application		Flight sample	Method of application
1. 0.002-in. silvered Teflon	Nickel powder filled acrylic adhesive	11.	Second surface mirror	RTV-566 adhesive
2. 0.005-in. aluminized Teflon	Nickel powder filled acrylic adhesive	12.	ZOT paint (8-10 mil thick)[a]	On AZ31B magnesium
3. Silica cloth No. 581	Heat laminated to 0.001 in aluminized Teflon	13.	Second surface mirror	RTV-566 adhesive
4. Silica cloth No. 581	Double faced tape with No. 585	14.	ZOT paint (10-12 mil thick)[a]	On 6061-T6 aluminum
5. 0.002-in. silvered Teflon	Nickel powder filled acrylic adhesive	15.	Second surface mirror	RTV-566 adhesive
6. Indium oxide front coated mirror	RTV-566 adhesive	16.	ZOT paint (6-7 mil thick)[a]	On 6061-T6 aluminum
7. 0.005-in. embossed silvered Teflon	With pressure sensitive P/223 tape	17.	Silver-alumina-silica	Vapor deposited on aluminum (Table 4)
8. Second surface mirror	RTV-566 adhesive	18.	ZOT paint (10-12 mil thick)[a]	On 6061-T6 aluminum
9. ZOT paint (8-10 mil thick)[a]	On 6061-T6 aluminum	19.	Second surface mirror	RTV-566 adhesive
10. ZOT paint (8-10 mil thick)[a]	On 6061-T6 aluminum	20.	ZOT paint (10-12 mil thick)[a]	On AZ31B magnesium

[a] Zinc orthotitanate pigment in potassium silicate binder YB-71 manufactured by IITRI.(10)

CALORIMETRIC MEASUREMENTS

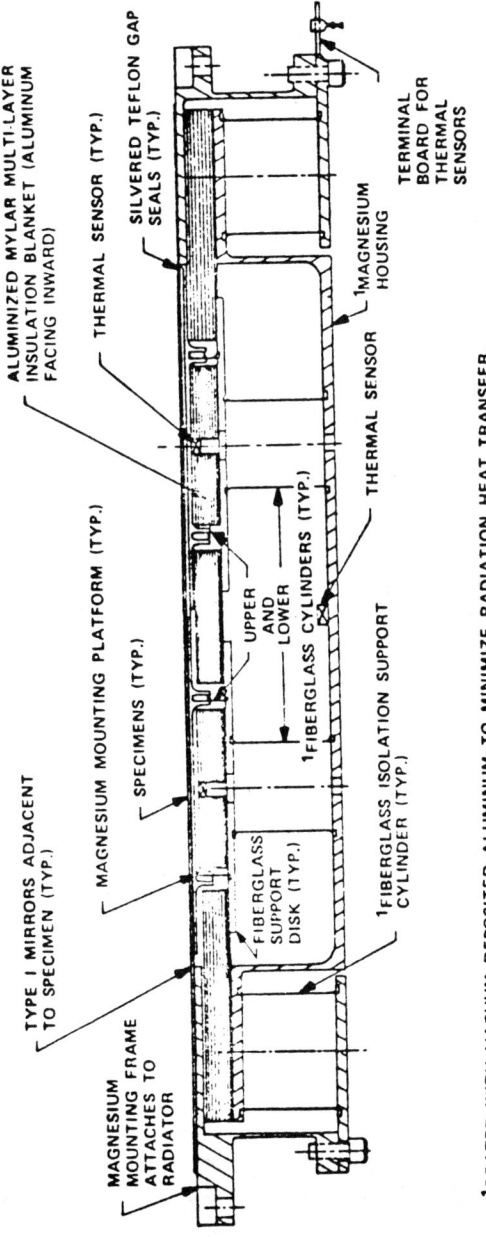

Fig. 2 Cutaway view of calorimeter design.

products from warm electronic and propulsion devices can not be avoided by design, adjustment to the clean data presented in this paper is required for satisfactory prediction of actual flight performance.

In the next section the calorimeter design is described including a description of the computer model for data reduction and an error analysis. This is followed by a discussion of the flight measurements with comparisons between the different types of surfaces. The data are developed into equation form as a function of equivalent sun hours for design application. Comparison of the results are made with other data from the literature.

Calorimeter Design and Analysis

Calorimeter Design

The calorimeter used in this program is shown in the photograph Fig. 1, and a cross section showing the general construction features is presented in Fig. 2. The basic design approach for the calorimeter was directed to maximizing the thermal isolation of the thermal control surfaces under test. This involved using minimum thickness, low-thermal-conducting fiberglass supports, multilayer insulation, and reflective aluminized surfaces. Further, the calorimeter was installed in a location chosen to minimize radiant heat transfer between the calorimeter samples and spacecraft external surfaces. A total view factor from caloric samples to spacecraft external surfaces of less than 2% was achieved. Finally, the design was configured to avoid complexity in computer thermal modeling so that reliable data evaluation could be achieved. Five calorimeters of this design were flown with no indication of structural failure since all temperature monitors operated throughout their life and the performance of similar reference samples on the different flights matched well. Figures 3 and 4 show the comparison between the reference samples from the five flight calorimeters.

Calorimeter Analysis

The node locations and the description of the nodes are given in Fig. 2 while Fig. 5 shows a typical thermal network for one specimen. Several of the nodes are external surfaces of the spacecraft that have a view of the test samples. The radiation coupling between the calorimeter samples and the spacecraft surfaces were obtained using an Aerojet-developed Monte Carlo program. Since the temperatures of these

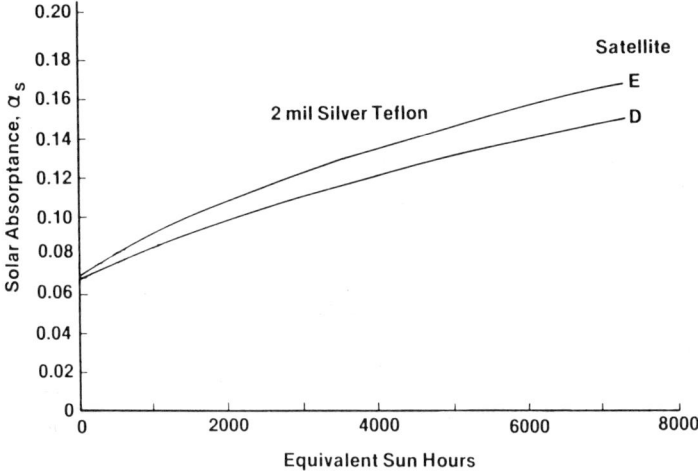

Fig. 3   Comparison of flight reference samples, satellites D and E.

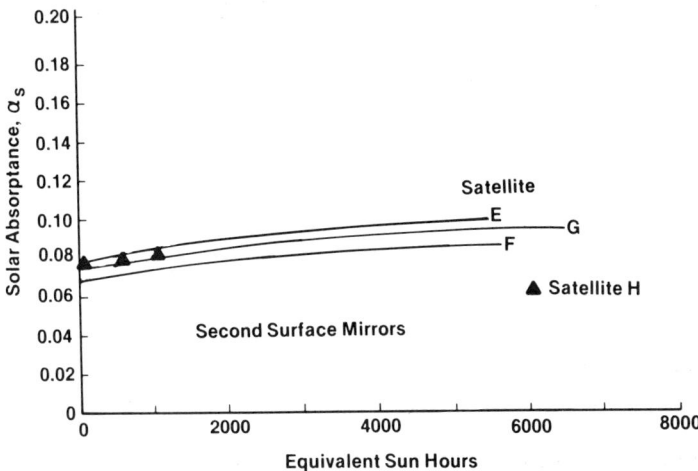

Fig. 4   Comparison of flight reference samples, satellites E, F, G, and H.

spacecraft surfaces were inputted in the computer program, only the i.r. radiation between the surfaces and the test samples was involved. Solar heating of the test samples was considered in the computer model by a diurnal shape factor table that was adjustable for β angle and solar heat flux automatically.

A flow chart detailing the algorithm by which the solar absorptance values of the flight test samples were evaluated is given in Fig. 6. For the day of evaluation the diurnal

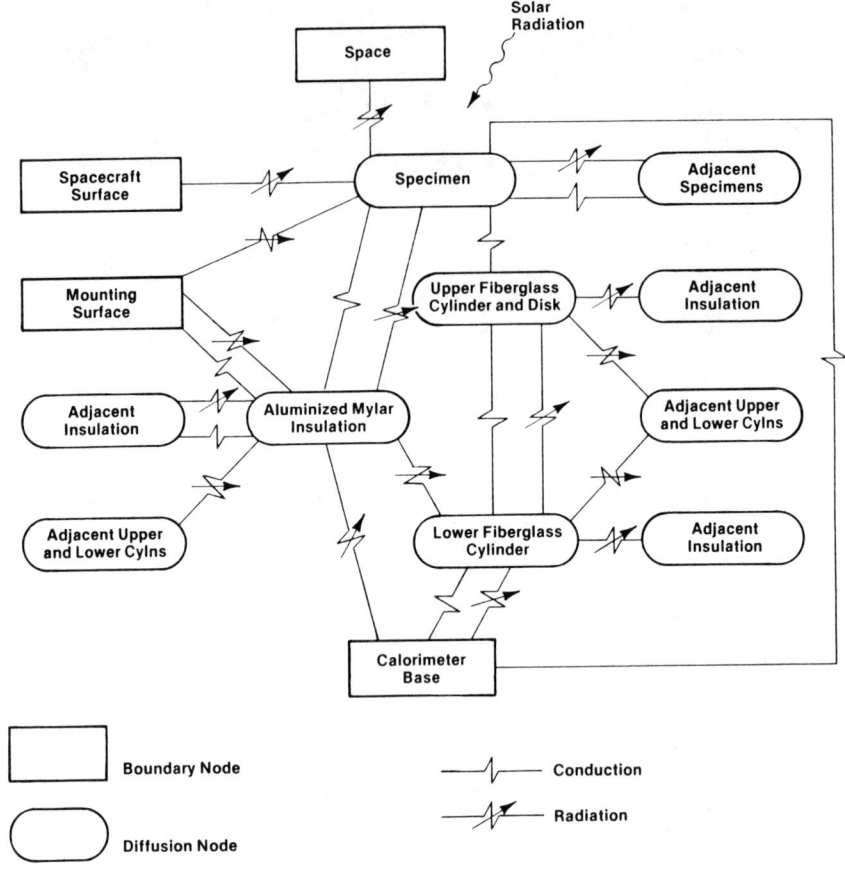

Fig. 5 Calorimeter thermal network for one specimen (typical).

temperatures (available at 15-min intervals) for the radiation linked spacecraft nodes and the calorimeter base node are inputted into the program. The previously calculated value of the solar absorptance (or an estimated value) for each test sample is entered into the computer program and the corresponding sample temperature diurnals are calculated. As noted in the flow chart Fig. 6 the calculated maximum temperature for each test sample is compared to the actual maximum flight temperatures that were inputted in the computer program. The computer will make adjustments to the solar absorptance values accordingly and iterate until the calculated temperature for the adjusted solar absorptance values matches the flight data within $0.1°F$ thus giving the solar absorptance value for that day.

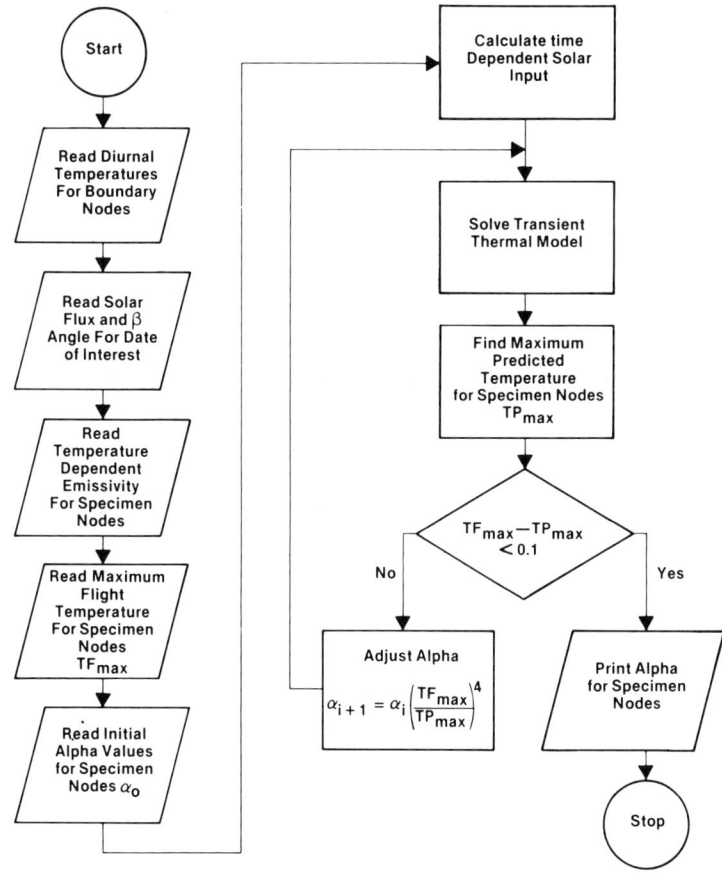

Fig. 6 Computer model flow chart.

The analysis of the calorimeter sample performance was done at varying intervals. In general, the time span between data reduction days was determined by the temperature rise rate since the accuracy of determining the solar absorptance value is strongly dependent on the telemetry count error as discussed later. In most cases the solar absorptance rise rates followed a roll-off trend so that early data were obtained at small time intervals on the order of one month and increased with time so that at three years data was taken at approximately three month intervals.

The data in this paper are given as a function of equivalent sun hours so direct design applications can be made. It also simplifies the comparison of this flight data with other flight and laboratory data when they are given

Table 2 Error analysis of α_s measurements

A. Calorimeter Design Error

Components of potential uncertainty in calorimeter design	Q heat leak, W	Uncertainty, %	Calculating error, Q'	$\left(\frac{d\alpha}{dQ}\Delta Q\right)^2$
Support conductance	0.003	30	0.0009	7.29×10^{-6}
Lead wire conductance	0.0044	25	0.0011	10.9
Multilayer insulation conductance	0.0003	10	0.0001	0.01
Solar absorptance in gap	0.0018	50	0.0009	7.29
Solar absorptance on Al Teflon	0.0003	50	0.0002	0.04
Temperature sensor power dissipation	0.0015	15	0.0002	0.004
				25.6×10^{-6}

From analysis $\frac{d\alpha}{dQ} \cong 3.0$

$$\alpha_{s_{Design}} = \left[\sum\left(\frac{d\alpha}{dQ}\Delta Q\right)^2\right]^{0.5} = \pm 0.005$$

RSS error in absolute α_s Due to calorimeter design uncertainty

B. Temperature Calibration Error
From flight data

$$\frac{d\alpha_s}{dT} = 0.0014$$

Calibration curves developed for a $\pm 1.0°F$ maximum error ($\Delta T = \pm 1.0$)
α_s Calibration $= \pm 0.0014$

C. Temperature monitor telemetry error
From telemetry quantization

$$\Delta T = +1.6°F, -0.0°F$$

From flight data

$$\frac{d\alpha_s}{dT} = 0.0014$$

$$\alpha_s \text{ Telemetry} = \left[\left(\frac{d\alpha_s}{dT}T\right)^2\right]^{0.5} = +0.0022, -0.0$$

D. Conclusion
The calorimeter design errors are essentially negligible when calculating the α_s change over a small temperature range

Absolute α_s Measurement Error	α_s Measurement Error
+ 0.009	+ 0.004
- 0.006	- 0.001

Table 3 Comparison of laboratory and initial flight measurements of solar absorptance α_s

Sample	Laboratory α_s	Initial flight α_s
1. 0.002-in. silver Teflon	0.068	0.068
2. 0.005-in. aluminized Teflon	0.144	0.140
3. Silica cloth No. 581	0.197	0.199
4. Silica cloth No. 581	0.165	0.186
5. 0.002-in. silver Teflon	0.066	0.076
6. Indium oxide coated mirror	0.082	0.091
7. 0.005-in. embossed silver Teflon	0.095	0.094
8. Second surface mirror	0.068	0.074
9. ZOT on aluminum (8-10 mil thick)	0.194	0.197
10. ZOT on aluminum (8-10 mil thick)	0.181	0.185
11. Second surface mirror	0.068	0.068
12. ZOT on magnesium (8-10 mil thick)	0.188	0.192
13. Second surface mirror	0.068	0.065
14. ZOT on aluminum (10-12 mil thick)	0.167	0.177
15. Second surface mirror	0.068	0.067
16. ZOT on aluminum (6-7 mil thick)	0.214	0.216
17. Silver-alumina-silica	0.158	0.158
18. ZOT on aluminum (10-12 mil thick)	0.167	0.190
19. Second surface mirror	0.068	0.076
20. ZOT on magnesium (10-12 mil thick)	0.167	0.199

as a function of equivalent sun hours. An exponential least-square curve fit for each sample material is made for values of solar absortance vs equivalent sun hours allowing easy insertion into thermal design computer programs.

During the process of reducing the flight data errors are introduced that must be considered in the application of the design data.

Error Analysis

The potential sources of errors in developing the solar absorptance values from flight temperature measurements are given in Table 2 along with the uncertainty associated with the source and the error impact on the results.

The three primary sources of error are in the calorimeter design, the temperature monitor calibration, and the telemetry quantization. The measurement of absolute solar absorptance involves the full calorimeter design errors, the temperature monitor calibration error, and the telemetry quantization error. As noted in Table 2 the estimated error for measuring the absolute solar absorptance is +0.009, -0.006. The actual initial flight measured values of α_s for all 20 samples are listed in Table 3 and compared to the laboratory values measured before flight. Except for two zinc orthotitanate (ZOT) paint samples and one silica cloth sample, all the flight measurements match the laboratory value within the error band. A general conclusion can be made from these results that the ground handling and launch conditions did not contribute significantly to performance degradation of the thermal control surface samples.

The primary interest in the calorimeter data is the variation of solar absorptance with time of environmental exposure. When the change in solar absorptance is considered, the uncertainties associated with the calorimeter design become negligible because they are essentially constant over small temperature changes. The temperature monitor calibration error and the telemetry quantization error are still fully involved in the α_s measurement so the final estimated error for measuring the change of solar absorptance with time is +0.004, -0.001 as noted in Table 2.

Flight Sample Descriptions

The thermal control surface samples evaluated on these calorimeters were selected for their potential application on future spacecraft. Consideration was given to parameters other than thermal performance in the selection process such as weight, cost, complexity of installation, and ability to maintain a clean surface. As a result, the materials tested included white paints, silica cloth, and metallized Teflon samples as well as the second surface mirrors tested in the early phase of the calorimeter experiment (Ref. 1). The detailed description of each of the 20 test samples is given in Table 1 including the method of attaching the sample to the holder.

The laboratory measured values of α_s of these samples are given in Table 3. The laboratory measurements of some sample materials show property variations with temperature and thickness. Fig. 7 shows the variation of emissivity with temperature for several of the materials tested in this

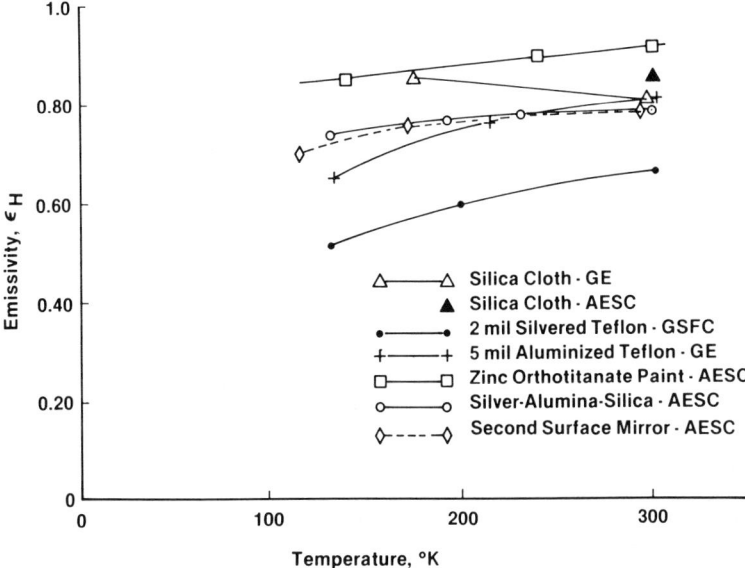

Fig. 7 Hemispherical emittance vs temperature for thermal control coatings.

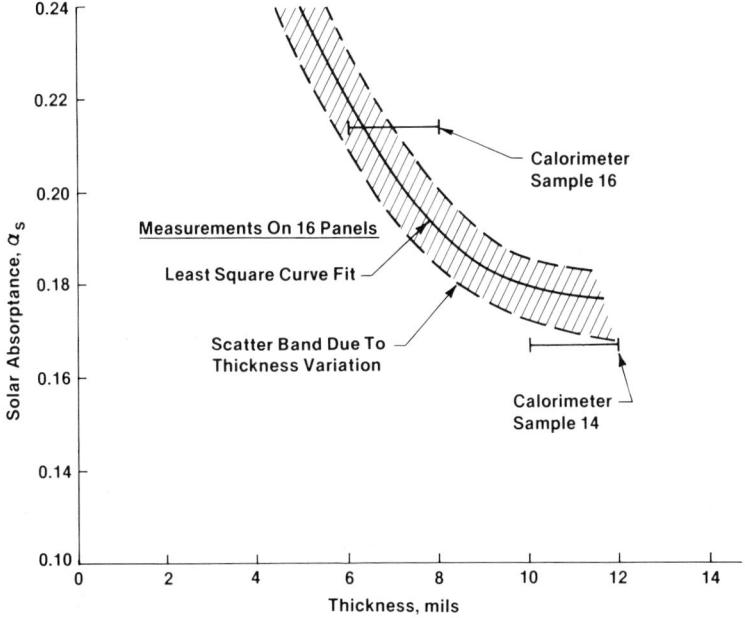

Fig. 8 α_s laboratory measurements of zinc-orthotitante paint (ZOT).

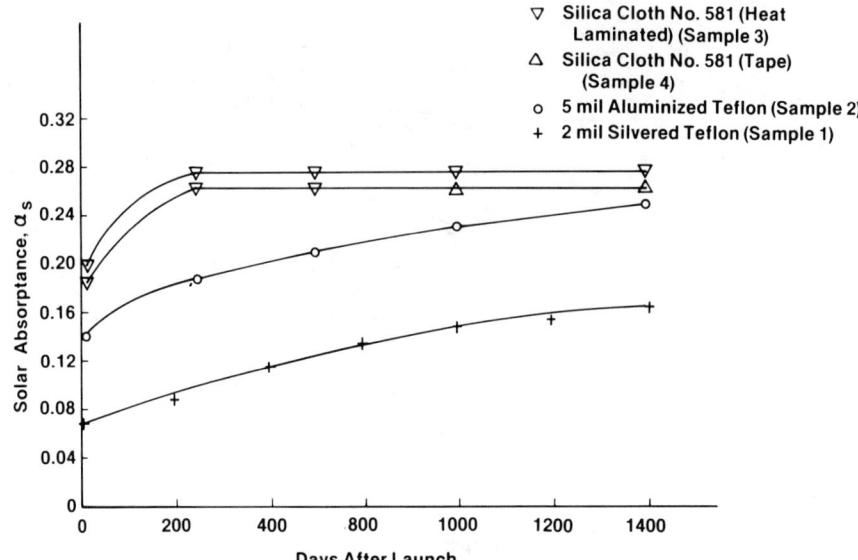

Fig. 9 Updated data for satellite D (Fig. 7 in ref. 1).

program. The solar absorptance value of the zinc orthotitanate (ZOT) white paint was shown to vary with thickness from laboratory measurements. Fig. 8 shows measurements taken on panels and calorimeter samples coated with ZOT paint of varying thickness.

Results of Flight Measurements

The flight data values for the solar absorptance of the samples were tabulated and plotted as a function of flight time for periodic comparison with existing data. The data from early flight calorimeters involving primarily second surface mirrors were presented in this manner in Ref. 1. Curves of the solar absorptance values of silvered and aluminized Teflon and silica cloth given in Ref. 1 are updated in Fig. 9.

The solar absorptance flight measurements of representative test samples are shown in Fig. 10 as a function of equivalent sun hours and years in orbit. Subsequent curves are given in equivalent sun hours only. The conversion factor for the satellites is 2400 equivalent sun hours per year The performance characteristics of each class of material is now analyzed and compared to other flight and laboratory data.

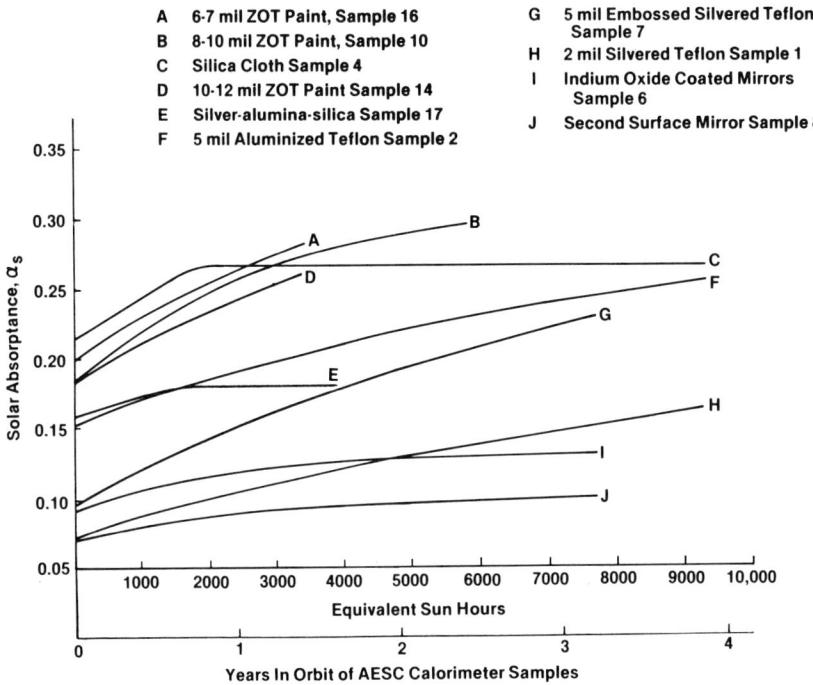

Fig. 10 Solar absorptance degradation with time for all test sample types.

Comparative Behavior of Test Samples

Metallized Teflon, α_s. The solar absorptance values for silvered and aluminized Teflon from these calorimeter experiments are plotted in Fig. 11 as a function of equivalent sun hours. These samples include two silvered 2-mil-thick Teflon samples, an aluminized 5-mil-thick Teflon sample, and an embossed silvered 5-mil Teflon sample. This latter sample can be seen in the lower right part of the calorimeter in Fig. 1. The aluminized and embossed samples show higher initial α_s values than the plain silvered Teflon samples and the embossed sample shows the highest α_s rise rate with time. The embossed sample was developed to provide flexibility to the surface when large temperature cycles are encountered, but it is apparent that a significant performance penalty is involved. AESC has used 2-mil-thick silvered Teflon on sunshades with 300°F (167°C) diurnal temperature swings for several years with no apparent failure.

Two samples from the NAVSTAR 5 satellite (Ref. 4) are shown in Fig. 11. The 5-mil silvered Teflon sample curve E has the same general characteristics as the plain AESC

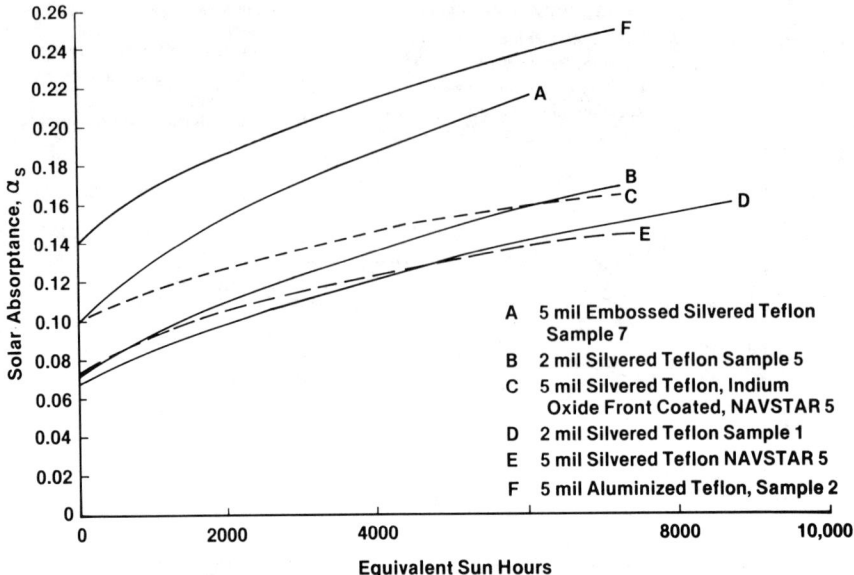

Fig. 11 Silvered and aluminized Teflon.

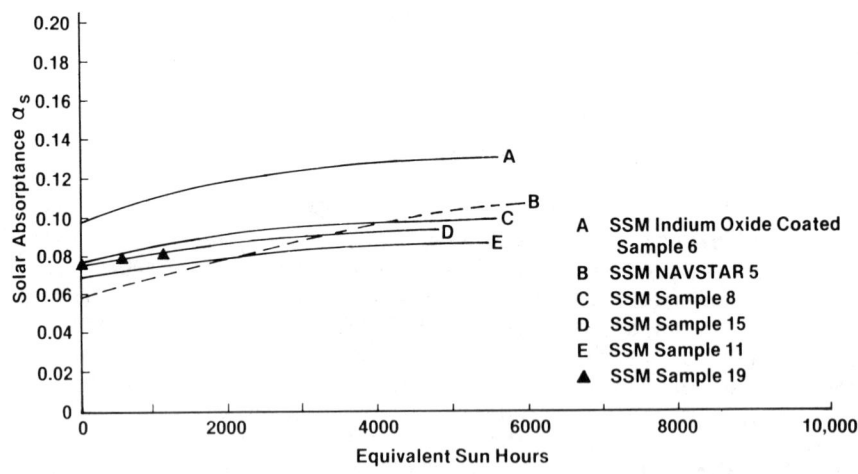

Fig. 12 Second surface mirrors (SSM).

silvered Teflon samples. The indium oxide front coated sample shows a higher initial α_s value than the plain silvered Teflon but has the same α_s rise rate with time. This behavior is consistent with the performance of the indium oxide front coated mirror shown in Fig. 12.

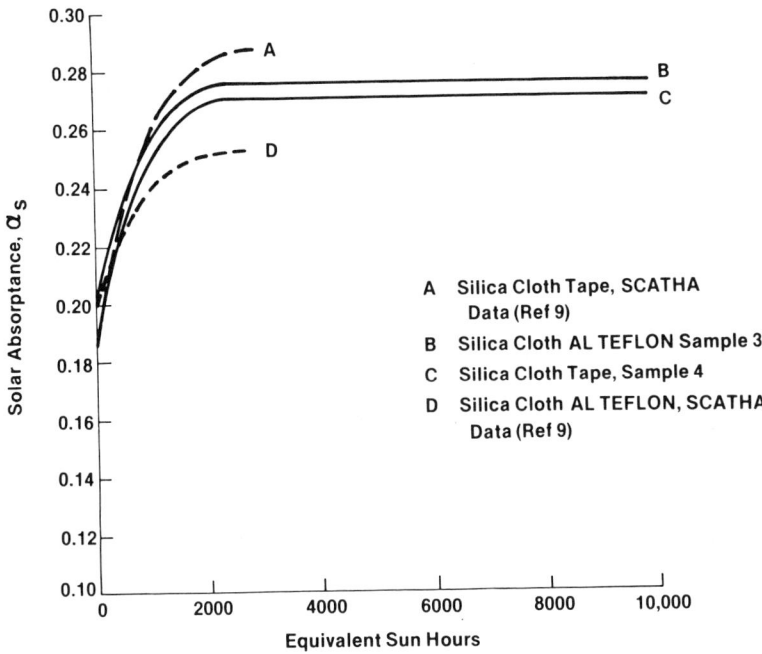

Fig. 13   Silica cloth α_s degradation with time.

Second surface mirrors α_s. The second surface mirror samples tested in this phase of the calorimeter experiment were selected primarily as reference surfaces since extensive performance data was obtained in the early phase and reported in Ref. 1. The α_s values plotted vs equivalent sun hours in Fig. 12 generally follow the data obtained previously for this calorimeter location. Also plotted in Fig. 12 for comparison is the second surface mirror data from NAVSTAR 5(4). The degradation rise rate of α_s as a function of exposure time are comparable for these samples located in clean spacecraft areas. The indium oxide front coated mirror sample shows a higher initial α_s value but its rise rate with exposure is similar to other mirror samples. This behavior is similar to that experienced for the indium oxide front coated silvered Teflon sample on NAVSTAR 5 shown in Fig. 11. The indium oxide coated mirror sample is seen as the lower left sample on the calorimeter shown in Fig. 1. An analysis of the sample indicates that a significant part of the increased initial α_s value over the conventional mirror samples can be attributed to the solder grounding tabs that can be seen in Fig. 1.

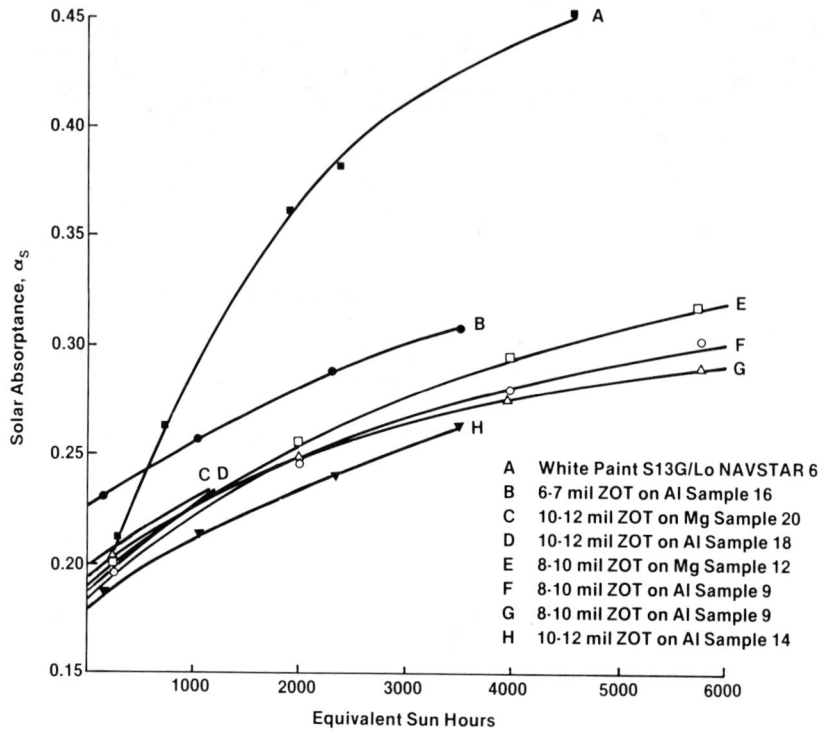

Fig. 14   White paint.

Silica cloth, α_s. The two test samples of silica cloth have an initial high α_s rise rate as shown in Fig. 13 but level off after about 2000 hours of sun exposure and remain constant through the rest of the measured flight. Comparative flight data from SCATHA(9) spacecraft are also shown in Fig. 13.

White paint, α_s. Curves of solar absorptance vs equivalent sun hours of exposure for the seven zinc orthotitanate samples are shown in Fig. 14. These samples have different thicknesses, varying from 6-7 up to 10-12 mil, and show the decrease in initial solar absorptance with increase in thickness as was demonstrated with the laboratory data in Fig. 8. The existence of the apparent discrepancy in the dependence on thickness of curves C and D (samples 18 and 20) is not yet determined as the data is preliminary. All samples do, however, show comparable rates of solar absorptance degradation indicating the lack of a dependence of the α_s degradation rate on sample thickness. Three 8-10-mil samples were tested, one on a magnesium substrate (curve

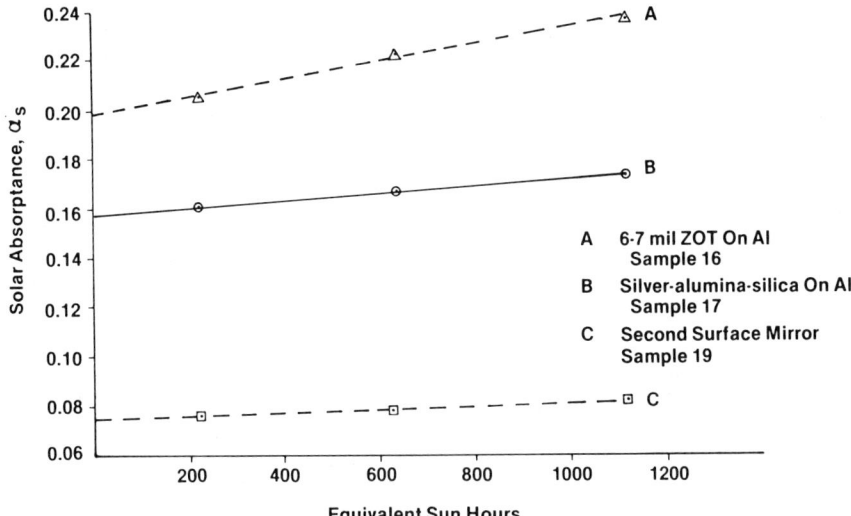

Fig. 15 Silver-alumina silica.

Table 4 Surface layers of silver-alumina-silica

Layer 4:	SiO_2, 825 + 84 nm,-0 nm
Layer 3:	AL_2O_3, 3025 + 275 nm,-0 nm
Layer 2:	Opaque silver, 900 $\pm$ 100 Å
	Deposition rate, 10-20 nm/sec
Layer 1:	Diffusion barrier
	AL_2O_3 100 $\pm$ 10 nm
Substrate:	Aluminum

E) and two on aluminum (curves F and G). Although the difference is not large, the magnesium substrate sample has a α_s degradation rate greater than those on aluminum. This discrepancy lies outside the error band for α_s. The reason for this difference is not yet known, but as all three samples were tested on the same flight it cannot be explained by differences in their space environment. In addition to the sub-strate difference, their surfaces were prepared differently but the effect of this has not been resolved.

Also shown in Fig. 14 are data on white paint S13G/Lo from NAVSTAR 6 (Ref. 4). The initial values of α_S for ZOT and S13G/LO are comparable, however the degradation rate and asymptotic α_S, for S13G/LO is substantially greater than for all ZOT samples.

Silver-alumina-silica, α_S. A sample of vapor deposited silver-alumina-silica (SAS) on aluminum was flight tested and the solar absorptance α_S change with time is shown in Fig. 15. Also shown for comparison are two representative curves for second surface mirrors and ZOT paint. The SAS sample was prepared following the specifications given in Table 4 which was developed from information provided in Refs. 5-7. The aluminum oxide overcoat was 22 quarter wavelengths thick while the silicon dioxide film was 6 quarter wavelengths thick. The initial solar absorptance value lies between those of second surface mirrors and white paint and similarly the initial data shows the α_S degradation rate for SAS is larger than for second surface mirrors but less than that of white paint indicating that it may be an acceptable thermal control coating for some applications. Lower α_S values may be expected from improved fabrication control in view of the information in the above references.

Since the initial presentation of this paper, the SAS degradation rate has changed significantly. The data, therefore, has been updated in Fig. 10 and Table 5 only to show performance up to 4000 equivalent sun hours. At 1500 ESH the α_S degradation rate decreased dramatically and is now approximately zero.

Measurements of emissivity. The data on solar absorptance presented above was obtained from flight measurements that were reduced assuming no change in the emissivity with time of the thermal control surface under investigation. Throughout the life of this calorimeter experiment program several evaluations were made on different samples to validate this assumption. This was done by comparing the minimum diurnal temperature when the sample was shaded from the sun with the computed predicted value over a one year period with an assumption of a constant emissivity. Fig. 16 shows that the measured change in minimum temperature with time over a 1-yr. period was matched well with the predicted temperature change assuming a constant emissivity.

The value of emissivity of some thermal control surfaces does vary with temperature. Laboratory measurements made on the thermal control surface materials used in this experimental program are plotted in Fig. 7.

Table 5 Solar absorptance (α_s) and emissivity (ϵ) data

$$\alpha_s = \alpha_o + (\alpha_m - \alpha_o)(1 - e^{-t/\tau})$$

Flight sample	Surface type	α_o	α_m	τ (ESH)	Emissivity[a] (ϵ)	Range (Max ESH)
1.	0.002-in. silvered Teflon	0.080	0.241	12123	0.66	9300
2.	0.005-in. aluminized Teflon	0.163	0.316	9525	0.80	9300
3.	Silica cloth No. 581	$\alpha_s = 0.224 + 3.03 \times 10^{-5} t$			0.86	1800
		$\alpha_s = 0.273$				1800
4.	Silica cloth No. 581	$\alpha_s = 0.213 + 2.97 \times 10^{-5} t$			0.86	1800
		$\alpha_s = 0.267$				1800
5.	0.002-in. silvered Teflon	0.085	0.246	8769	0.66	7700
6.	Indium oxide coated mirrors	0.096	0.133	2265	0.79	7700
7.	0.005-in. embossed Ag Teflon	0.108	0.297	7467	0.80	7700
8.	Second surface mirrors	0.077	0.103	3138	0.79	7700
9.	ZOT paint (8-10 mil thick)	0.198	0.301	2863	0.91	5800
10.	ZOT paint (8-10 mil thick)	0.185	0.314	2933	0.91	5800
11.	Second surface mirrors	$\alpha_s = 0.069 + 1.5 \times 10^{-6} t$			0.79	5800
12.	ZOT paint (8-10 mil thick)	0.192	0.350	3836	0.91	5800
13.	Second surface mirror	$\alpha_s = 0.075 + 4.7 \times 10^{-6} t$			0.79	3400
14.	ZOT paint (10-12 mil thick)	0.188	0.358	5920	0.91	3400
15.	Second surface mirror	0.079	0.109	6259	0.79	3400
16.	ZOT paint (6-7 mil thick)	0.230	0.405	5757	0.91	3400
17.	Silver-alumina-silica	$\alpha_s = 0.156 + 1.50 \times 10^{-5} t$[b]			0.79	1500
		$\alpha_s = 0.178$				1500
18.	ZOT paint (10-12 mil thick)	$\alpha_s = 0.190 + 3.65 \times 10^{-5} t$[b]			0.91	1100
19.	Second surface mirrors	$\alpha_s = 0.075 + 6.1 \times 10^{-6} t$[b]			0.79	1100
20.	ZOT paint (10-12 mil thick)	$\alpha_s = 0.199 + 3.09 \times 10^{-5} t$[b]			0.91	1100

[a] Emissivity values at 295 K (see Fig. 7 for ϵ vs temp.).
[b] Preliminary data. Sample 17 has been updated to reflect the change in the degradation rates since the presentation of the paper.

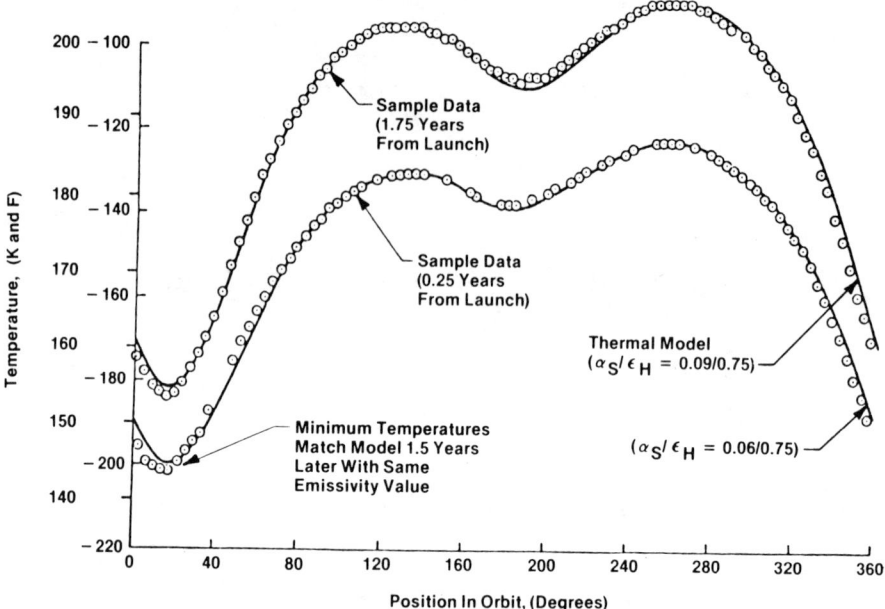

Fig. 16 Comparison of minimum diurnal temperatures 1.5 yrs apart with computer model having constant emissivity.[1]

Development of Design Working Equations

Data on the degradation of thermal control surfaces on several orbital spacecraft have been presented in the literature. However, because of the diverse spacecraft configurations and the different thermal test surface locations relative to contamination sources, it is difficult to compile these data into a usable design source for spacecraft thermal control systems. The data developed in this calorimeter experimental program evaluated several types of thermal control surfaces in a clean location on a synchronous orbiting satellite. These data can provide a reliable design baseline for thermal control radiator surfaces on long-life satellites.

To be most useful for design applications the measured solar absorptance data were converted into equivalent sun hours of exposure. The data used to plot the flight measurement results given in Figs. 10-14 were entered into a computer program that produced the coefficients of the exponential equation

$$\alpha_s = \alpha_0 + (\alpha_m - \alpha_0)(1 - e^{-t/\tau}) \tag{1}$$

which gives the value of solar absorptance α_s as a function of equivalent sun hours. These coefficents are given in Table 5. To achieve a suitable curve fit over the longlife performance of most of these surfaces it was necessary to eliminate the first flight measurement. This was caused by the generally sharp change in degradation rate after several months of flight. The curves in Figs. 10-14 have been drawn with the initial values fitted in but the equations without the first point will give a greater error of up to +0.005 for the first 400 equivalent sun hours of exposure.

Application of Data

Passive radiators are the primary method of temperature control for operating equipment and experiments on orbiting spacecraft. Most satellite experiments and operating equipment must be maintained within a given temperature range over its operating life. These temperatures are determined by qualification tests and by reliability derating values when longlife is involved. These temperature specifications along with the heat loads and available radiator surface orientation are required to begin the radiator design analysis.

In addition to the heat generated by the equipment, the heat absorbed by the radiator from the Ssn, from earth emission, and from the spacecraft itself must be rejected. This aspect of passive radiator design and performance is controlled only by the selection of the thermal control surface properties. (Louvers and other devices to block the sun from the radiator surface are considered to be active devices.) To minimize the influence of solar heat on the radiator performance, a low value of solar absorptance is desired. To minimize the size and weight of a radiator, a high value of emissivity is desired. Thermal control surfaces developed for spacecraft application have had, as their goal, a low value of solar absorptance and a high value of emissivity in the range of infrared wavelength.

The longlife aspects of the thermal performance analysis is dependent on the degraded values of the radiator surface properties. The thermal control surface properties in this experiment were evaluated in a clean location. In the presence of contamination however, higher degradation rates will be experienced and adjustments must be made to the "clean" degradation rates. At present, the influence of contamination has not been adequately quantified for design purposes but efforts are being intensified in this area as a result of the Shuttle Transportation System studies. (11,12)

The performance analysis of a passive radiator for a spacecraft application initially involves two steps. The first step is to screen the potential thermal control surfaces and select a few that can be examined in more detail. This will establish the predicted performance diurnally and with time in orbit. The screening is accomplished under steady-state conditions to ascertain the average operating temperature, and to approximate the the maximum and minimum diurnal temperatures.

The equation for calculating the radiator temperature under steady-state conditions is

$$T = \left(\left(\frac{1}{\epsilon \sigma A}(\alpha_s E_s + \alpha_s E_a + E_e + Q)\right)\right)^{1/4} \quad (2)$$

where direct solar heat load is given by

$$E_s = S_0 A (\cos \theta) \quad (3)$$

and θ is the angle between surface normal and sun line. Albedo heat load is given by

$$E_a = a S_0 F_a \quad (4)$$

Earthshine heat load is given by

$$E_e = H_e F_e \quad (5)$$

and Q = is the heat load from component or experiment.

The geometrical radiator orientation factors in the above equations F_a, F_e, and the heat loads from Earthshine and albedo can be obtained from spacecraft design manuals and handbooks such as Ref. 8.

The prediction of diurnal temperatures of a passively cooled device on a spacecraft requires that a transient analysis be made to account for the heat capacitance in the system. The basic equation for conducting a transient analysis is

$$T_2 = T_1 + \frac{\Delta t}{C}(\alpha_s E_s + \alpha_s E_a + E_e + Q_1 - \sigma \epsilon A T_1^4) \quad (6)$$

The radiator surface values of α_s and ϵ play a major role in predicting the radiator temperature and in the design of spacecraft thermal control systems as seen from Eqs. (2) and (6). The life degraded values of α_s are required to predict realistic longlife thermal performance characteristics of satellites.

Conclusions

The performance data presented above is comprised of a unique group of different types of thermal control surfaces exposed to the same longlife orbital environment. Direct comparison of the tested surface samples to each other is made practical by the use of similar reference samples on each of the satellites. The use of an efficient calorimeter that thermally isolates the test samples from the spacecraft provides accurate and reliable results. As a consequence the data compilation presented above forms a reliable design baseline for thermal control of satellite systems.

This baseline data is referenced to a clean (noncontaminated) surface in an geosynchronous orbit. Where surface contamination is expected from the spacecraft, adjustments must be made to the "clean" degradation rates presented in this paper. The technology for selection of contamination factors as a function of surface material, temperature, location, and time is only now being seriously investigated. When the contamination correction factor is adequately developed quantitatively, its application in conjunction with the above "clean" baseline data will provide engineers the complete tools necessary for design of reliable, longlife thermal control systems for satellites.

References

[1] Curran, D. G. T., and Millard, J. M., "Results of Contamination/Degradation Measurements on Thermal Control Surfaces of an Operational Satellite," 12th Thermophysics Conference, Albuquerque, N. Mex., AIAA Paper No. 77-740, June 1977.

[2] Rajagopalan, R. and Willson, V. J., "Thermal Performance of Anik-B Satellite in Orbit," AIAA Paper 80-1498, 15th Thermophysics Conference, Snowmass, Colo., July 1980.

[3] Bouchez, J. P. and Howle, D., "The Orbital Test Satellite (OTS) Thermal Experience After 3.5 Years in Orbit," AIAA Paper 82-0831, 3rd Joint Thermophysics Fluids, Plasma and Heat Transfer Conference, St. Louis, Mo., June 1982.

[4] Pence, W. R. and Grant, T. J., "α_s Measurements of Thermal Control Coatings on NAVSTAR Global Positioning System Spacecraft," AIAA Paper 81- 16th Thermal Physics Conference, Palo Alto, Calif., June 1981.

[5] Haas, G., Heaney, J. B., and Triolo, J. J., "Evaporated Ag Coated with Double Layers of Al_2O_3 and Silicon Oxide to Produce Surface Films with Low Solar Absorptivity and High Emissivity," <u>Optics Communications</u>, Vol. 8, July 1973, p. 183.

[6] Haas, G., Ramsey, J. B., Heaney, J. B., and Triolo, J. J., "Reflectance, Solar Absorptivity and Thermal Emissivity of S_1O_2-Coated Aluminum," <u>Applied Optics</u>, Vol. 8, Feb. 1969, P. 275.

[7] Haas, G., Ramsey, J. B., Heaney, J. B., and Triolo, J. J., Thermal Emissivity and Solar Absorptivity of Aluminum Coated with Double Layer of Aluminum Oxide and Silicon Oxide ," <u>Applied Optics</u>, Vol. 10, June 1971, P. 1296.

[8] Donabedian, M., <u>The Infrared Handbook</u>, Office of Naval Research, 1978, Chap. 15.

[9] Hall, D. F. and Fote, A. A., "α_S/ε Measurements of Thermal Control Coatings on the P78-2 (SCATHA) Spacecraft," AIAA Paper 80-1530, 15th Thermophysics Conference, Snowmass, Colo., July 1980.

[10] Lockerby, S. C., Barsh, M. K., and Mossman, D. L., "ZOT-A White Thermal Control Coating for Space Environment: Considerations, "<u>National SAMPE Technical Conference Series</u>, Vol. 14, pp. 49-51, 1982.

[11] Barnhart, B. J. and Balse, J. C., "Shuttle Contamination and Experimentation: DOD Implications," <u>SPIE Proceedings</u>, Vol. 280, 1981, pg. 127.

[12] Various authors, Spacecraft Contamination Environment session, <u>SPIE Proceedings</u>, Vol. 338, May 1982.

Experimental Investigation of Bipropellant Exhaust Plume Flowfield, Heating, and Contamination and Comparison with the CONTAM Computer Model Predictions

H. Trinks*
Technical University of Hamburg-Harburg
Hamburg, Federal Republic of Germany
and
R. J. Hoffman[†]
Science Applications, Inc., Los Angeles, California

Abstract

Contamination of satellite surfaces and components by thruster exhaust products is receiving increased attention from the satellite design community. This concern is attributed to the development of more sophisticated satellite systems, designed to perform multiple and more complex missions for longer periods of time in space.

This paper describes two independent efforts to characterize the transient operation of a small bipropellant thruster, and to determine its contamination potential relative to a specific satellite system: the GALILEO Spacecraft. The analytical study used an early version of the CONTAM III computer code for prediction of transient combustion processes, multiphase flows, and contamination effects. The vacuum chamber experimental study measured gas species mass flux, momentum distribution, droplet outflow, composition, and heat flux. The engine used in both studies was the Messerschmitt-Bolkow-Blohm GmbH (MBB) 10-N bipropellant vortex injection thruster.

The experimental results confirmed the ability of the

Presented as Paper 83-1447 at the AIAA 18th Thermophysics Conference, Montreal, Canada, June 1-3, 1983. Copyright © American Institute of Aeronautics and Astronautics, Inc., 1983. All rights reserved.
*Prof. Dr., Technical University.
†Manager, Plume Technology and Spacecraft Contamination Division.

CONTAM III model to accurately predict the important transient combustion events (even for an unconventional injector) and to quantitatively predict the multiphase flowfield and contaminant species produced.

Background

The CONTAM computer code for the prediction of combustion dynamics, multiphase nozzle and plume flows, chemical kinetics, and contamination of sensitive spacecraft surfaces, has been under development by the U.S. Air Force Rocket Propulsion Laboratory since 1969. The latest version of the code, CONTAM III, has recently been released to U.S. aerospace industry and government users, and is documented in Refs. 1, 2, and 3. Technical summaries are included in Refs. 4 and 5.

Two years prior to the completion of the CONTAM III code development by the prime contractor, Science Applicatons, Inc. (SAI), a preliminary version of the code was used to analyze the transient operation of the MBB 10-N bipropellant engine, which will be used on NASA/JPL's Project GALILEO. Estimates of contaminant species production, type and amount of contaminant exiting the engine, and plume flowfield properties were generated. The results of this JPL sponsored study, conducted by SAI, have been reported in Ref. 5.

About a year after the completion of the analytical study, the authors of this paper met and were encouraged by a superfical comparison of the experimental vacuum chamber results with the prior CONTAM analysis. The MBB 10-N vacuum chamber experimental program was conducted under sponsorship of the European Space Agency (ESA), by Prof. Trinks in West Germany.

This paper is the result of an effort to compare the experimental results and the analytical modeling effort in more detail. In addition, certain improvements to the computer code (CONTAM III) were completed after the MBB 10-N analytical study. The effect of these improvements were included in the comparison.

PART I: Analytical Computations
- CONTAM Analysis

MBB 10-N Thruster Description

The MBB 10-N thruster was developed and qualified for the SYMPHONIE Program. With the launches of SYMPHONIE I in December 1974 and of SYMPHONIE II in August 1975, 14 thrusters have successfully operated in space. A total of 62 thrusters were fabricated. SYMPHONIE I and SYMPHONIE II flight data indicates an achieved (vacuum performance-I_{sp}) of between 287.0 and 291.9 s.

The engine employs a coaxial vortex injector system, generating two concentric liquid cones which impinge approximately 1 mm. from the injector face. Some of the details of the vortex injector are proprietary to MBB and were not available at the time of the study. Figure 1 schematically represents the MBB vortex injection system.

Fuel, which regeneratively cools the nozzle throat region, is injected tangentially into the central plenum of the injector, forming a fuel vortex. The oxidizer is tangentially injected into an annulus surrounding the fuel plenum, forming a vortex sheet, with an angular rotation in the same direction as the fuel vortex. Both the fuel and oxidizer exit the plenum with outboard momentum and angle due to the centrifugal force created by the angular swirl velocity.

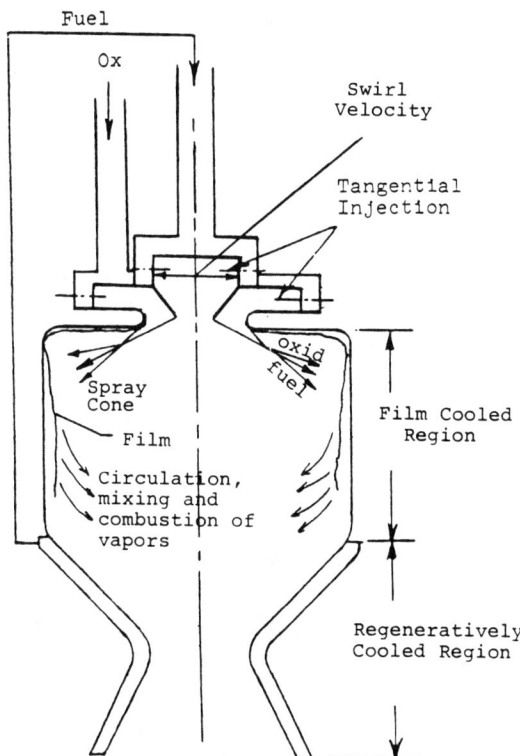

Fig. 1 Schematic of vortex injector system.

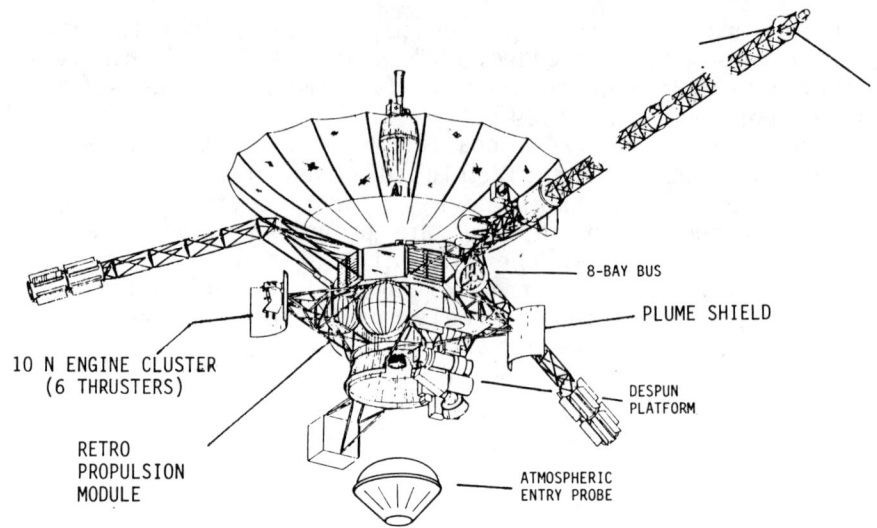

Fig. 2　GALILEO Spacecraft with 10-N engine clusters.

Based on cold flow injector spray patterns provided by MBB, a fuel spray cone angle of 100 deg. and an oxidizer cone angle of 120 deg. was assumed. Conceptually, the oxidizer sheet is deposited on the combustion chamber wall and begins traveling towards the throat. The fuel droplet spray impinges on the oxidizer wall film causing vaporization and combustion, with flow generally directed towards the center of the combustion chamber. Part of the oxidizer remains near the wall to provide film cooling.

Under nominal conditions, the engine operates with a MON-1/MMH mixture ratio of 1.645, and a chamber pressure of approximately 128 psia.

MBB 10-N Engines on the GALILEO Spacecraft

Project GALILEO, the next U.S. planetary mission, will be launched by the Shuttle/Centaur in the late spring of 1986, to perform a comprehensive investigation of the Jupiter system, as in Astronautics & Aeronautics.[6] Figure 2 depicts the Galileo Spacecraft and the location of the 10-N engine clusters and plume shields designed to protect sensitive instruments from plume contamination.

Identification of GALILEO Sensitive Surfaces

As part of a JPL in house effort, surfaces which might be sensitive to direct or scattered plume flux from the 10-

Table 1 GALILEO sensitive surfaces

RCT-NIMS	Radiometric calibration target (NIMS)	
RCT-PPR	Radiometric calibration target (PPR)	
SSI radiator		
SSI aperture	Solid-state imaging sub-system	Scan platform
SSI lens		
NIMS cooler shade (inside)		
NIMS radiator		
NIMS aperture	Near infrared mapping spectrometer	Scan platform
NIMS mirror		
UVS aperture	Ultraviolet spectrometer	Scan platform
PPR aperture	Photo polarimeter radiometer	Scan platform
RRH antenna	Probe relay antenna	
PCM	Photometric calibration mirror	Science boom
PCT	Photometric calibration target	Science boom
DDS sensor	Dust detector subsystem	Science boom
EPD sensor	Energetic particle detector	Science boom
PLS Sensor	Plasma subsystem	Science boom

N engine were identified. These surfaces, either on the scan platform or science boom, are presented in Table 1. The analysis was to consider deposition, and the effects of contamination, on those surfaces which were determined to receive significant plume contaminant mass flux.

CONTAM Analysis

Study Approach. An early version of the CONTAM III code was used for the analysis of the MBB 10-N engine. The study approach concentrated first on a detailed evaluation of the production of potential contaminant species during the transient operation of the 10-N engine. Unburned propellant droplets and vapors, wall film mixtures of propellants and MMH nitrate, and certain combustion products were considered as potential contaminant sources.

After determining the time-dependent "production" of potentially contaminating species in the combustion chamber, the multiphase nozzle and plume flowfields were generated. The flowfield contains both propellant droplets and vapor/product gases. A computation of nonequilibrium chemical kinetics, along computed gaseous streamlines, completed the flowfield description.

Next, plume species flux to the location of specific sensitive surfaces were evaluated. Engineering judgments were made as to the significance of the potential contamination problem relative to those surfaces. The study approach is quite representive of the approach to be taken for any bipropellant plume contamination analysis involving a small, pulsing thruster.

Analysis. Modeling of a vortex injector with the CONTAM III/TPP subprogram provided a difficult challenge. TPP was developed for orifice injector elements with impinging liquid streams - not for impinging propellant sheets with angular velocity.

To simulate the injector, a pair of unimpinging rings were employed. Fuel was injected from near the center of the injector face in two cones having half-angles of 45 and 55 deg. 50% of the fuel was contained in each cone. By conserving angular momentum, the fuel injection velocity was set to 9.5 m/s.

75% of the oxidizer was assumed to be injected at a point just slightly outward (radially) from the fuel injectors, with a half-cone angle of 60 deg. and the same assumed swirl as the fuel. The total oxidizer steady-state injection velocity was approximately 7.4 m/s. The remaining 25% of the oxidizer was injected in such a way as

Table 2 CONTAM III / TPP runs
(MBB 10-N engine)

Pulse length:	9, 20, 40 ms (to valve close)
Tail-off:	30-ms computation time
Injector Simulation:	
Fuel:	1 pair of unimpinging rings 45-deg half-cone - 50% of fuel 55-deg half-cone - 50% of fuel Injection velocity - 9.5 m/s (including swirl component)
Oxidizer:	60 deg half-cone - 75% of oxidizer Direct wall impingement - 25% of oxidizer Injection velocity - 7.4 m/s (including swirl component)

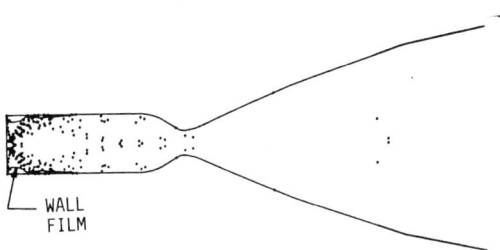

Fig. 3 CONTAM III computer graphics of the MBB 10-N engine showing droplet injection and liquid wall film.

to immediately hit the wall and begin flowing along the chamber.

Combustion gas properties were computed by the TPP subprogram assuming chemical equilibrium within the combustion chamber.

Standard TPP fan and showerhead break-up lengths were employed. Standard TPP droplet size distributions, and injected size and velocity formulation were used, but modified for the vortex injector flow as follows:
1) fuel droplet size factored by 0.25, 2) oxidizer droplet size factored to 0.35, and 3) fuel and oxidizer droplet injection velocities factored by 0.3.

Table 2 provides the important input parameters for the CONTAM III transient engine analysis using the TPP (transient performance program) subprogram.

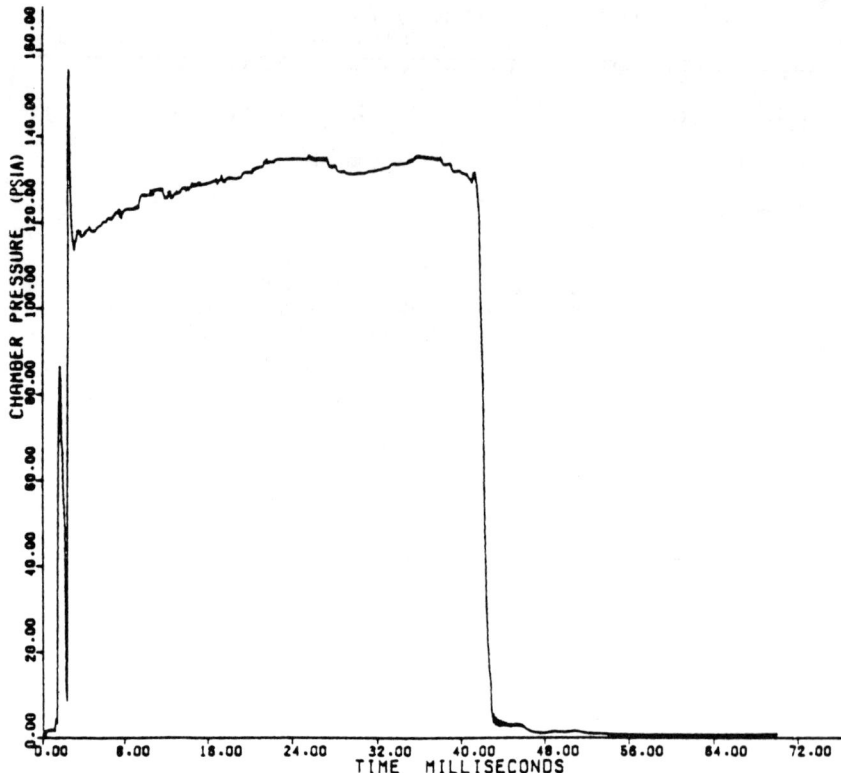

Fig. 4 CONTAM III computer-generated plot of time-dependent chamber pressure: MBB 10-N engine.

Figure 3 is a computer-generated graphic of the 10-N engine combustion chamber and nozzle, showing the injection of fuel and oxidizer droplets and the formation of liquid wall film near the injector end of the chamber, as simulated by the CONTAM III/TPP model. A careful examination of the figure shows the continuation of wall film from the injector end to midway down the combustion chamber wall. Our interpretation is that the liquid wall film is burned off the combustor wall at a point midway to the nozzle throat, as a result of film boiling, combustion, and droplet entrainment by the accelerating combustion chamber gases. Additional pulses will continue to deposit wall film on the combustion chamber wall and the depletion of wall film, as a result of film boiling and droplet entrainment, will be a function of combustion gasdynamics, and wall temperature.

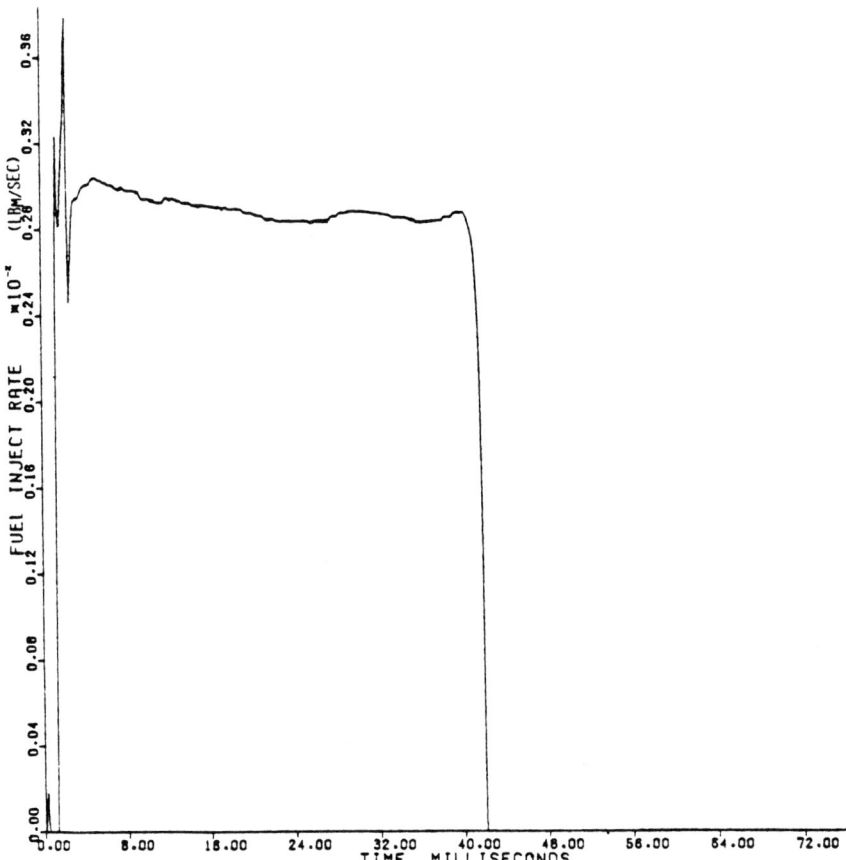

Fig. 5 CONTAM III computer-generated plot of time-dependent fuel injection rate: MBB 10-N engine.

Result of CONTAM Analysis

Detailed printed and graphical information is provided by the CONTAM III/TPP subprogram to describe the transient combustion and gasdynamic processes, and the production of contaminant species, in a bipropellant engine. Pulse trains, as well as single pulses, may be considered, including engine off-time between pulses during which wall film is evaporated from the chamber and nozzle wall, and propellant continues to enter the combustion chamber as a result of void volume injection.

The computed transient chamber pressure profile for the MBB 10-N engine is shown in Fig. 4. The initial pressure spike is typical of bipropellant engine starts. During start-up, there appeared to be some fuel and

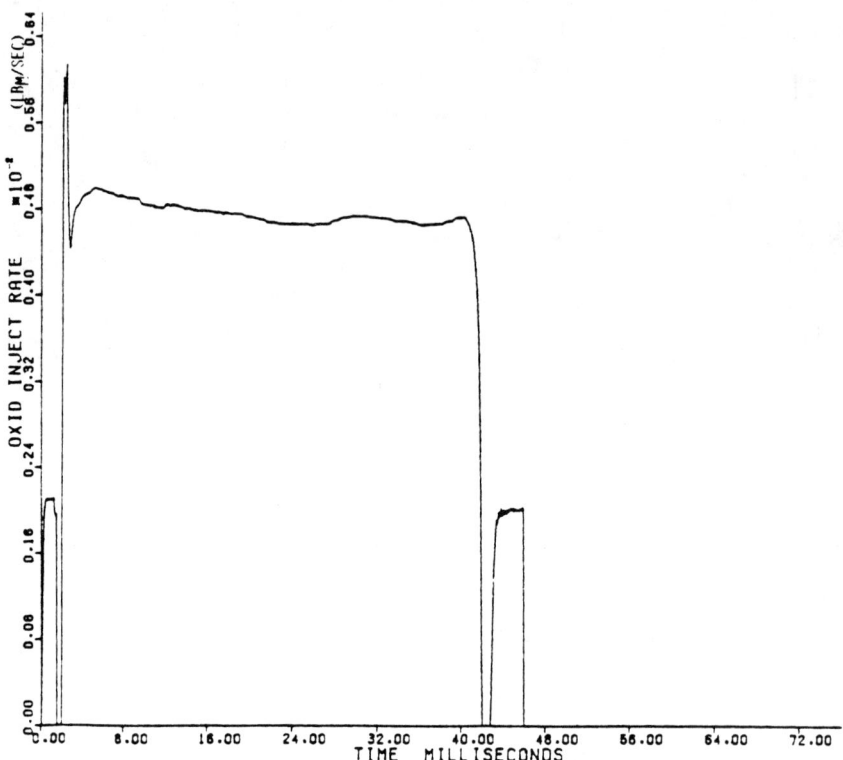

Fig. 6 CONTAM III computer-generated plot of time-dependent oxidizer injection rate: MBB 10-N engine.

oxidizer boiling in the injector and some propellant flashing. Ignition occurs at 2.4 ms. During the first 9 ms of operation, an average fuel droplet outflow of 2.6 x 10^{-5} lbs/s, and an average oxidizer droplet outflow of 2.3 x 10^{-5} lbs/s occurred. Very small amounts of unburned oxidizer vapor and wall film is expelled during start-up. At about 9 ms a wall film has been established which is oxygen rich close to the injector face and increases in fuel fraction with distance downstream. An adduct mixture exists where both species are present.

The 10-N engine required about 18 ms to reach steady-state pressure, although pressure was close to steady-state in only 5 ms. Other indicators of steady-state, e.g., chamber gas temperature, propellant injection rate, propellant droplet outflow (out of engine), droplet evaporation rates, etc., behaved similarly. During steady-state operation, approximately 1% of the injected fuel mass and 0.8% of the injected oxidizer mass is expelled. The wall film axial fuel fraction distribution and thickness is similar to that at 9 ms.

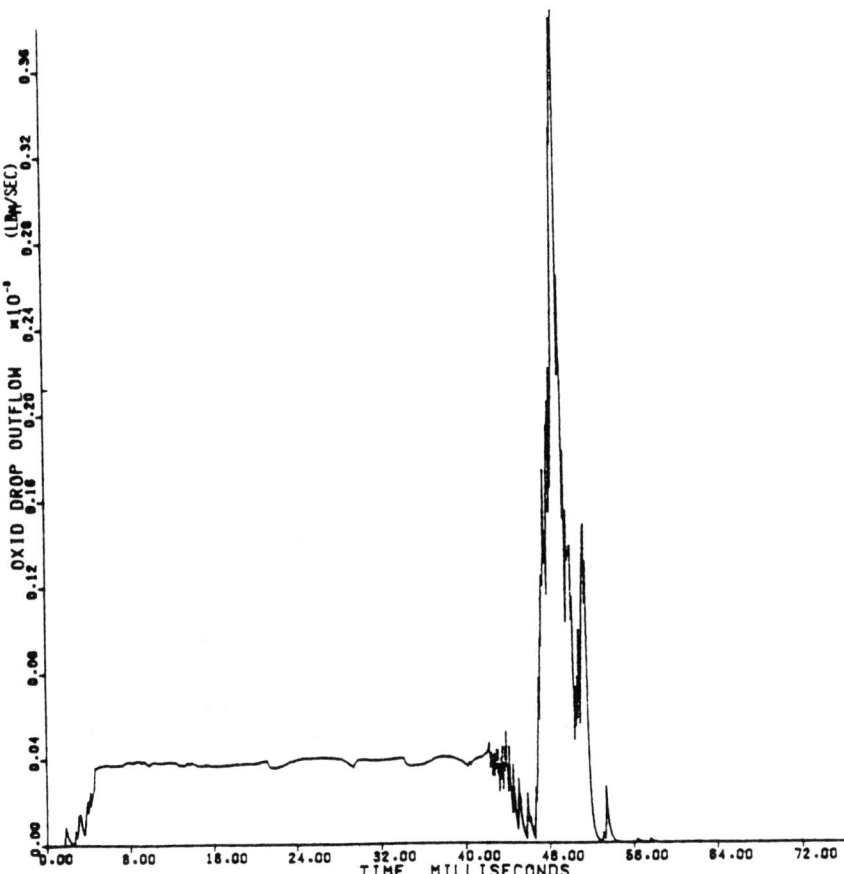

Fig. 7 CONTAM III computer-generated plot of time-dependent droplet outflow rate: MBB 10-N engine.

After the start of valve closure at 40 ms the pressure rapidly decreases but does not vanish. The continued small, but finite, pressure for several milliseconds is a result of oxidizer void volume "dribble" after the valves are closed. This may be clearly seen in Fig. 5, 6 which present the computed fuel and oxidizer injection rates as a function of time, respectively. Fuel injection ceases almost immediately after valve closure. Oxidizer injection, also ceases within 2 ms of the start of valve closure but immediately increases to approximately 1/3 of the steady-state injection value for an additional 3 ms. This is due to boiling of the oxidizer in the injector as the pressure drops, forcing additional liquid out of the injector. The much lower vapor pressure of the fuel prevents this from happening in the fuel injector.

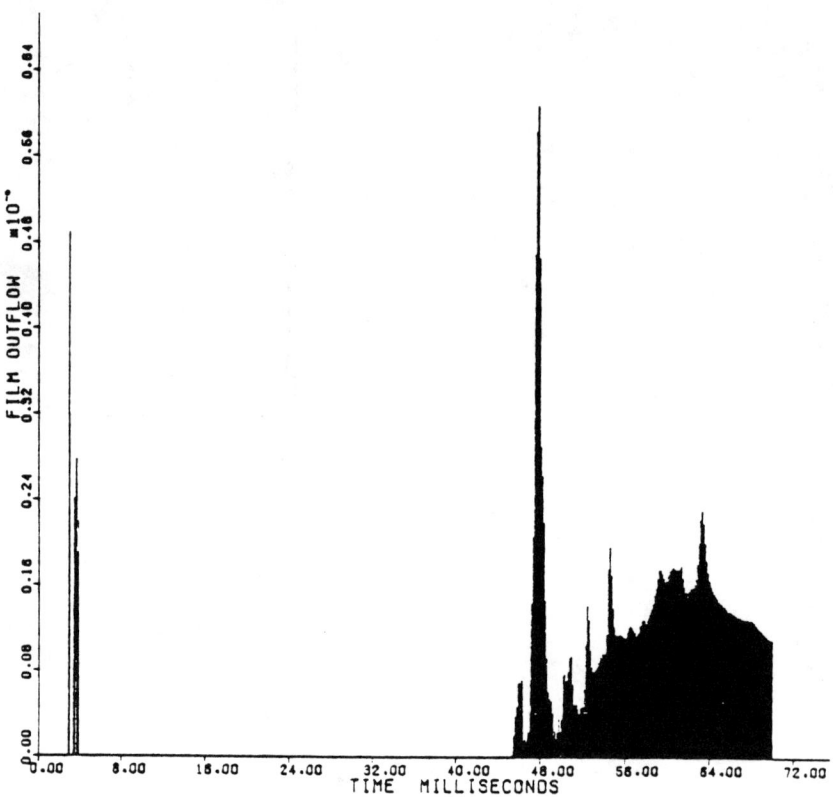

Fig. 8 CONTAM III computer-generated plot of time-dependent wall film outflow.

The surge of liquid oxidizer after valve closure causes a relatively large accumulation of liquid oxidizer in the combustor. These droplets were originally thought to be ejected from the chamber due to the chamber pressure and flow maintained by the evaporating oxidizer droplets and the fuel/oxidizer wall film. For the vortex injector, however, oxidizer droplets injected after valve closure are directed towards the wall and immediately become part of the wall film or produce a mixture of wall film adduct droplets as a result of the impingement, shear, and gas flow entrainment. Unfortunately, the CONTAM/TPP analysis, not being specifically designed for vortex injectors, did not distinguish between oxidizer drops and wall film adduct drops ejected during tail-off, and identified all ejected drops as oxidizer. Figure 7 depicts the time-dependent droplet outflow rate for the MBB 10-N engine. Figure 8 is the computed film outflow for the mixture of fuel, oxidizer, and fuel-derived nitrates.

Table 3   Results of CONTAM III analysis
of the MBB 10-N engine

1) During transient start-up:
   a) 0.7% fuel expelled as unburned drops
   b) 1.2% oxid expelled as unburned drops
   c) no wall film expelled

2) During steady-state operation:
   a) steady-state nearly attained at 9 ms
   b) wall film extends to approx. 1/2 of chamber length
   c) 0.8% fuel expelled as unburned drops
   d) 1.0% oxid expelled as unburned drops
   e) no wall film expelled

3) During shut-down:
   a) additional oxidizer is injected after valve
      closes for 3 ms
   b) considerable unburned oxid vapor is expelled
      (18%, 3.34 mlb)
   c) considerable unburned fuel, oxid, and MMH nitrate
      drops are expelled (4%, 1.1 mlb)
   d) negligible wall film reaches nozzle lip and is
      expelled.

4) The plume flowfield is characterized by:
   a) 5-150 m droplets of propellant and MMH nitrate
      near the plume axis
      (5 m less than 13 deg off-axis)
      (10 m less than 6 deg off-axis)
      (100 m less than 2 deg off-axis)

   b) smaller droplets, extending to greater angles
      off-axis during the transient start-up and
      tail-off

   c) gaseous chemical species (mass fraction)

N_2	0.41796
H_2O	0.29830
CO	0.17383
CO_2	0.08818
H_2	0.01480
NO	0.00287
H	0.00142
O_2	0.00138
OH	0.00077
O	0.00048

(Table 3 continued on next page.)

Table 3 (cont.) Results of CONTAM II analysis
of the MBB 10-N engine

---

5) Relative to the contaminant potential of the
MBB engine, it was concluded that:
   a) The MBB engine is a relatively "clean" engine
   b) Potential for wall film to reach the nozzle
      lip, and splatter into the plume backflow
      region, is near zero
   c) Expelled droplet outflow is moderate (1%), but
      confined to an extremely small area near the
      nozzle axis
   d) The MBB 10-N engine should not be pulsed for
      less than 20 ms, or risk contamination from
      wall film and unburned vapor
   e) Expelled liquid droplets should stay close to
      the plume axis and not pose a threat to
      GALILEO instruments
   f) The GALILEO 10-N engine plume shield should
      be effective.

It is clear from the discussion above that the details of the injection, vaporization, and gasdynamic processes occurring after valve closure are often important to the analysis of contaminant species production.

These results provided input to the multiphase nozzle and exhaust plume computation using the MULTRAN subprogram of CONTAM III. The MULTRAN subprogram computed the coupled multiphase flow of combustion products, vapors, and unburned droplets through the nozzle and exhaust plume. Of significance to the GALILEO configuration is the computation of droplet limiting streamlines, beyond which no droplets of a given size are expected to be found. The model predicted that 50+ m size drops would be confined to 2 deg. from the nozzle axis, 10+ m drops to 6 degrees, and 5+ m drops to 13 deg.

Table 3 summarizes the significant results and conclusions of the analytical study.

Part II: Experimental Program and Results

Experimental Program

The MBB 10-N bipropellant thruster, which is being used on the US/German GALILEO spacecraft, has been investigated systematically for exhaust plume formation and contamination effects. The investigated thruster was manufactured by MBB Space Division (Germany) and equipped with a valve which is under development at MBB. This valve

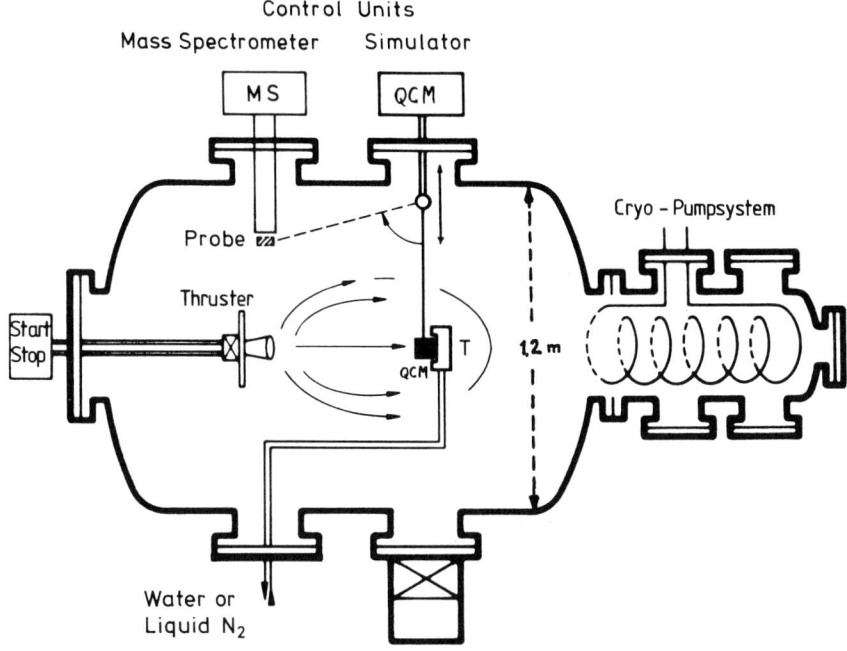

Fig. 9 Test facility.

permits a pulse operation mode with a pulse duration of less than 10 ms.

The experimental research program concentrated on the transient mode to investigate the nonsteady processes of small bipropellant thrusters. The HAMBURG test facility, utilized in this study, is particularly suitable for investigating transient nonsteady processes.

The test facility consists of a high-vacuum chamber, different pumping systems and a number of special test instrumentations. The stainless steel chamber with a length of 2 m and a diameter of 1.2 m is well equipped with numerous flanges, feed through connections, and windows for all kinds of observations. The MBB 10-N thruster is installed axially in the centerline of the cylindrical vessel.[7]

The vacuum chamber is pumped by oil diffusion pumps, liquid N_2 cryosystems, and liquid He cryosystems. The pump capacity of all pumping systems is about 14.000 l/s. The pressure in the chamber before thruster operation is about 10^{-6} mb. The test facility is equipped with instrumentation concerning high speed photography, mass spectrometry, measurement device for density, velocity, heat flux, momentum distribution, and electric charging

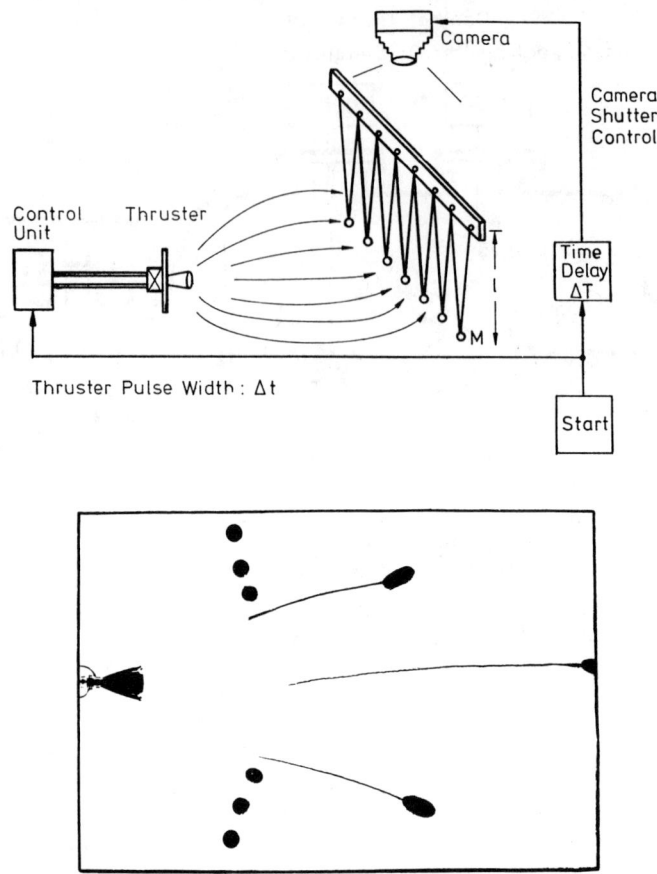

Fig. 10 Momentum distribution measurement.

effects. Most of the measuring techniques were developed especially for the small thruster plume investigations. They are suitable to measure simultaneously transient parameters concerning gaseous as well as condensed reaction products with high time resolution. The test facility is shown in Fig. 9.

During this investigation program the following activities were performed: high speed photography for visualization of plume formation, gaseous and condensed reaction products analysis, plume velocity measurements, mass flux and heat flux measurements, charging effects investigations, and finally a test result evaluation and comparison with theoretical computations.

All investigations were made by varying the operational parameters for the thruster within the limits indicated:

BIPROPELLANT EXHAUST PLUME FLOWFIELD

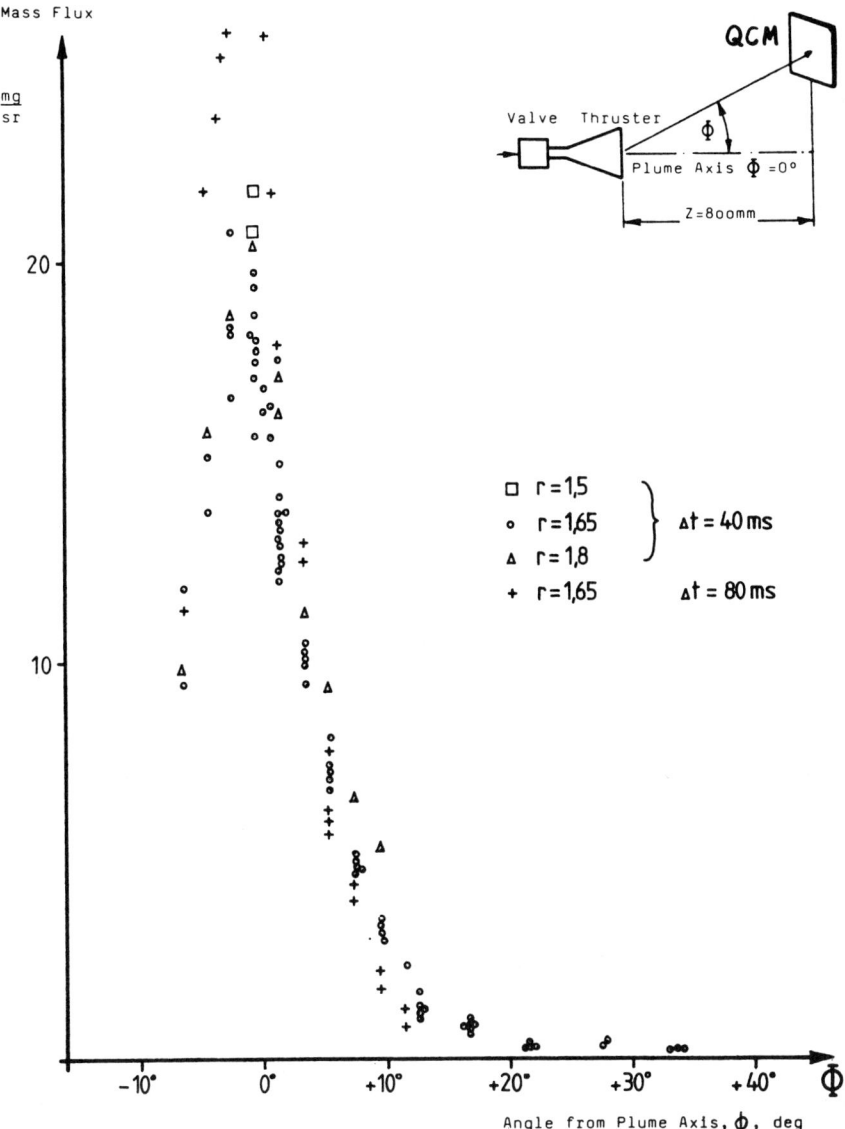

Fig. 11 Plume deposition distribution.

1) nominal total propellant flow rate, m_N=3.5g/s; 2) mixture ratio $m_{N_2O_4}/m_{MMH}$, r=1.5 - 1.8; 3) electrical pulse width, Δt(EPW=7-200ms; 4) off-time between pulses, t_{off} = ∞ -10ms and 5) number of pulses in a train, depending on EPW, n=1-10.

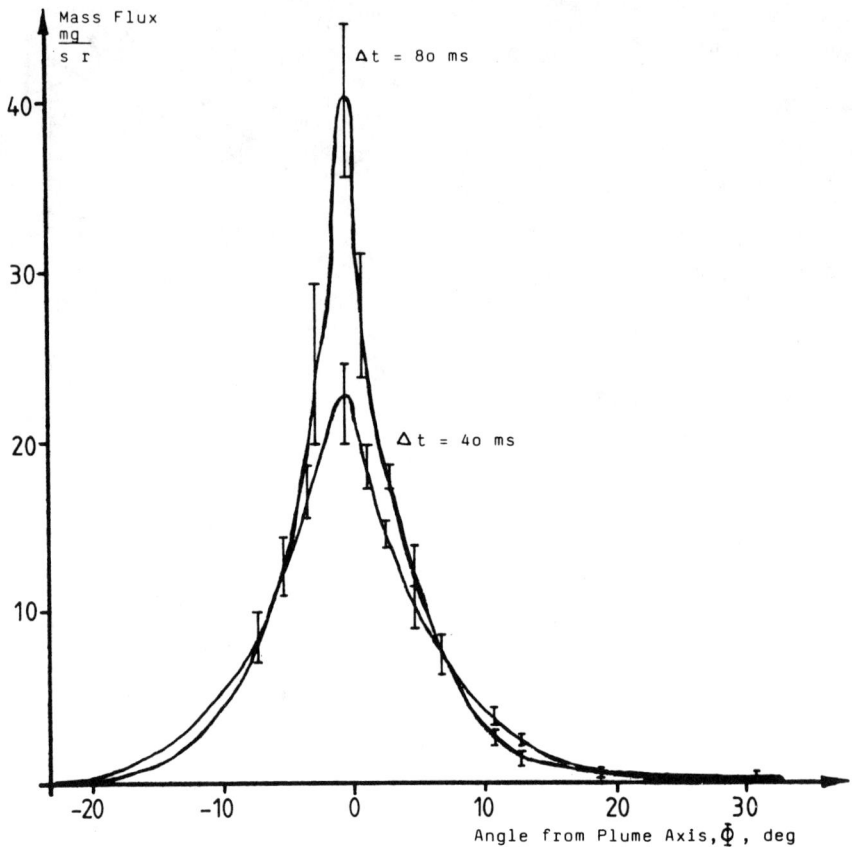

Fig. 12 Plume deposition distribution.

Experimental Results

Momentum Distribution. The momentum distribution device and typical results are shown in Fig. 10 illustrating the number of pendula installed in the vacuum chamber. The thruster plume interacts during the short single pulse operation time EPW with each pendulum according to the specific momentum. After the thruster operation is stopped, each pendulum swings with a specific amplitude, which is measured photographically. The measured swing amplitude gives the specific momentum in the plume flowfield. The momentum distribution corresponds to the mass flux distribution because the velocity of the plume flow is nearly constant and practically independent of the location in the plume flowfield.

Contamination Measurements. Contamination measurements were performed using quartz crystal microbalance techniques (QCM) and mass spectrometers to analyze the contaminant

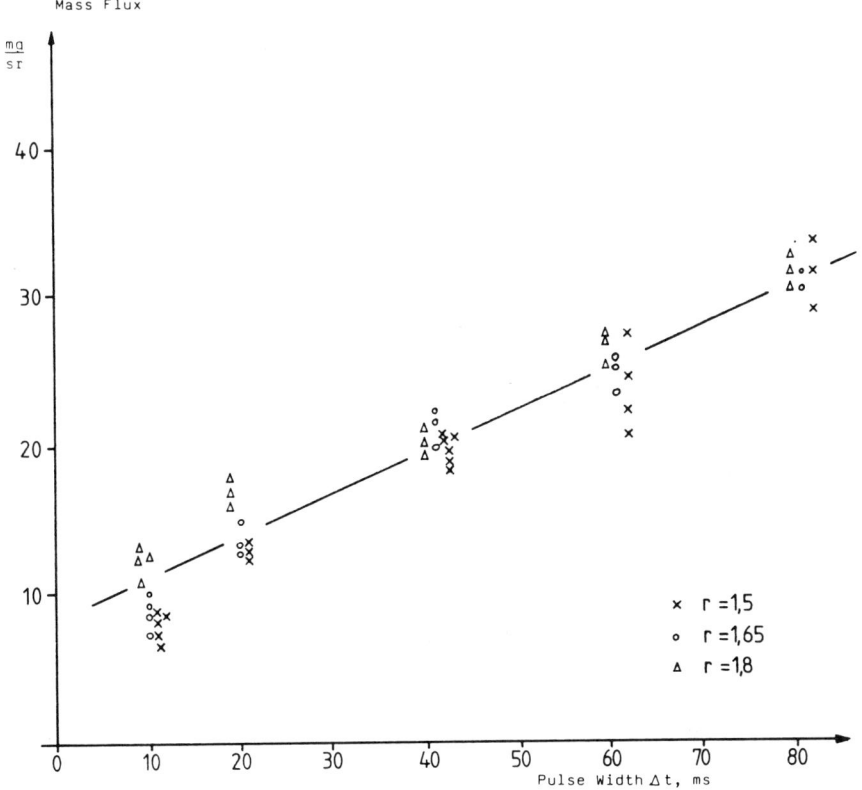

Fig. 13 Plume deposition in plume axis o = 0 deg.

composition. The QCM sensor was installed in the vacuum chamber according to Fig. 9. By a mechanical simulator it was possible to vary the sensor location in the plume flowfield from one thruster pulse to the next. In this way all measurements were performed using the same QCM sensor. The deposited material was analyzed by mass spectrometer within a few seconds after the thruster operation. Thus the influence of the evaporation rate of the deposited material in the time range of minutes was diminished. Fig. 11 shows typical results measured by the QCM. The mixture ratio was varied.

It was concluded that:

1) The amount of deposited material in the plume strongly decreases with the angle from the plume axis (see Fig. 11).

2) The deposition distribution is narrower for a long pulse duration EPW (t = 80 ms) than for short pulse duration (see Fig. 12).

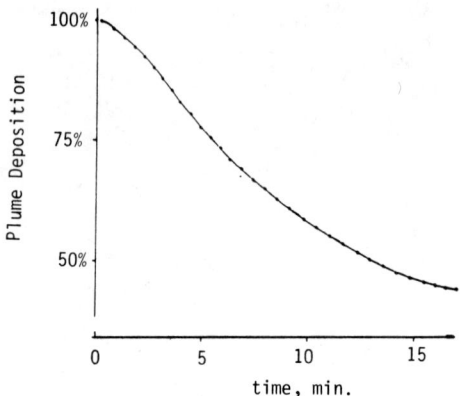

Fig. 14 Plume deposition evaporation rate.

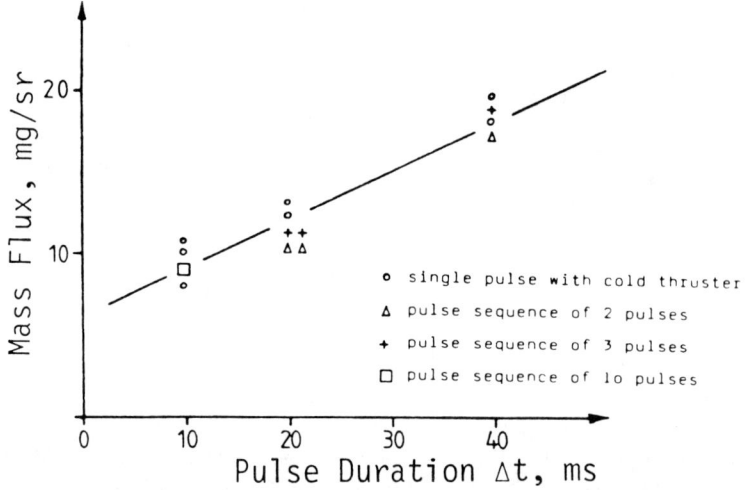

Fig. 15 Plume deposition on plume axis.

3) The plume deposition seems to be nearly independent of the mixture ratio r in the range r = 1.5 - 1.8 (see Fig. 13).

4) The deposited material evaporates within the first 15 min. (at temperature T - 300 K). The evaporating material is identified by mass spectrometer as H_2O and MMH. The MMH-nitrate droplets remain on the surface (see Fig. 14).

5) The thruster operation mode (single pulse, pulse sequence) seems to have no essential influence on the amount of deposited material, within the plume region measured (see Fig. 15).

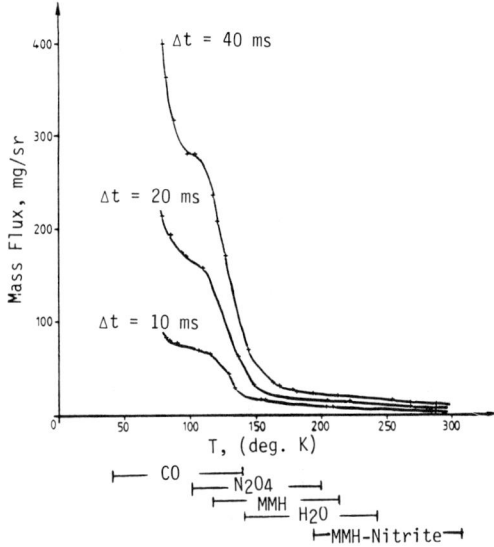

Fig. 16 Plume deposition on plume axis o = 0 deg. dependent on QCM temperature T.

Table 4 Condensation vs target temperature

Specie	Target Temp (K)
MMH nitrate	200-300
H_2O	140-250
MMH	120-210
N_2H_4	100-200
CO	50-140

6) The plume deposition is strongly dependent on the QCM crystal temperature (see Fig. 16).

7) The composition of the deposited material is independent of the angle to the plume axis during steady-state operation of the thruster.

8) During steady-state operation, the main constituent of the plume deposition is MMH-nitrate with MMH and H_2O dissolved in the nitrate.

9) During the start phase (first 2 ms) mainly N_2O_4 is expelled. Later on MMH and N_2O_4, and finally (after 8-10 ms) mostly MMH-nitrate is expelled.

10) During the tail-off phase mainly MMH-nitrate, with a large amount of dissolved MMH, is expelled.

11) In the case of the cooled target plate, successive condensation can be observed in Table 4.

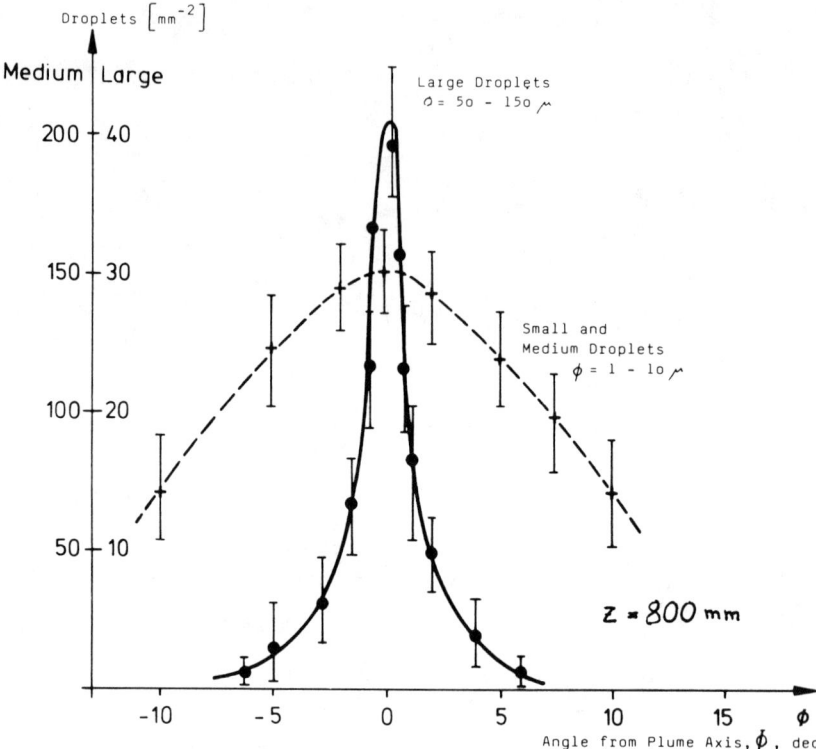

Fig. 17 Droplet distribution in the far field.

Droplet Distribution. Droplet distribution measurements were performed by numerous glass plates installed in the vacuum chamber in the plume flowfield. The deposited droplets were evaluated by mass spectrometer, microscopic and photographic methods, immediately after the thruster operation. The composition, size and angular distribution were derived as shown in Fig. 17.

Droplet Outflow and Plume Velocity. Droplet outflow and plume velocity were investigated by electric field sensors installed in the vacuum chamber.[8] The electrically charged particles are identified either via influence effects when flying past field sensors, or when impinging on electrical sensor plates. When striking, they release their electrical charge to the sensor and every droplet produces an electric pulse according to its size. By the installation of these rather simple electric field sensors and by the evaluation of the strong electric sensor signals it is possible to get signficant information about plume flow field structure, droplet outflow rate, and plume velocity.

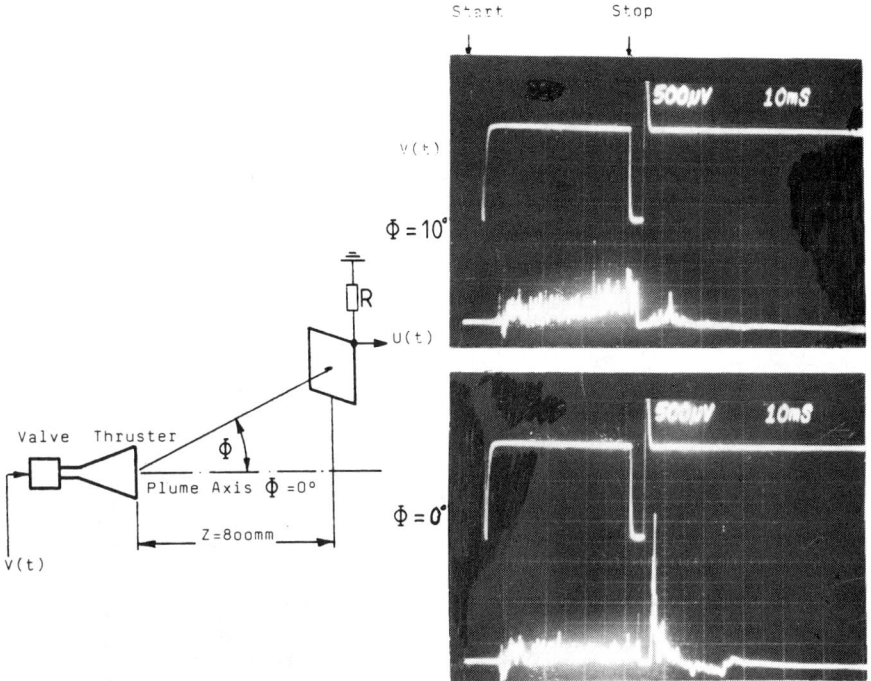

Fig. 18 Droplet outflow measurement.

Fig. 18 provides a typical example of droplet outflow measurements with the electric field sensors. In Fig. 19 typical results are shown for the plume velocity measurement using the electric field sensors. The velocity values demonstrate that the plume velocity, even during the steady-state operation of the thruster, is not constant. Strong oscillations and density fluctuations in the plume flowfield, corresponding to rapid velocity fluctuations, were observed.

Heat Flux Measurements. Heat flux measurements were performed using temperature sensor and thermo paper arranged in the plume flow field. Typical results for the 10-N thruster plume heat flux distribution are given in Figure 20. The heat flux in the plume axis region can be seen to be relatively low when compared with the region immediately adjacent to the axis. This effect is particularly strong in the case of thruster operation with small pulse widths.

Plume Composition Measurements. The plume composition of gaseous reaction products were analyzed with a number of

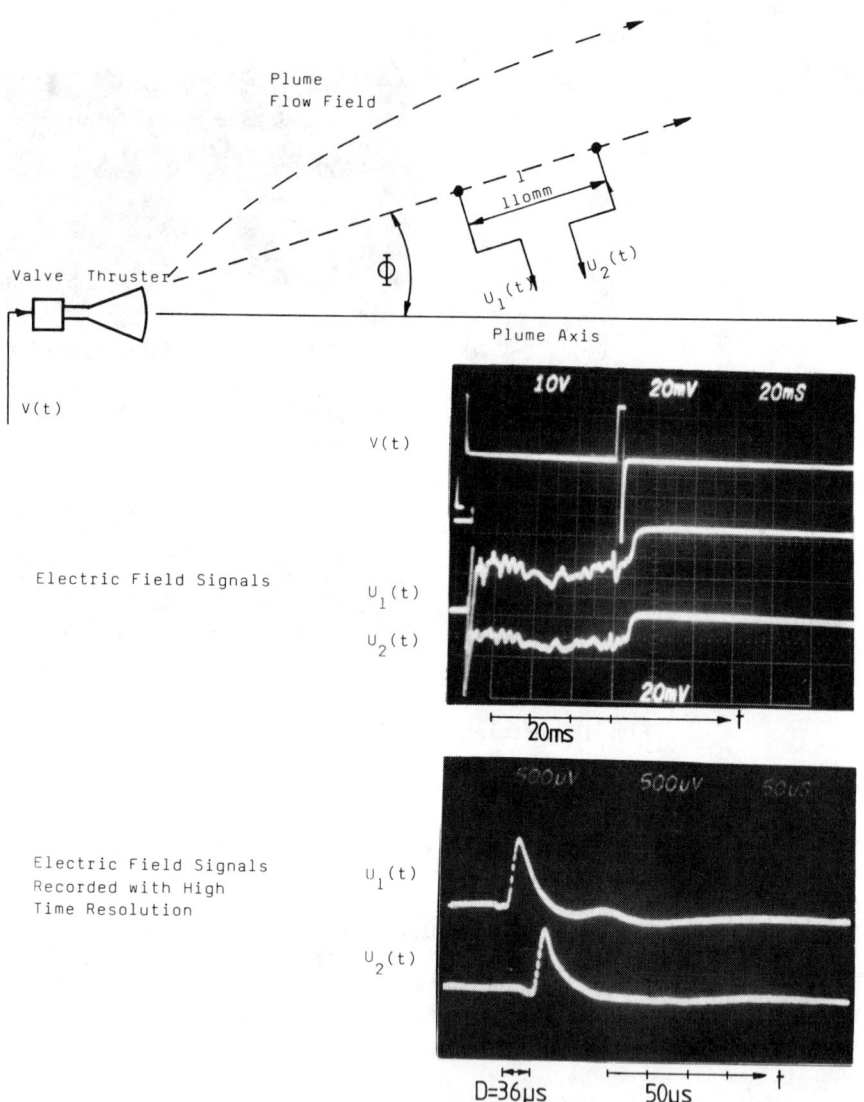

Fig. 19 Plume velocity measurement.

mass spectrometers. Different mass spectrometers were located in the test chamber. Through multichannel mass spectrometer it was possible to analyze the thruster plume composition with high time resolution. These highly dynamic in situ measurements are suitable to determine the time dependent development of the the plume structure.[9-11]

In Fig. 21 gas concentration distributions are shown, derived from mass spectrometric results. Quite similar to

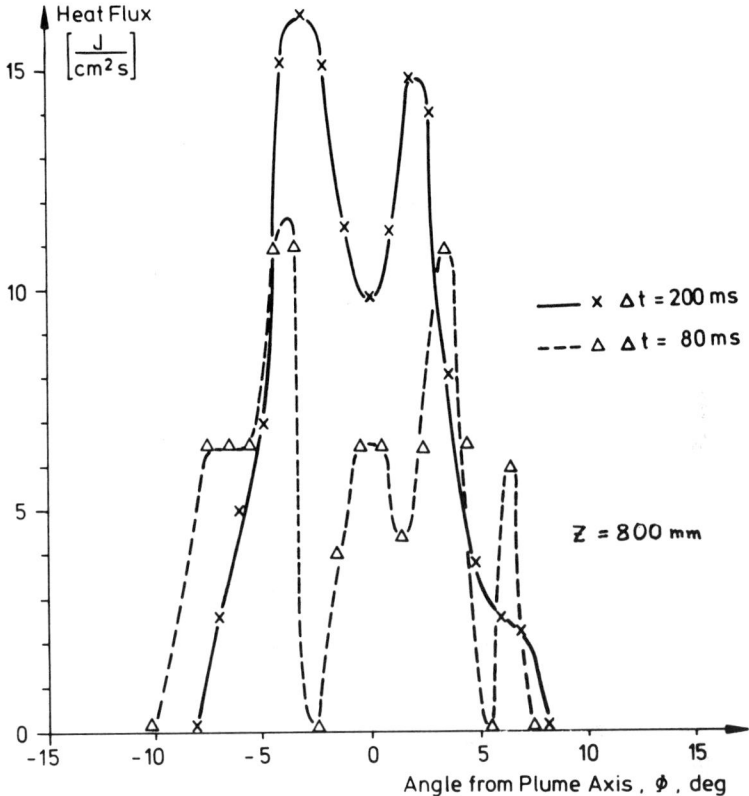

Fig. 20 Heat flux measurement.

the heat flux measurement results, the gas density is relatively low very near the plume axis. The increasing amount of NO with increasing angle from the plume axis demonstrates the increased chemical nonequilibrium in the lower density portions of the plume. Figure 21 also indicates the nonequilibrium state of the start-up and tail-off phase of the pulse.

Compilation of Results. Experimental results have been compiled for angular distribution of momentum (corresponding to the mass flux), plume deposition, gas density, and heat flux as shown in Fig. 22. The momentum distribution is a superposition of gas density distribution and the plume deposition distribution corresponding to the droplets distribution.

Figure 23 shows gas density contours derived from mass spectrometric and momentum measurement results. The gas density contours were registered in high vacuum with and

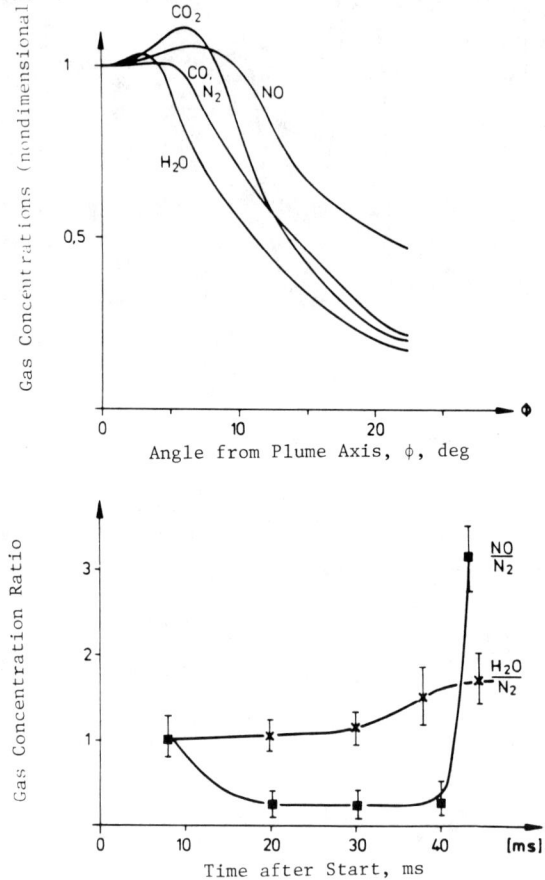

Fig. 21 Gas concentration in the plume.

without the plume shield model. This plume shield model was manufactured similar to the planned plume shields for the GALILEO Project spacecraft. The results shown in Fig. 23 demonstrate that the plume shields should indeed be effective.

All experimental results were evaluated, compiled, and compared. A conceptualization model for the formation of the 10-N thruster plume flowfield was derived as follows. During the start phase (1-2 ms) a wide spread cloud of unburned fuel and oxidizer vapor is expelled mainly from the thruster inner wall. During the transient start phase larger droplets were expelled, more or less focused in the axis zone. The steady-state phase is depicted n Fig. 24 with all essential values concerning plume angle and droplets distribution. During the tail-off phase some

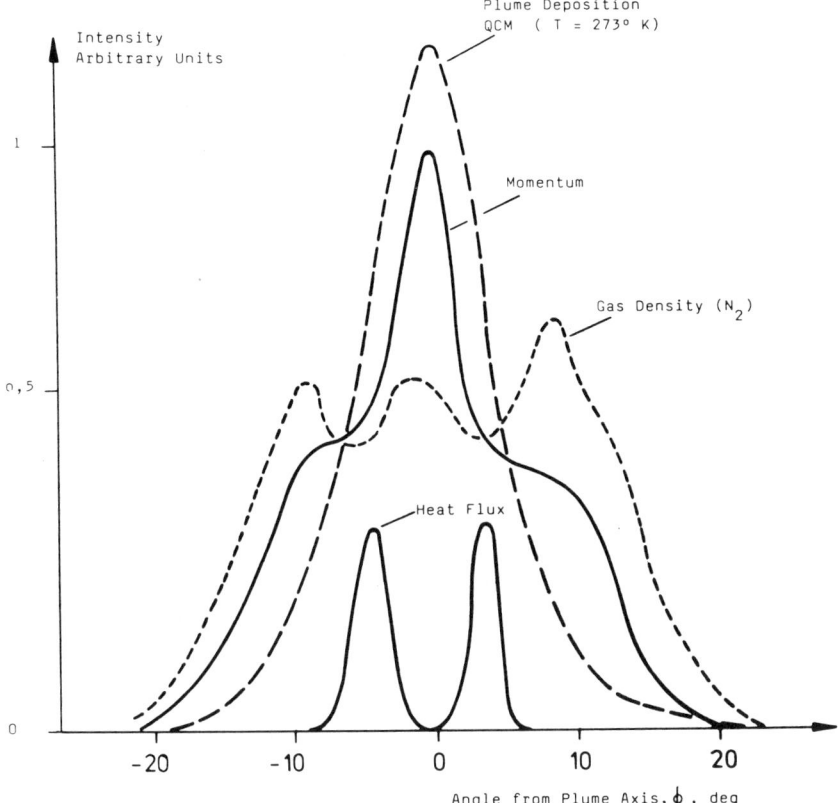

Fig. 22 Compilation of experimental results.

droplets from the thruster inner wall were expelled consisting of MMH-nitrate with MMH dissolved.

Part III: Comparison of Experimental Results with the CONTAM III Analytical Model

Although the two engines investigated had different propellant valves, a comparison of the analytical and experimental results is very encouraging. The experimental results confirm the analytical prediction that steady-state is attained in about 9 ms based on chamber pressure, gas temperature, injecton rate, and droplet outflow. During steady-state operation the CONTAM III analysis predicted 1% of the fuel mass expelled as droplets, with experimental values reaching 1.25%. The analysis predicted large quantities of droplets ejected from the engine during shutdown (within 30 ms after valve closure) which is consistent with the experimentally measured results.

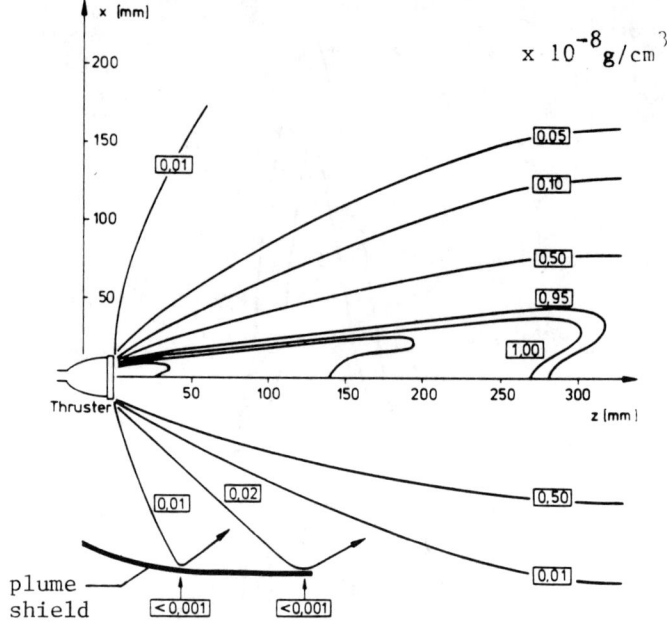

Fig. 23 Gas density contours.

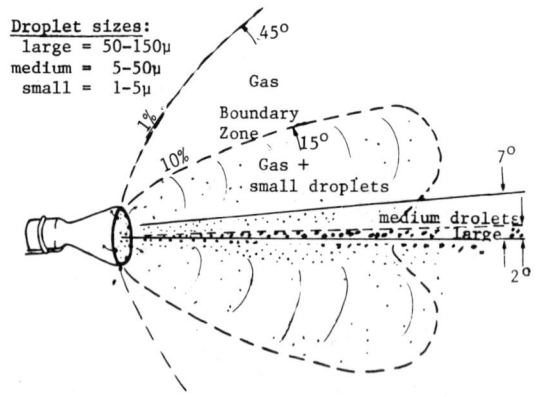

Fig. 24 Model conception: 10-N thruster plume.

Figure 25 provides a comparison between the CONTAM III predicted time-dependent droplet outflow rate with the experimentally derived values measured by electric field sensor.

Analytical results indicate that no wall film should be expelled from the MBB engine if the engine is pulsed for

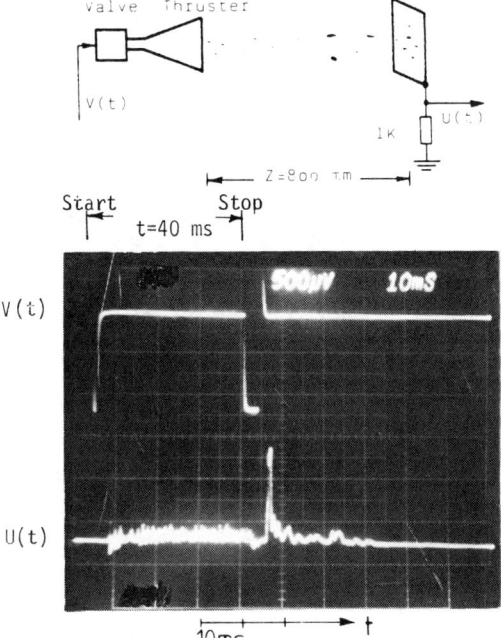

Measured Droplet Outflow Rate

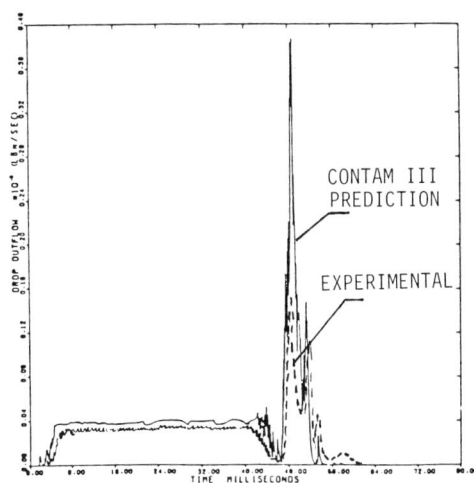

CONTAM III Computer-Generated Plot of Time-Dependent Droplet Outflow Rate: GALILEO MBB 10N ENGINE.

Fig. 25 Droplet outflow rate: experimental and theoretical results.

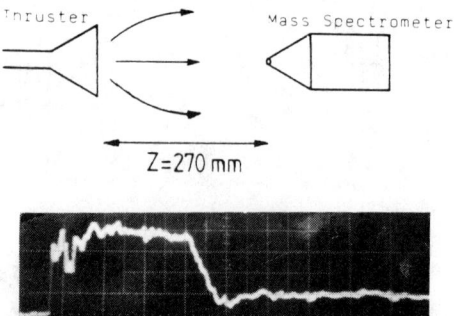

Experimental Result: Time-Dependent Concentration of N_2/CO in the Thruster Plume Flow Field

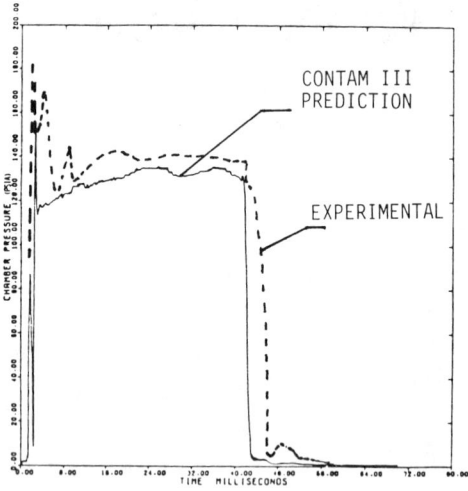

CONTAM III Computer-Generated Plot of Time-Dependent Chamber Pressure: GALILEO MMB 10N Engine

Fig. 26 Measured plume gas density and computed chamber pressure.

at least 20 ms. Pulsing as low as 9 ms, however, produced an analytical prediction of 0.5 - 1.0% wall film expelled, which is consistent with the experimentally measured value.

Figure 26 shows the CONTAM III computer generated plot of time-dependent chamber pressure and, in comparison, the experimentally derived gas density of N_2 and CO in the thruster plume flowfield by mass spectrometric measurements. This example elucidates the satisfactory agreement between the theoretical and experimental results. Even the oscillating plume flowfield behavior during the start phase is theoretically predicted nearly quantitatively.

Additional comparisons between the experimental values and analytical results concerning the gas density contours in the plume flowfield also show good agreement. Further investigations are in progress.

Acknowledgement

The experimental research described in this paper was carried out as part of a contract sponsored by the European Space Agency - European Space Technology Center (ESA-ESTEC). The responsibility for the contents resides with the author.

References

[1] Hoffman, R. J. et. al., "The CONTAM III Computer Program: Vol I - Contaminant Production and Combustion Processes," Science Applications, Inc. AFRPL-TR-82-33, June 1982.

[2] Hoffman, R. J. et. al., "The CONTAM III Computer Program: Vol II - Contaminant Transport and Chemical Kinetics," Science Applications, Inc. AFRPL-TR-82-33, March 1983.

[3] Hoffman, R. J. et. al., "The CONTAM III Computer Program: Vol. III - Contaminant Deposition and Surface Effects," Science Applications, Inc., AFRPL-TR-82-33, April 1983.

[4] Hoffman, R. J., Hetrick, M. A. Jr., and Nickerson, G. R., "CONTAM III MULTRAN-KINCON Computer Model Improvements (Flow Fields)," 12th JANNAF Plume Technology Meeting, U.S. Air Force Academy, Colorado Springs, Colo., CPIA Pub. 332, Nov. 1980.

[5] Hoffman, R. J., "Plume Contamination Effects Prediction for Several Real Satellite Systems Using the CONTAM III Computer Code," Science Applicaations, Inc., 13th JANNAF Plume Technology Meeting, Houston, Texas (NASA/JSC), April 1982.

[6] Diaz, A. I., and Casani, J. R., "Galileo 1986 on Centaur," Astronautics & Aeronautics, Vol. 21, Feb. 1983, pp. 24-33.

[7] Trinks, H. "Experimental Exhaust Plume Investigation with MBB 10-N Thruster - V Experimental Results," ESA Contract No. 4888/81/NL/AK (SC), 1982.

[8] Trinks, H. "Aerospace Vehicles Charging By Thruster Plumes," Int. Conf. Lightning and Static Electricity, Fort Worth, Texas, June 1983.

[9]Trinks, H., "Mass Spectroscopic Investigation of Dissociation and Ionization in a Simulated Re-Entry Plasma," Planet-Space Science, Vol. 21, 1973, pp. 291-296.

[10]Schilf, N. and Trinks, H. "Four-Channel Magnetic Mass Spectrometer for Dynamic Measurement of Fast Reactions," International Journal of Mass Spectrometry Ion Physics, Vol. 37, 1981.

[11]Trinks, H. and Schilf, N., "Gasdynamic Investigation of Lead Azide Lead Styphnate Detonation Processes in Vacuum by Multichannel-Mass-Spectrometry," Gasdynamics of Detonations and Explosions: AIAA Progress in Astronautics and Aeronautics, Vol. 75, edited by Bowen, Manson, Oppenheim, and Sokukhin, AIAA, New York, 1981, pp. 242-252.

Particle Sampling of Solid Rocket Motor Exhausts in High-Altitude Test Cells

P. T. Girata Jr.* and W. K. McGregor[†]
Sverdrup Technology, Inc. / AEDC Group
Arnold Air Force Station, Tennessee

Abstract

A technique has been developed by which particle samples were obtained in the near-field exhaust plume during solid-propellant rocket motor firings at a simulated altitude of 120,000 ft. A tungsten-tipped, ablatively cooled sampling probe was used. The particles, collected on witness stubs and in water samples, were analyzed by a scanning electron microscope and x-ray spectroscopy. Samples were collected at three radial locations in the plume. Analysis revealed aluminum oxide (Al_2O_3) exclusively within the core region and Al_2O_3 with solid carbon in the plume boundary. The particles were of submicron unit size with agglomerations up to a few microns. The Al_2O_3 particle size distribution was not dependent on radial position.

Introduction

In composite solid-propellant rocket motors the final product of combustion that arises from the combustion of aluminum is aluminum oxide (Al_2O_3) in the form of solid particulates. The properties of the Al_2O_3 particles - size, density, temperature, physical state, velocity - play an ever increasing role in motor performance, nozzle design, operational aspects of spacecraft, Space Transportation System (Shuttle) deployment, base heat transfer, infrared signature, environmental fallout from launches, and pos-

---

Presented as Paper 83-0245 at the AIAA 21st Aerospace Sciences Meeting, Reno, Nev., Jan. 10-13, 1983. Copyright © American Institute of Aeronautics and Astronautics, Inc., 1983. All rights reserved.
 *Engineer, Propulsion Diagnostics.
 †Sr. Technical Specialist, Propulsion Diagnostics.

sibly other functions. Many studies have been performed in recent years to shed light on these elusive properties. Most have centered on the measurement of size distribution.[1-5] The reported size distribution is often biased either by measurement technique[3] or by the investigator's principal interest.[5] Hermsen[5] presents a summary of the known data and uses the results of measurements made on samples taken from the exhaust of 66 different rocket motors to arrive at a data correlation function for use in making performance predictions. His correlation function, however, is for the mass-weighted average diameter which accents large ($\sim$10-μm) particles. The environmental fallout problem, on the other hand, is more concerned with the smaller particles so that Strand et al.,[4] used methods that would detect and catalog the smaller particles and in this way de-emphasized the larger ones.

The mechanism of formation of the particles in the combustion and expansion process is not well understood, although many theories have been presented.[6-9] Most theories are based upon a liquid breakup model for the condensed droplets near and just downstream of the nozzle throat. Most of these models rely on a critical Weber number[9] (ratio of inertial forces to surface tension) which is determined largely by the acceleration in the nozzle. The models, however, do a relatively poor job of predicting what is observed, leading to the conclusion that the physical mechanism of formation and breakup are not well understood.

The Space Transportation System (STS) will be used to carry many payloads into low Earth orbit. Upper stages which utilize solid rocket motors will place satellites into high-energy trajectories. The launch from the vicinity of the STS imposes the possibility of damage to the thermal protection system (tiles) and the crew observation windows either by the impact of single large particles or by the erosion caused by a high-velocity cloud of small particles.[10] The solid-propellant rocket motors employed for the space applications are different from those previously sampled for particle properties in that they exhaust into a vacuum and use a high-expansion-ratio carbon-carbon nozzle. Therefore, the sampling must be done under conditions where the nozzle is flowing full in order to observe the complete effects of the expansion process on the particulates. Furthermore, the particulates resulting from the erosion of the carbon-carbon nozzle are equally important from the STS damage standpoint. The emphasis for the STS damage problem must be placed on an unbiased sample and on any radial dependence in the size distribution.

One of the principal users of the STS will be the spacecraft powered by the inertial upper stage (IUS) rocket motors. These motors have undergone tests in the J-5 test cell[11] at the Arnold Engineering Development Center (AEDC), and advantage was taken of these tests to sample the exhaust and attempt to determine the particle size distribution as a function of radial position. The IUS propulsion system[12] consists of two solid-propellant rocket motors, SRM-1 first stage with an average thrust of 44,000 lbf and SRM-2 second stage at 17,300 lbf of average thrust with standard nozzles. The SRM-2 as used on STS employs an extendible exit cone (EEC) capable of delivering 18,200-lbf average thrust. Both use the same propellant; the ingredients are HTPB:AP:Al in the approximate ratio 13%:68%:18% by weight. The nozzle expansion ratios are 63.8:1 for SRM-1 and 49.3:1 for SRM-2 (181.1:1 with EEC).

The approach taken to the sampling problem was restricted by the test environment to one of direct sampling of the exhaust in the near-field core flow. Thus, the main problem was to design a probing technique that would survive the environment long enough to obtain a representative sample. This was done by use of a tungsten-tipped probe supported by ablation struts; these probes were considered expendable. This paper will describe the results of sampling the effluents from two SRM-2 motor firings. The results are similar to those which can be expected from the SRM-1 (first stage) motor.

Fig. 1 Photograph of IUS solid-propellant rocket motor firing in J-5 test cell.

IUS Test Configurations

The solid-propellant rocket motor development program for the IUS includes a number of altitude tests at AEDC. The motors were tested in AEDC rocket development test cell (J-5) at a simulated altitude of approximately 120,000 ft. All core flow samples were taken during two tests in which the carbon-carbon nozzle extended into the diffuser section (Fig. 1). The probe locations for the two tests are given in Table 1.

The 87-in. axial location was chosen to position the probe upstream of the intersection of reflected shocks off the diffuser wall, and as far as possible from the nozzle exit to minimize heat transfer. The two radial locations

Fig. 2 Probe installation for IUS-DS8C motor in J-5 diffuser.

Table 1 Schedule of probe sampling on IUS motors

IUS Motor	Date tested	No. of probes	Location
DS4A	13 February 1981	1	10 in. Off ℄, 87 in. Downstream
DS8C	30 June 1981	3	1) ℄, 87 in. Downstream 2) 10 in. Off ℄, 87 in. Downstream 3) Tangent to Nozzle Exit Angle, 51 in. Downstream

were chosen to meet the need for determining the spatial variation in the aluminum-oxide particle size distribution. Because of a restricted IUS testing schedule, probes could not be positioned at other locations. The tangent location was chosen in order to determine the size distribution of carbon particles that erode from the nozzle surface.

Sampling Technique

The sampling technique used in acquiring the core flow samples was to locate fixed probes (Fig. 2) at accurately known distances downstream of the nozzle exit at the different radial positions. The total pressure behind the bow shock ranged from 15-25 psia for the chosen location. The probe bodies consisted of a 1-in. stainless steel tube surrounded by a carbon steel water jacket that was, in turn, covered with a carbon steel plate which served as an ablation coating (Fig. 3). Since the tip of the probe was to be immersed in a high-temperature ($\sim$4000 K) environment with a large concentration of high-velocity aluminum oxide particles, a tungsten tip was fabricated and attached to the stainless steel tube sampling line. The probes were considered to be expendable. The water jacket was not intended to cool the outside carbon steel jacket but to keep the sample tube as cool as possible. The water was turned off before motor burnout in order to ensure that water would not be sprayed on the carbon-carbon nozzle. After the firings the probe hardware located in the core region were found to be severely eroded, but the tungsten tips were not appreciably damaged.

Each core flow probe was connected through a 1-in.-diam sample transfer line to a sample station. Flow was maintained through the probe by a sample return line which was routed back to the low-pressure ($\sim$20 Torr) test cell. The

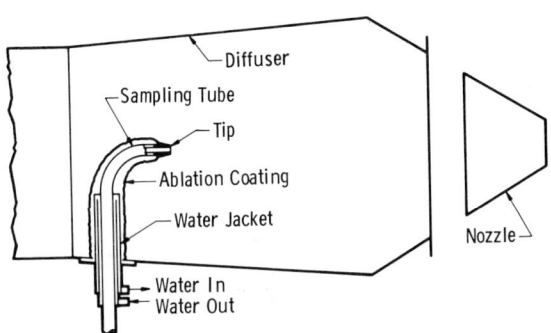

Fig. 3 Cross section of core flow probe installed in J-5 diffuser.

system is shown schematically in Fig. 4. Contained in the sample station are three sample bottles of the type shown in Fig. 5. The sample bottles contained a holder for mounting a witness stub to gather particles for analysis using a scanning electron microscope (SEM) combined with an x-ray energy dispersive spectrometer (XES). The 15-mm-diam stubs were made of either copper, carbon, or beryllium. Filter paper, to divide the sample bottle into two compartments (particle/gas), is also held in place by the SEM stub holder as shown in Fig. 5. The gas compartment should then be relatively clear of particles so that the gas can be removed

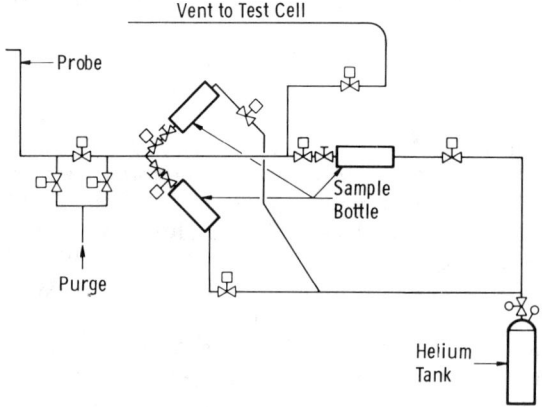

Fig. 4 Schematic of gas/particle sampling system.

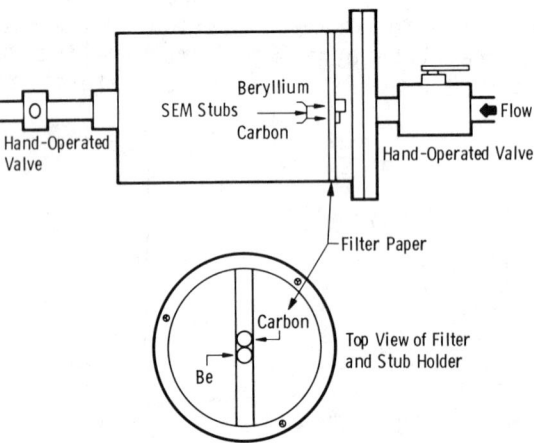

Fig. 5 Cross section of gas/particle sampling bottle and SEM stub location.

through a line attached to the end opposite the entrance to the sample bottle for gas analysis. The gaseous species collected in the core region by the hot tungsten-tipped probe suffers continued surface reactions and are not representative of the freestream gases. Therefore, the gas analysis results will not be discussed further.

The sample station was designed to minimize the number of turns the gas/particle flow has to make before a sample is captured in order to minimize wall collision effects. Gas/particle dynamic calculations show that particles less than 7 μm in diameter will not impact the rear wall of the first turn in the probe, and particles as large as 15 μm will probably survive such a collision. Particles of 16 μm and smaller can be collected by the 45-deg turns into sample bottles 1 and 2, and all particles should be collected by the head-on orientation of bottle 3. For motor DS4A (see Table 1) the three samples were taken during the first three stationary periods of the nozzle gimbaling program (a part of the test plan) in order to ensure sample acquisition at known nozzle positions. For motor DS8C the first two samples were taken at the null position while the third was taken with the nozzle gimballed 4.5 deg away from the tangent probe.

A large sample of plume particulates (about 100 g) was collected by modifying the core flow probe, as shown in Fig. 6, by injecting high-pressure demineralized and filtered water into the probe at a point about 4 in. from the probe inlet. A shower was created in which the plume gas/particle

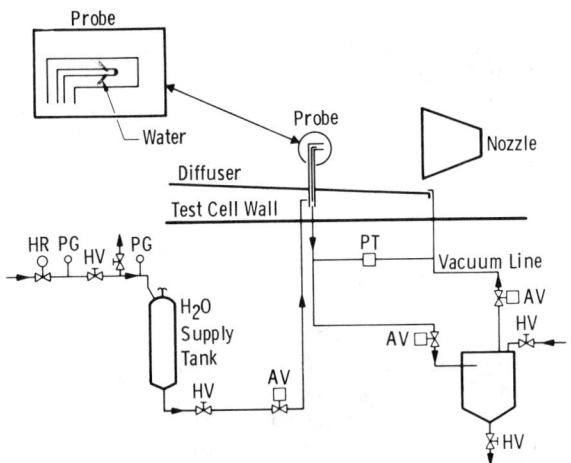

Fig. 6 Schematic of water ingestion probe.

flows can be ingested and thus capture the particulate matter in the water. The water/particle slurry was then captured in a tank, and the solid material was isolated by a filter process for analysis. This method of collection avoids the probe wall collisions by the particles.

Particulate Analysis

The sample bottles (Fig. 5) were configured to direct sample flows over the SEM stub where some of the particles were deposited. The remainder of the solid material was collected on the filter paper, except for some very small particles which might pass through the paper.

The resolution limit of the SEM used for the particle analyses was about 0.01 μm. The XES detector was capable of analyzing elements having atomic numbers greater than 12 (sodium). The XES detector was also used as the sensing element for an automatic scanning/sizing system used here at a lower resolution limit of 0.25 μm. The SEM stubs and portions of the filter paper were scanned first to search for large particles, and then representative regions of the stubs, or filter paper, were selected and photographs were made at several magnifications. Selected areas of the stubs were subjected to automatic counting/sizing and the data were recorded. In the case of the water samples the analyses were either made from small droplets placed on a stub and dried, or from the final dried sample.

Results

The analysis of the particulate matter found on the SEM stubs and filter paper taken from the sample bottles is presented in this section. Data presented include example photographs of the electron microscope image, size histograms, and x-ray spectral analyses of different areas on the stubs and filters. The procedures used in sampling and data analysis for the two motors, DS4A and DS8C, were sufficiently different that the results will be reported separately. Even though the procedures were different, some important observations were made that should be discussed before the test data are presented.

It was observed that, with sufficient magnification, all particles (evaluated by the SEM/x-ray unit) were made up of smaller particles (an agglomerate) (Fig. 7). This observation is very important when reviewing the histograms presented, since the mechanism of particle formation and agglomeration has not been conclusively defined.

DS4A Test Data

For the DS4A test, a single probe was installed in the diffuser at a position 10-in. off centerline at a station 87-in. downstream of the nozzle exit plane. The sampling times for each bottle (see Fig. 5) were as follows: bottle 1, T = 1.5 s; bottle 2, T = 13.5 s; and bottle 3, T = 23.8 s. During sample acquisition the thrust vector control system was stationary at the null position and the nozzle flow was aligned with the diffuser.

<u>Sample Bottle 1 (1.5 s)</u>. Analysis of the copper witness stub using the SEM/x-ray unit (Fig. 8) revealed only particles containing aluminum, which were assumed to be aluminum oxide. Most of the particle sizes ranged from 1-40 μm (Fig. 9) with only four particles larger than 4 μm. All particles observed were agglomerations of smaller particles and had different shapes; they were not spherical. Analysis for tungsten and/or iron was negative, indicating that the probe was still intact when the sample was taken. The filter paper was also scanned for large particles to ensure that the procedure for dispersing the particles onto the stub did not bias the sample. The analysis of the particles on the filter paper revealed the same size range as on the stub.

<u>Sample Bottle 2 (13.5 s)</u>. Analysis of the witness stub from this bottle was accomplished by the Aerospace Corpora-

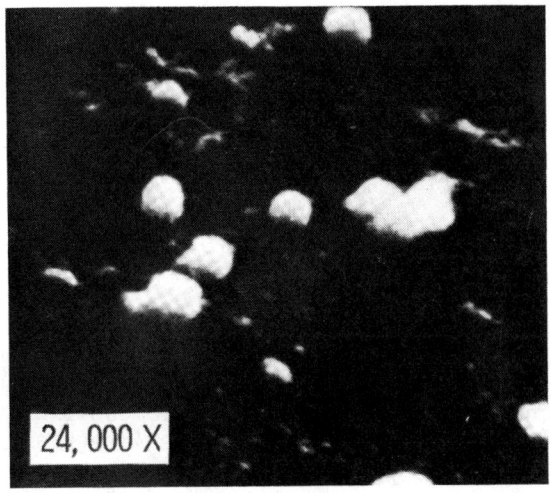

Fig. 7 High-magnification photograph of Al_2O_3 particles.

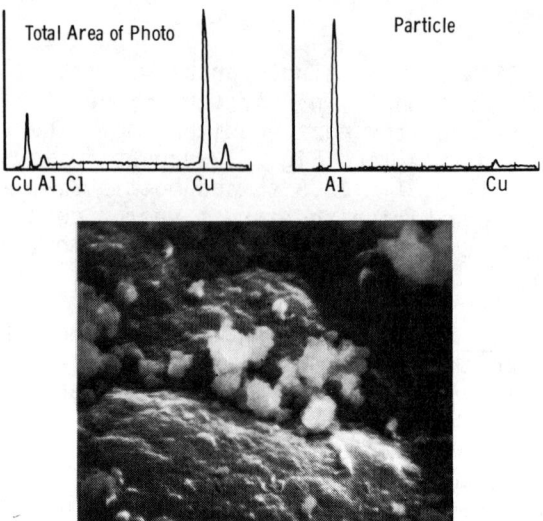

Fig. 8 SEM photograph of one area of SEM stub from sample No. 1 with x-ray analysis.

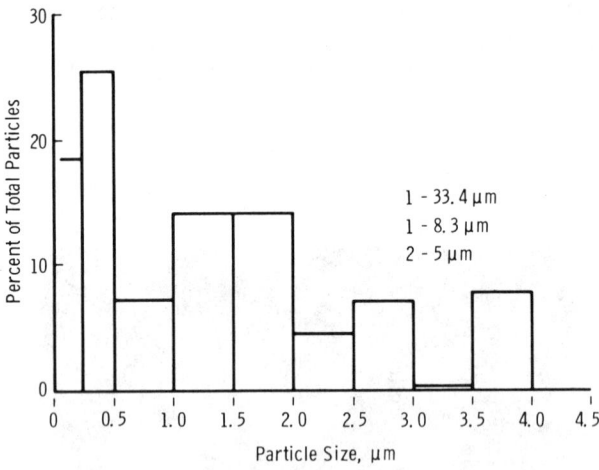

Fig. 9 Histogram of Al_2O_3 particles from IUS motor DS4A, sample 1.

tion using similar SEM/x-ray equipment. A close spacing of agglomerated smaller particles was found and were identified as aluminum oxide. No appreciable amounts of other elements were found. The particles were irregular in size and ranged from 0.6-10 μm, with an average size of 1.8 μm. A diffraction analysis was also made, and the pattern corresponded closely with "alpha" Al_2O_3.

<u>Sample Bottle 3 (28.3 s)</u>. Particle analysis of the witness stub in sample bottle 3 revealed that the stub was completely coated with a grayish material containing some large particles. Large particles ranging from 50-100 μm, and a few as large as 1 mm, were found in the particle collection chamber. X-ray analysis of the large particles showed that they were made of stainless steel and/or tungsten coated with aluminum oxide. The SEM stub was coated with a fused agglomeration of small Al_2O_3 particles. It was concluded that the probe had deteriorated by the time this sample was taken.

<u>DS8C Test Data</u>

For the DS8C test, two core flow probes and the water ingestion probe were installed in the J-5 diffuser (Fig. 2) as described previously. Three samples were acquired from each probe at 1.5, 13.5, and 21.5 s into the firing. The water ingestion sample was collected continuously from 1.5 until 45 s after ignition. The total motor burn time was about 72 s.

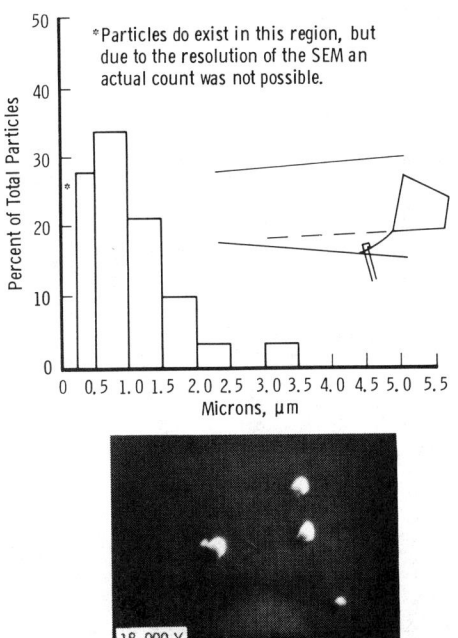

Fig. 10 Histogram and photograph of Al_2O_3 particles from tangent probe bottle 3, IUS motor DS8C.

The thrust vector control system was operating during the firing and was stationary at the null position for the 1.5- and 13.5-s samples. During the sampling at 21.5 s the nozzle was stationary but pointed away from the tangent probe at an angle of 4.5 deg with the axis (see Fig. 10). The centerline probe was thus positioned approximately 13 in. off the nozzle axis for the 21.5-s sample. The tangent probe position was located between the particle core flow and the plume boundary, as indicated by the sample bottle pressure (44 Torr compared to about 10-Torr test cell ambient pressure). Thus, the tangent probe should provide a good sample of particles in the boundary layer.

Centerline Probe Samples. The particle samples obtained from bottles 1 and 2 at 1.5 and 13.5 s, respectively, by the bottles oriented 45 deg from the sample line axis contained only aluminum and were assumed to be Al_2O_3. The particle size histograms of these samples were almost identical; a photograph of an area on the SEM stub from bottle 2 and the corresponding histogram for samples 1, 2, and 3 ap-

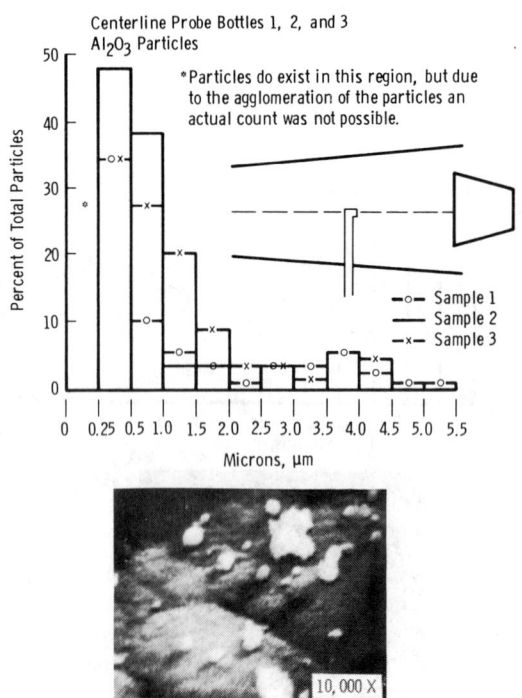

Fig. 11 Histogram and photograph from centerline probe, IUS motor DS8C.

pears in Fig. 11. Note that the distribution is shifted even farther toward smaller particles than on previously shown data.

The location of the probe for sample 3 (21.5 s) was approximately 13 in. from the centerline because of the nozzle gimbaling. However, only a slight difference in the size histogram of the Al_2O_3 particles was detected. Note also that sample bottle 3 was oriented directly into the sample line.

Tangent Probe Samples. The first two samples collected by the tangent probe were from a region of the plume just inside the visible core flow of the plume with the nozzle at the null position. The particles collected in sample bottles 1 and 2 (1.5 and 13.5 s, respectively) were very similar in appearance on the SEM micrograph as the particles found at other locations in the plume. Example photographs from the witness stub taken from bottle 2 are shown in Fig. 7.

It should be recalled that the XES detector used had a lower atomic number limit of 12 so that particles containing Al (A = 13) could be identified, but those containing only carbon (A = 6) could not. In the x-ray analysis of these samples only two kinds of particles could be identified - those containing aluminum (assumed to be Al_2O_3) and those giving no identification (assumed to be carbon, C_S). In bottle 1 about 40% of the particles were found to be C_S, and in bottle 2 about 20% were C_S. The size histograms for Al_2O_3 and C_S are given in Fig. 12. Note that particles less than 0.25 µm were not counted because of the instrument limit, but optical observation indicates that very few Al_2O_3 particles smaller than 0.25 µm were not agglomerated, and the carbon particles were predominantly in the 0.25-0.5-µm range.

Fewer particles were collected in sample bottle 3 because the probe was located outside the particle core flow, but within the plume boundary. The major difference between sample 3 and samples 1 and 2 was that about 60 percent of the particles analyzed were carbon (C_S). Histograms of the C_S and the Al_2O_3 (Fig. 10) particle sizes compared very well with samples 1 and 2. No large particles of either C_S or Al_2O_3 were found on the witness stubs or the filter paper.

Water Ingestion Probe. The water ingestion probe was designed to collect a large sample of rocket exhaust particles; this was done by injecting high-pressure demineralized and filtered water into the probe and creating a shower effect to entrap the particles of the flow. A total of 1.5

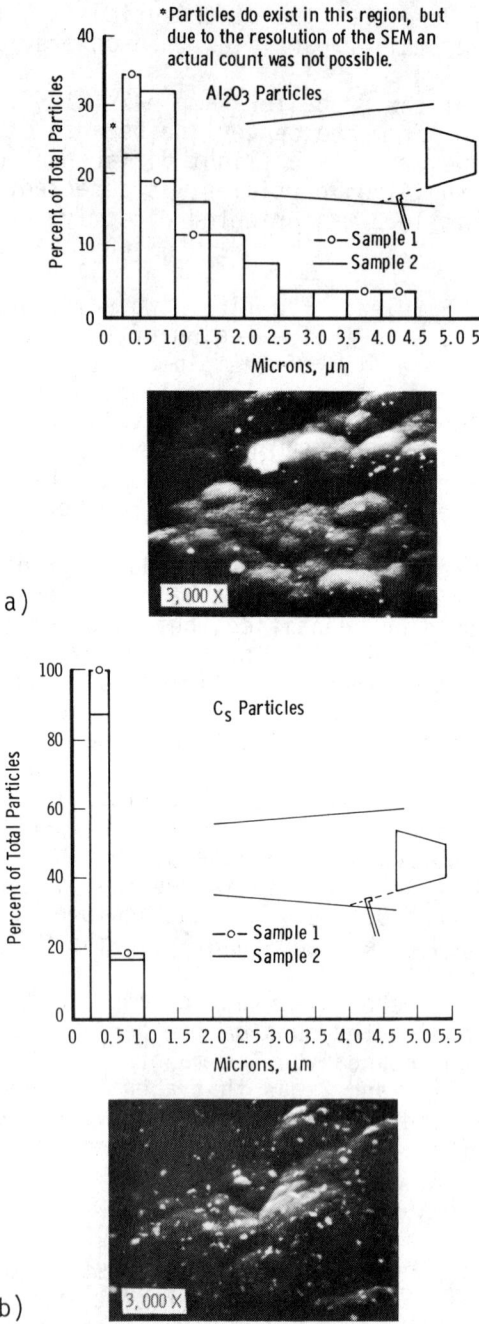

Fig. 12 Histogram and photograph from tangent probe; a) Al_2O_3 particles; b) C_s particles.

gallons of water, which contained the particulate matter, was collected over approximately 44 s of the burn. The method used to extract the solids from the water consisted of a series of dilution and filtering exercises which eventually yielded about 45 g of very pure Al_2O_3. A typical SEM analysis of the particles is shown in Fig. 13; note that the size distribution is not appreciably different than that found by the dry sampling method. Samples obtained directly from the water by drying droplets on a witness stub showed about the same size distribution (see Fig. 13).

A small sample of the particles collected in the water was also analyzed at the Aerospace Corporation laboratories. Again, the size distribution was very similar to Fig. 13. However, a more complete x-ray diffraction analysis than was possible for the small dry sample gave evidence of both α and γ crystalline structure of the Al_2O_3 particles.

Discussion of Sampling Biases

The main difficulty with all probe sampling techniques is in relating the properties of the sample to the proper-

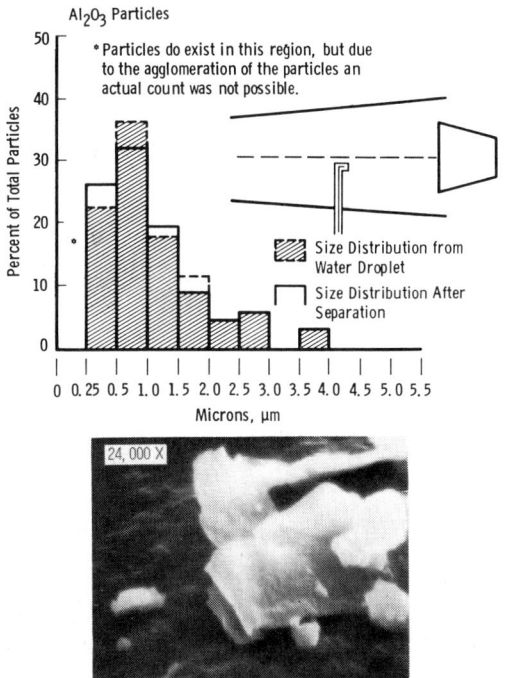

Fig. 13 Histogram and photograph of water ingestion probe from IUS motor DS8C.

ties in the undisturbed flow. With regard to capturing particle samples from a highly reactive, high-temperature, supersonic gas/particle flow, the following biases might be expected:

1) The bow shock created by the probe in the flow would tend to divert smaller particles around the probe, thus biasing the sample collected to larger particles. However, agglomerates might be sheared apart by the stresses created in passing through the shock wave, thus biasing to smaller particles.

2) Adherence of material to the probe inlet and sample line and agglomeration of small particles in the low velocity flows would appear to bias the sample to larger particles.

3) Sharp turns into the sampling bottles would tend to bias the sample toward smaller particles. However, a careful cleaning of the sample lines for the tangent probe did not reveal any large particles.

4) Impact of agglomerated particles on surfaces near the probe inlet would tend to break the cluster into smaller particles, thus biasing the sample toward the smaller particles. Those, however, would probably be swept around the probe in the reaccelerating flow.

5) Particles striking the witness stubs inside the sample bottles might either break up or bounce off the surface, again possibly biasing the sample toward smaller particles. However, the size distribution of the particles found on the filter paper was not appreciably different than on the SEM stubs.

All of these possibilities were considered during this study and are worthy of further investigation. However, such a study would be difficult because of the necessity to analyze particles made up of agglomerations of the smaller, more fundamental Al_2O_3 particles.

Summary of Results

The results obtained in this study of the particle emission from IUS rocket exhausts may be summarized as follows:

1) The method of sampling (by the use of tungsten-tipped probes with a carbon steel ablation protected stainless steel sample transfer line leading to sample bottles equipped with particle witness stubs) proved to be an effective means to gather samples from solid-propellant rocket motor exhausts.

2) Within the core flow region of the exhaust plume of IUS motors, only Al_2O_3 solid particles were found. Under

scanning electron microscope observation the particles were found to consist of very fine unit particles (less than 0.1 μm) which agglomerated either in the exhaust flow or in the sampling system to form larger particles.

3) In the boundary region tangent to the nozzle exit angle (just inside visible coreflow) the particulate samples were made up of about 60-80% Al_2O_3 and 20-40% solid carbon C_s. The size distribution of the Al_2O_3 particles was very nearly the same as for the inner core. The largest C_s agglomerates were 0.5 μm with most less than 0.25 μm.

4) Particulates were also found in the boundary region between the core flow (tangent to nozzle exit) and the underexpanded plume boundary. The particles were a mixture of Al_2O_3 and C_s in about a 40/60% ratio. The size distribution was approximately the same as found in the inner core near the boundary, but the density of particles collected was much less.

5) Particles collected in a water ingestion sample did not appreciable differ in composition from the dry sampling method.

6) No conclusive reasons have been found to indicate a biasing of the particle sizes by the sample-handling procedure, but neither breakup within the probe bow shock or agglomeration within the sampline lines can be ruled out as possibilities.

It is not possible to generalize these results to other solid-propellant rocket motor configurations. The carbon-carbon nozzle used in the IUS motors evidently does not permit condensation or melting of the Al_2O_3 in the throat region since the nozzle is clean and uniform after the firing, with no apparent Al_2O_3 deposits. The larger particles observed previously[5] from Minuteman class motors with ablative nozzles are possibly the result of this condensation or melting effect. Also, the large carbon particles observed from other carbon-carbon nozzle designs did not appear here. It is thus recommended that sampling of the exhausts of other solid-propellant rocket motors of interest be undertaken before general conclusions are drawn.

If these results are accurate, then the previously used particle size distribution for performance calculations[5] is inaccurate and other sources must be found to explain performance losses. In the STS damage assessment problem, there appears to be little danger from the impact of large particles on heat shield titles, but erosion caused by a large number of small particles must still be assessed. The IUS will be properly distanced and oriented during SRM-1 burns to minimize/negate this possibility.

Acknowledgments

The research reported herein was performed by the Arnold Engineering Development Center, Air Force Systems Command. Work and analysis for this research were done by personnel of Sverdrup Technology, Inc./AEDC Group, operating contractor of the propulsion test facilities at AEDC. Further reproduction is authorized to satisfy needs of the U.S. Government.

The authors wish to acknowledge Mr. Randy Quinn and Mr. Linn Weller of Sverdrup Technology, Inc./AEDC Group, Mr. R. J. Bryson of Arvin/Calspan, and Mr. Randy Johnson of Pan Am, MSSP, for their contributions to the success of the program. Mr. Ron Peterson was responsible for the independent analyses carried out in the Physical Sciences Laboratory at the Aerospace Corporation, El Segundo, California.

References

[1] Geisler, R. L., Beckman, C. W., and Kinkead, S. A., "The Relationship Between Solid-Propellant Formulation Variables and Motor Performance," AIAA Paper 75-1199, Oct. 1975.

[2] Coats, D. E. et al., "A Computer Program for the Prediction of Solid Propellant Rocket Motor Performance," Ultrasystems, Inc., Irvine, Calif., AFRPL-TR-75-36 (AD-A015140), July 1975.

[3] Dawbarn, R. and Kinslow, M., "Studies of the Exhaust Products from Solid Propellant Rocket Motors," AEDC-TR-76-49 (AD-A029569), Sept. 1976.

[4] Strand, L. D. et al. "Characterization of Particulates in the Exhaust Plume of Large Solid-Propellant Rockets," Journal of Spacecraft and Rockets, Vol. 18, July-August 1981, p. 297.

[5] Hermsen, R. W., "Aluminum Oxide Particle Size for Solid Rocket Motor Performance Prediction," AIAA Paper 81-0035, Jan. 1981.

[6] Fein, H. L., "A Theoretical Model for Predicting Aluminum Oxide Particle Size Distribution in Rocket Exhausts," AIAA Journal, Vol. 4, Jan. 1966, p. 92.

[7] Jenkins, R. M. and Hoglund, F., "A Unified Theory of Particle Growth in Rocket Chambers and Nozzles," AIAA Paper 69-541, June 1969.

[8] Cheung, H. and Cohen, N. S., "Performance of Solid Propellants Containing Metal Additives," AIAA Journal, Vol. 3, Feb. 1965, p. 250.

[9] Bartlett, R. W. and Delaney, L. J. "Effect of Liquid Surface Tension on Maximum Particle Size in Two-Phase Nozzle Flow," Pyrodynamics, Vol. 4, Oct. 1966, p. 337.

[10] McDonnell Douglas Technical Services Corporation, "Orbiter Surface Damage Due to SRM Plume Impingement," TM-1.4-MAB-314, March 30, 1979.

[11] Test Facilities Handbook (11th Edition), Arnold Engineering Development Center, Arnold Air Force Station, Tenn., April 1981.

[12] Chase, C. A., "Development Status of the Inertial Upper Stage Solid Motors," AIAA Paper 80-1267, 30 June - 2 July 1980.

Postfire Sampling of Solid Rocket Motors for Contamination Sources in High-Altitude Test Cells

P. T. Girata Jr.* and W. K. McGregor†
Sverdrup Technology, Inc./AEDC Group
Arnold Air Force Station, Tennessee

Abstract

Various sampling techniques were used to analyze the effluents being emitted from inertial upper stage (IUS) solid rocket motors (SRM's) after burnout, at a simulated altitude of about 120,000 ft. A 1-in. stainless steel, heated probe moved into position on nozzle centerline after motor burnout to sample chamber effluents. A sampling system acquired three batch samples for gas/particle analysis and a gas chromatograph monitored carbon dioxide and total hydrocarbons semicontinuously (every 15 s). Another probe located near the boot region (where the nozzle attaches to the motor case) was used to acquire batch samples for gas/particle analysis, and a flame ionization detector monitored total hydrocarbons. The study was carried out for 4 second-stage IUS motors and 3 first-stage IUS motors. The chief source of contamination was found to result from rupture of various components of the flexible nozzle emitting oil and grease vapors into the surrounding region.

Introduction

It has been shown in previous space flight and vacuum chamber measurement programs that the effluents from small mono- and bipropellant attitude control thrusters contaminate optical and thermal control surfaces and thereby contribute to deterioration of performance and the useful life

---

Presented as Paper 83-1448 at the AIAA 18th Thermophysics Conference, Montreal, Canada, June 1-3, 1983. This paper is declared a work of the U.S. Government and therefore is in the public domain.
*Engineer, Advanced Technology, AEDC Group.
†Staff Scientist, AEDC Group.

of the spacecraft.[1,2] The current use of large solid rocket motors (SRM's) to propel spacecraft into orbit from either standard launchers or the Space Transportation System (STS) has extended this problem immeasurably. The contaminating effluents come not only from the exhaust plume itself during the firing of the rockets but also from the effluents being emitted from the nozzle during the postfire period as well as outgassing from the SRM auxiliary hardware that may occur over a long period of time after burnout due to thermal heating.

Several serious concerns about possible contamination of payloads during operation of SRM's in space stem from the inability to predict contamination levels. The inadequate plume models for describing the expansion into a high vacuum, especially for two-phase flow (particles and gasses) led to the need to sample SRM plumes during simulated altitude tests and to identify the gas/particle effluents being emitted. Sampling of two inertial upper stage (IUS) SRM-1[3] plumes (18% aluminum as an additive in the propellant) at the Arnold Engineering Development Center (AEDC) showed an aluminum oxide particle size distribution across the core region of less than 0.25-5.5 μm.[4] In the plume boundary, particles ranged from less than 0.25-3.5 μm, with an appreciable amount of submicron solid carbon from erosion of the carbon-carbon nozzle. The effect of these particles is seen mostly as impingement damage on the STS or other spacecraft. The details of the gas expansion and scattering of particles around the nozzle lip are still unknown.

Other concerns about contamination come from the postfire period in which the low velocity effluents being emitted from the nozzle continue for a long period, possibly as long as 20-35 min. Currently there are no data identifying the effluents being emitted from the SRM during the postfire period, even though there have been laboratory studies of the outgassing of the propellant, liner, insulation, chamber casing, and nozzle materials (e.g., Ref. 5). When the propellant has burned out, the chamber liner starts to smoulder and the case continues to heat up due to thermal lag, giving off gaseous and particulate effluents. The effluents being emitted from the nozzle during this postfire period are at a low velocity (SRM chamber pressure is less than 0.5 psia) and have a greater probability of scattering back onto the spacecraft than do the high velocity plume gases during full thrust. These gases lead to an unknown contamination source. Other unknown contaminants come from outgassing of adjacent materials outside the nozzle because of the heat radiating from the SRM nozzle to surrounding surfaces; more importantly, emissions from auxiliary hardware such as actuators, lubricated seals, etc., may occur over a several minute period after burnout due to the thermal heating.

Solid-propellant rocket motors intended for space application are tested in moderately high altitude (100-140 kft) test cells. The ambient pressures that can be simulated are on the order of 10 Torr before and after the burn but perhaps as low as 3 Torr during the burn due to the rocket plumes acting as its own jet pump. These pressures are sufficient to maintain underexpended plumes for nozzles with area ratios up to about 200 so that the inner core of the plume is not affected by the larger than space pressures. However, the expansion in the nozzle lip boundary region does not behave in the same manner as in space, and the postfire emissions must be examined in light of the possible effects of the test cell environment. Nevertheless, much insight into the possible sources of contamination can be gained by observing the emissions from the nozzle as well as other parts of the motors during the postfire period.

The IUS motors are to provide the principal means of propelling large spacecraft into final orbit from the STS and booster launchers over the next few years. They consist of two stages, a large (SRM-1) and a small (SRM-2) motor; their characteristics are given in Table 1.[3] The extensive test program of these motors at AEDC provides an opportunity to study the plume[4] and postfire emissions from the motors. Although some of the specific findings are peculiar to IUS motors, the general class of emissions can be expected from all SRM's considered for space applications.

Included in this paper are the identification of the particulate matter and gases being emitted from IUS motors through the nozzle and from the area where the nozzle attaches to the motor (the so called "boot" area). The data

Table 1 Characteristics of IUS motors

	SRM-1	SRM-2
Thrust (average)	44,000 lb	17,300 lb
Weight	21,400 lb	6,000 lb
Nozzle expansion ratio	63.8:1	49.3:1 with EEC (181.1:1)
Propellant	HTPB:AP:Al	HTPB:AP:al
Nozzle material		Carbon-Carbon
Liner material	(EPDM) Kirkhill Ethylene Propylene Diene Monomer	
Casing Material	Kevlar epoxy	

presented are meant only to identify the effluents and the location/time at which they are emitted. Prediction of their effect on spacecraft performance requires more in-depth consideration.

Description of Tests

Test Configuration

A typical altitude test cell for testing large solid-propellant rocket motors is the Rocket Development Test Cell (J-5) at AEDC.[6] It consists of a test cell connected to a diffuser duct which has a coaxial steam ejector mounted in the aft end. The configuration is depicted in Fig. 1 with a motor having the volume envelope typical of an IUS motor. For very large expansion ratios, such as may be afforded by storable extended exit cones (EEC), the nozzle exit plane may extend inside the diffuser.

The test procedure usually calls for an environmental conditioning period of a few days prior to the test. All test hardware, transducers, electrical lines, etc., must be protected from especially cold environments; large quantities of protective insulation and other materials may be employed. This introduces a "test peculiar" source of effluents during the firings which must be taken into account. During the test itself, the cell is reduced in pressure to about 25 Torr by large mechanical exhausters. Then, just prior to ignition, the steam ejector is energized, reducing the pressure to about 8-10 Torr. When the motor is fired, the jet pump action of the plume/diffuser reduces the cell to its lowest pressure, about 2-5 Torr. Upon completion of the firing, the plume pumping action ceases and the steam ejector again takes over. When the steam capacity is depleted (approximately 15 min after motor burnout), the

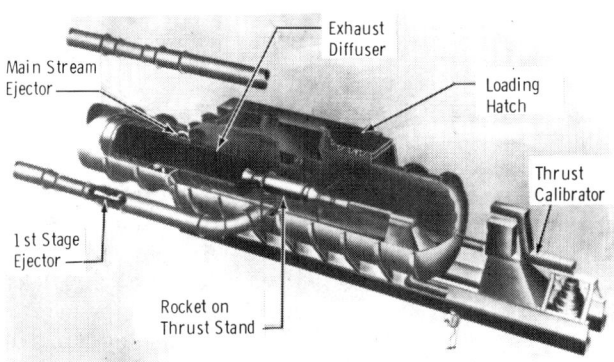

Fig. 1 Test cell configuration.

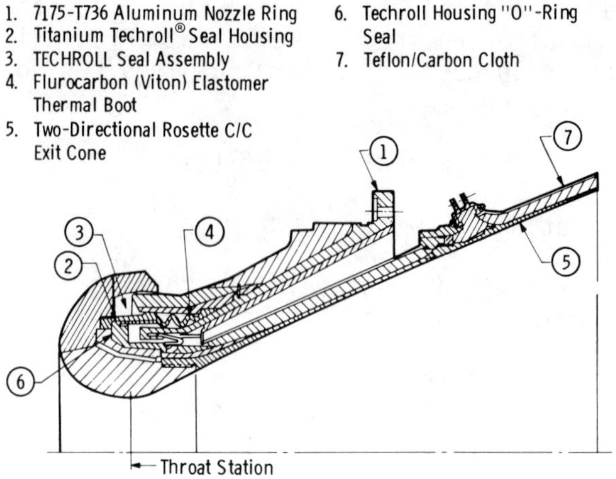

1. 7175-T736 Aluminum Nozzle Ring
2. Titanium Techroll® Seal Housing
3. TECHROLL Seal Assembly
4. Flurocarbon (Viton) Elastomer Thermal Boot
5. Two-Directional Rosette C/C Exit Cone
6. Techroll Housing "O"-Ring Seal
7. Teflon/Carbon Cloth

←— Throat Station

Fig. 2   Illustration of IUS nozzle/chamber seal.

pressure again rises to a level which can be supported by the mechanical pumps. It is the pump-back transients that occur when the motor burnout and the steam flow are exhausted that make studies during the postfire period difficult. Steam and water, carrying a variety of debris, suddenly flow back into the cell, covering windows and lenses used by the photographic instrumentation. This difficulty has recently been eliminated by reverting to pinhole camera tactics with a nitrogen purge through the pinhole. This method is used to permit viewing the motors during the postfire period prior to steam ejector shutdown. After that, there is usually too much vapor in the cell for sampling and the pressures are larger than the residual rocket chamber pressure; thus the sampling is limited to the approximately 15-min period.

IUS Tests

The IUS SRM development program includes a number of altitude tests at AEDC consisting of two phases for the large (SRM-1) and two for the small (SRM-2) motors: a "development" (D) phase and a "qualification" (Q) phase. This has led to the designation DS- and DL- for the development program and QS- and QL- for the quality assurance tests of small and large motors, respectively. Seven of the SRM's, tested at a simulated altitude of approximately 120 kft, were subjected to postfire investigations. A peculiarity of the IUS motors is the way in which the immersed nozzle is sealed to the rocket case.[3] The seal (designated "Techroll®

Table 2 Schedule of postfire sampling on IUS motors

SRM-I.D. No.	Date tested	Sampling technique
DL-3A	5-13-81	Postfire probe (batch samples), witness plates.
QS-5	2-10-82	Postfire probe (batch samples, on-line HC and CO_2).
QS-3	2-19-82	Postfire probe (batch samples, on-line HC and CO_2), Boot probe (ATR plates, batch samples), motion picture coverage, on-line HC and drip pan.
QL-1	4-30-82	Postfire probe (batch samples, on-line HC), Boot probe (ATR plates, on-line HC, motion picture coverage).
QL-3	7-17-82	Postfire probe (batch samples, on-line HC), Boot probe (on-line HC, motion picture coverage).
QS-1	8-24-82	Postfire probe (batch samples, on-line HC), Boot probe (on-line HC, motion picture coverage).
QS-4	9-10-82	Postfire probe (batch samples, on-line HC), Boot probe (on-line HC, motion picture coverage), witness plates attached to boot probe.

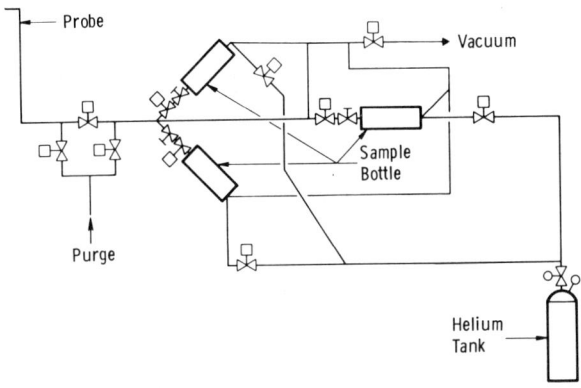

Fig. 3 Schematic of gas/particle sampling system.

seal," shown schematically in Fig. 2) contains lubricating grease and hydraulic oil. As the temperature continues to rise after burnout, the vapor pressure of these materials is exceeded and the seal can be ruptured. Thus, the area around this seal is of interest from the contamination standpoint, and a sample collection probe was configured and is designated as the "boot" probe in this study. The designation "postfire" probe in Table 2 refers to the nozzle exit probe, which drops to the nozzle centerline at the exit immediately after burnout. The various types of probes and analysis equipment are explained in detail in the following section.

Table 2 shows the motors tested and identifies which sampling technique was used on a particular test; not all sampling techniques were used on each motor.

Sampling Techniques

The sampling probe used in acquiring the postfire rocket chamber effluents was located 6 in. downstream of the nozzle exit. The probe body consisted of a 1-in. stainless steel tube wrapped with heater tape. After completion of the motor burn, the probe was inserted into position at the nozzle centerline. The probe was connected through a 1-in.-diam sample transfer line to a sample station (Fig. 3). Flow was maintained through the probe and sample station by a vacuum line which was routed to a large vacuum tank held at a pressure of about 2.5 Torr. Contained in the sample station are three sample bottles of the type shown in Fig. 4. The sample bottles contained a holder for mounting a witness stub to gather particles for analysis using a scanning electron microscope (SEM) combined with an x-ray energy dispersive spectrometer (XES). The 15-mm-diam stubs were made of either copper, carbon, or beryllium. Filter paper, used to divide the sample bottle into two compartments (particle/gas), is also held in place by the SEM stub holder as shown in Fig. 4. The gas compartment should then be relatively

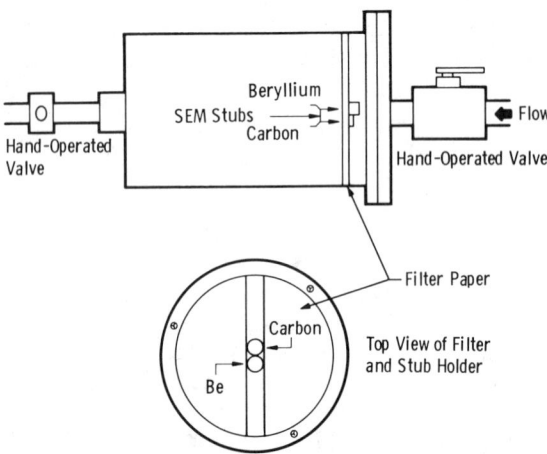

Fig. 4 Location of SEM stubs in sample bottles.

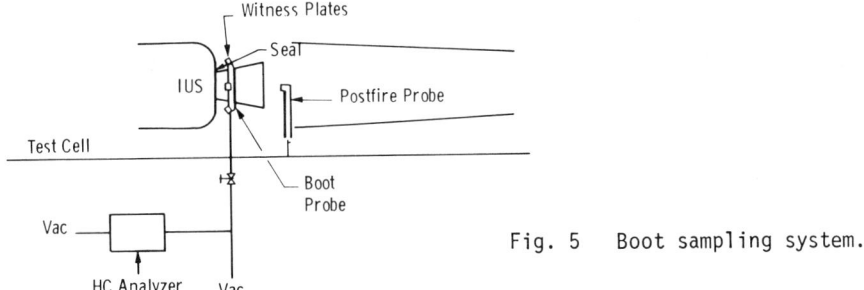

Fig. 5   Boot sampling system.

clear of particles so that the gas can be removed through a line attached to the end opposite the entrance to the sample bottle. Also attached to the postfire probe was a gas chromatograph used to monitor the HC and CO_2 concentrations on-line (semicontinuously) for the duration of the postfire period.

The boot probe (Fig. 5) was designed to sample the boot area using another sample station similar to that used for the postfire probe. Also used was a flame ionization detector to monitor total HC (every 15 s) from ignition to completion of the postfire period.

Particulate Analysis

The sample bottles (Fig. 4) were configured to direct sample flows over the SEM stub upon which particles were deposited. The remainder of the solid material was collected on the filter paper, except for some very small particles which might pass through the paper.

The resolution limit of the SEM used for the particle analyses was about 0.01 μm. The XES detector was capable of analyzing elements having atomic numbers greater than 12 (sodium). The XES detector was also used as the sensing element for an automatic scanning/sizing system used here at a lower resolution limit of 0.25 μm. The SEM stubs and portions of the filter paper were scanned first to search for large particles, and then representative regions of the stubs, or filter paper, were selected and photographs were made at several magnifications. Selected areas of the stubs were subjected to automatic counting/sizing, and the data were recorded.

Gas Analysis

The test cell pressure during the postfire period was approximately 15 Torr, and to ensure that air would not leak into the bottles after sample acquisition the bottles were automatically filled with helium (He) to 900 Torr. After completion of the test the gases were removed through the sample gas line (see Fig. 4) and a gas chromatograph was used to determine the species concentrations as mole percent.

A gas chromatograph and a hydrocarbon analyzer were used as an on-line analysis system to monitor the carbon dioxide and total hydrocarbons. Due to the low pressure inside the test cell, a continuous analysis by the analyzers was not possible. Using a carrier gas (air) made it possible to sample every 15 s with both analyzers.

Results

As stated in the previous sections the postfire emissions for the IUS SRM's may last for several minutes. However, due to test cell limitations all the samples were taken within the first 15 min after burnout.

At the beginning, the major goal of the IUS sampling was to identify the effluents being emitted from the nozzle during the postfire period. After completion of sampling the DL-3A motor, it was noted that the Techroll seal ruptures sometime during the postfire period.[5] Thus, a detailed study of the Techroll seal rupture was made along with the nozzle outgassing on the other six IUS tests.

The results presented are grouped into two categories: SRM-1, the large IUS motors, which include the DL-3A, QL-1, and QL-3 motors; and SRM-2, the small IUS motors, which in-

Table 3 Gas chromatograph data for IUS motor DL-3A

	Sample No. 1[a] (31.5 s)	Sample No. 2[a] (93.5 s)	Sample No. 3[a] (156.5 s)
Pressure (Torr)[b]	792.00	801.40	801.90
H_2 (mole %)	3.92	4.85	1.12
N_2 (mole %)	37.89	30.72	6.62
O_2 (mole %)	15.83	16.11	24.61
CO (mole %)	2.46	2.48	0.65
CO_2 (mole %)	35.69	32.12	0.94
C_xH_y (mole %)	3.45	12.56	65.98
Totals	99.24	98.84	99.92

[a] Corrected for He diluent.
[b] Pressure measured at time of analysis.

clude the QS-5, QS-3, QS-1, and QS-4 motors. All the sampling techniques were not used on every test (see Table 2).

Motion Picture Coverage

Several motion picture cameras were used during testing of the IUS motors. The cameras were used to monitor the nozzle exit and boot area for the entire 15-min postfire sampling period. The films showed that for all SRM's a heavy vapor was emitted from the nozzle during the first few minutes of the postfire period and then puffs of vapor were noticed at different times for the remaining 15 min.

For the SRM-1 motors there were no signs of emissions from the boot area, but it was observed that at times the vapors from the nozzle would flow back into the boot area.

The film of the boot area of SRM-2 showed the same backflow behavior from the nozzle, but there were also definite emissions of vapor from the boot area at different locations around the nozzle, which occurred in bursts of one to several minutes. On all SRM-2 tests vapors were being emitted from the boot area, even before burnout for some cases.

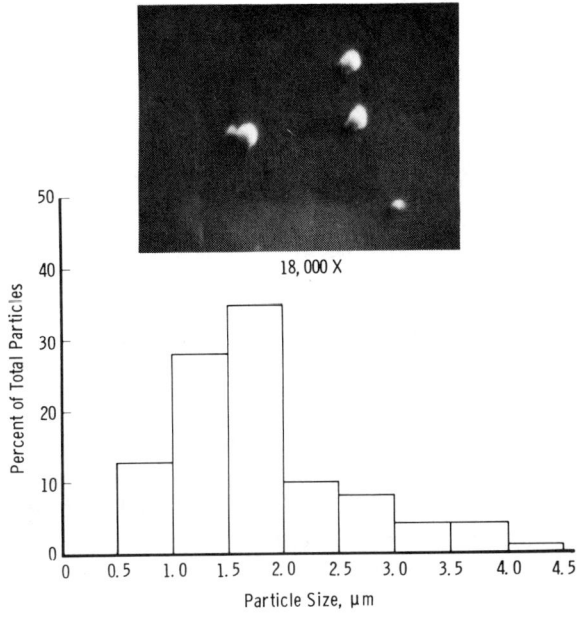

Fig. 6   Histogram of Al_2O_3 particles for DL-3A motor.

SRM-1

On the DL-3A motor, only the postfire probe (with sample bottles) and the witness plates (Fig. 3) located on the diffuser lip and exposed briefly during the postfire period were used. Table 3 shows the analysis of the gasses and reveals that for the first few minutes after burnout the gaseous products contained mostly carbon dioxide and nitrogen, with a smaller amount of hydrocarbons and oxygen, indicating that combustion was still in progress. The sample taken at 156.5 s (about 2.5 min) shows that nitrogen and carbon dioxide concentrations significantly decreased while the oxygen and hydrocarbon concentrations significantly increased. This behavior indicates that combustion has been completed before the 2.5-min sample and that thereafter the liner is merely smoldering and outgassing; or, the excessive hydrocarbon emission may indicate rupture of the seal, releasing oil into the combustion chamber. The source of increased O_2 is possibly the decomposition of remaining ammonium perchlorate. The particle analysis of the stubs and filter paper from the bottles revealed aluminum oxide (Al_2O_3) particles having diameters ranging from 0.5-4.5 μm (Fig. 6). There were also particles ranging in size 0.25-0.5 μm that could not be identified by the SEM, and they were assumed to be carbon (the SEM cannot analyze particles below atomic No. 11, sodium). Also found on the stubs was a coating containing Potassium (K), Phosphorous (P), Chlorine (Cl), Sodium (Na), and Calcium (Ca), Fig. 7. SEM analysis of the witness plates located on the diffuser lip was characteristic of rust and dirt, which are part of the test cell environment. The use of the diffuser witness plates was abandoned after the DL-3A test.

Postfire Probe. SEM analysis of the stubs taken from the bottles compared very well with the results from the DL-3A stubs (see Figs. 6 and 7). The Al_2O_3 and C_s particles had the same size distribution with the same coating of K, P, Cl, Na, and Ca on the stubs. The semicontinuous gas analysis from the FID is shown in Fig. 8, and the plots represent relative THC from time after ignition. Due to sample system peculiarities the first data point varied from test to test. Thus the curve starts at the first data point and does not indicate that the hydrocarbon emissions start at that point. High levels exist for the first 3 min and then drop off; then at different times (after 3 min) short bursts of hydrocarbon are seen, sometimes at a high level.

Boot Probe. SEM analysis of the stubs taken from the bottles showed no particles or coating, and the gas analysis

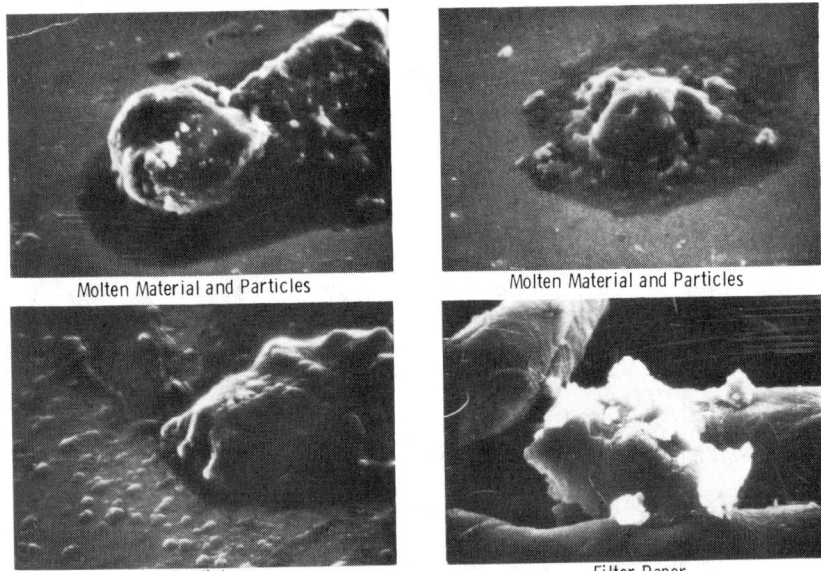

Fig. 7 Representative SEM photographs.

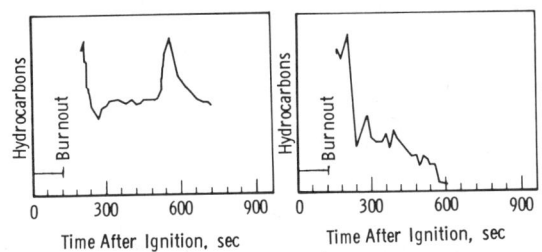

Fig. 8 Plots for QL-1 and QL-3 postfire probes (IUS test data): motors QL-1 (left) and QL-3 (right).

gave only air. Also, there was no indication of hydrocarbon emission from the semicontinuous sampling system and FID detector.

SRM-2

Visual inspection after an SRM-2 test that was not sampled showed a puddle of oil on the test cell floor, and indications were that the Techroll seal had ruptured at some time during the postfire period. A drip pan was used during the QS-5 test, and a sample of the oil was collected. An IR spectrometer was used to make a comparison between the used and unused oil which revealed that an ester was missing from the used oil. The oil is a hydrocarbon-based silicon compound that, when vaporized and then condensed in the labora-

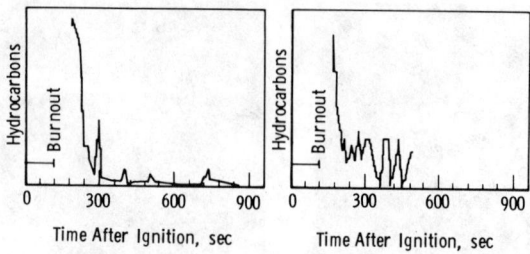

Fig. 9 Plots for QS-1 and QS-4 postfire probes (IUS test data): motors QS-1 (left) and QS-4 (right).

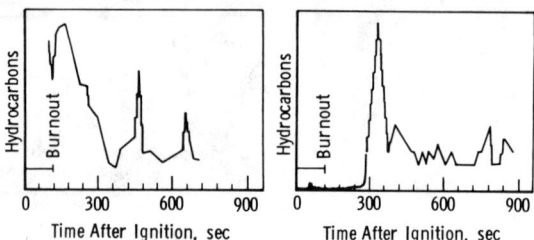

Fig. 10 Plots for QS-1 and QS-4 boot probes (IUS test data): motors QS-1 (left) and QS-4 (right).

tory, also loses the ester. The oil was present on the other 3 QS- motors tested, but no other samples of the oil were taken.

Postfire Probe. The FID and the sampling system used on the QS-5 and QS-3 tests suffered several equipment failures, and incomplete data were obtained on those firings.

SEM analysis of the QS-1 and QS-4 stubs taken from the bottles compared very well with the results from the SRM-1 tests.

The semicontinuous gas analysis for total hydrocarbons (Fig. 9) shows high levels during the first 3 min and then a decrease with minor peaks for the remainder of the test.

Boot Probe. The FID system used on the QS-5 motor failed, but a system was operational on the other tests. The analysis for the other three tests showed a significant rise in HC at various times. Figure 10 is an example of the hydrocarbon data taken, and the only difference between the three motors was the time at which the emission appeared. For the QS-1 motor, the emission started at burnout, for the QS-3 motor (not shown) it was at 150 s, and for the QS-4 motor (Fig. 10b) it was at 290 s.

Summary of Results

The approach used in sampling the IUS SRM's proved to be an effective means to gather samples during the postfire period. Various sampling techniques were used to identify the effluents, as well as the location and times at which they are emitted. Prediction of their effect on spacecraft performance requires more in-depth consideration.

The results may be summarized as follows:

1) For the SRM-1 and SRM-2 the solid material being emitted from the nozzle consisted of K, P, Cl, Na, and Ca in a molten state, Al_2O_3 particles ranging from 0.5-4.5 µm in size, and C_s particles ranging less than 0.25-0.5 µm in size.

2) The gas analysis of the nozzle batch sample for SRM-1 during the first 2 min showed principally CO_2 and N_2 with smaller amounts of O_2 and HC, and the sample taken at about 2.5 min showed a decrease in N_2 and CO_2 while the O_2 and HC increased. This behavior was interpreted to mean either that combustion (oxidation) ceases or the Techroll seal ruptures before 2.5 min after burnout. The high O_2 level may be produced by a ·omposition of remaining ammonium perchlorate.

3) The semicontinuous sampling of hydrocarbons from the nozzle showed a rise for the first 3 min for both SRM's (1 and 2), but major differences occurred after 3 min. For the SRM-1 motors the emission decreased after 3 min and then at different times a major burst was noted. For the SRM-2 after the initial decrease minor bursts were observed.

4) Semicontinuous sampling of hydrocarbons from the boot area showed major differences for the SRM-1 and SRM-2 motors. For the SRM-1 motors there was no indication of HC emissions. For the SRM-2 significant bursts at various times were observed and on some tests large concentrations were observed at burnout.

5) A liquid residue was found in the boot area of the SRM-2 motors, but not in the SRM-1 motors. The liquid was identified as the silicon oil emitted as a result of seal rupture.

6) Motion picture coverage of the nozzle showed no visual difference in gasses being emitted from the two SRM's. However, there were major differences in the film covering the boot area. Although the SRM-1 emitted no visible vapors, the SRM-2 emitted vapors in bursts lasting 1-3 min over the period from burnout to the end of the sampling period.

Conclusion

A conclusion to be reached from these tests is that for the IUS motors the major source of contaminant emission during the postfire period is the oil and grease released upon rupture of the nozzle flexible seal components. This event is characterized by large emissions during the first three minutes after burnout, followed by occasional bursts of vapors. It appears that the emission for the large motor (SRM-1) is always into the rocket chamber, while for the small motor (SRM-2) the emission is both into the chamber and outside into the area surrounding the nozzle attachment ring.

Although the investigation reported here has direct application to the IUS solid rocket motor emission problem, the implications are that similar investigations for all orbital transfer propulsion systems would be advisable. Outgassing behavior, lubricant leakage, and other emissions are all potential spacecraft contaminants candidates.

Acknowledgments

The research reported herein was performed by the Arnold Engineering Development Center, Air Force Systems Command. Work and analysis for this research were done by personnel of Sverdrup Technology, Inc./AEDC Group, operating contractor of the propulsion test facilities at AEDC. Further reproduction is authorized to satisfy needs of the U.S. Government.

The authors wish to acknowledge Mr. D. G. Gardner, Mr. Randy Quinn, and Mr. Linn Weller of Sverdrup Technology, Inc./AEDC Group, Mr. R. J. Bryson of Calspan Field Services, Inc./AEDC Division, and Mr. Randy Johnson of Pan Am World Services, MSSP, for their contributions to the success of this program. The work was sponsored by Headquarters Space Division (AFSC) (Capt. E. Bartchichat) and the AF Rocket Propulsion Laboratory (Capt. R. Fursteneau). Capt. K. Leners was the AF-AEDC Project Manager.

References

[1] Williams, W. D. et al., "Experimental Study of the Plume Characteristics of an Aged Monopropellant Hydrazine Thruster," AEDC-TR-79-2, April 1979.

[2] Alt, R. et al., "Phase I, Bipropellant Engine Plume Contamination Program," AEDC-TR-79-28, Dec. 1979.

[3]Chase, C. A., "Development Status of the Inertial Upper Stage Solid Motors," AIAA Paper 80-1267, June 30 - July 2, 1980.

[4]Girata, P. T. and McGregor, W. K., "Particle Sampling of Solid Rocket Motor (SRM) Exhauster in High Altitude Test Cells," AIAA-83-0245, Jan. 1983.

[5]Mullen, C. R. et al., "Prevention of Payload Contamination from IUS Propulsion Sources (U)," JANNAF Propulsion Meeting, Feb. 1983.

[6]Test Facilities Handbook (11th Edition), Arnold Engineering Development Center, Arnold Air Force Station, Tenn., April 1981.

Author Index for Volume 91

Ahern, J. E. 96, 235
Bareiss, L. E. 73
Belcher, R. L. 96
Calia, V. S. 186
del Casal, E. P. 39
Fote, A. A. 215
Girata, P. T. 293, 312
Glassford, A. P. M. 3
Hall, D. F. 215
Hitchcock, M. 3
Hoffman, R. J. 261
Jarossy, F. J. 73
Karperos, K. 235
Konopka, W. L. 180

Lee, A. L. 54
Liu, C. K. 3
McGregor, W. K. 293, 312
Moss, R. G. 29
Osiecki, R. A. 3
Palmer, K. F. 162
Reed, R. A. 180
Roux, J. A. 139, 162
Ruff, R. D. 96
Scialdone, J. J. 108
Smith, S. D. 197
Trinks, H. 261
Wood, B. E. 139, 162

PROGRESS IN ASTRONAUTICS AND AERONAUTICS SERIES VOLUMES

VOLUME TITLE/EDITORS

*1. **Solid Propellant Rocket Research** (1960)
Martin Summerfield
Princeton University

*2. **Liquid Rockets and Propellants** (1960)
Loren E. Bollinger
The Ohio State University
Martin Goldsmith
The Rand Corporation
Alexis W. Lemmon Jr.
Battelle Memorial Institute

*3. **Energy Conversion for Space Power** (1961)
Nathan W. Snyder
Institute for Defense Analyses

*4. **Space Power Systems** (1961)
Nathan W. Snyder
Institute for Defense Analyses

*5. **Electrostatic Propulsion** (1961)
David B. Langmuir
Space Technology Laboratories, Inc.
Ernst Stuhlinger
NASA George C. Marshall Space Flight Center
J.M. Sellen Jr.
Space Technology Laboratories, Inc.

*6. **Detonation and Two-Phase Flow** (1962)
S.S. Penner
California Institute of Technology
F.A. Williams
Harvard University

*7. **Hypersonic Flow Research** (1962)
Frederick R. Riddell
AVCO Corporation

*8. **Guidance and Control** (1962)
Robert E. Roberson
Consultant
James S. Farrior
Lockheed Missiles and Space Company

*9. **Electric Propulsion Development** (1963)
Ernst Stuhlinger
NASA George C. Marshall Space Flight Center

*10. **Technology of Lunar Exploration** (1963)
Clifford I. Cummings and Harold R. Lawrence
Jet Propulsion Laboratory

*11. **Power Systems for Space Flight** (1963)
Morris A. Zipkin and Russell N. Edwards
General Electric Company

*12. **Ionization in High-Temperature Gases** (1963)
Kurt E. Shuler, Editor
National Bureau of Standards
John B. Fenn, Associate Editor
Princeton University

*13. **Guidance and Control—II** (1964)
Robert C. Langford
General Precision Inc.
Charles J. Mundo
Institute of Naval Studies

*14. **Celestial Mechanics and Astrodynamics** (1964)
Victor G. Szebehely
Yale University Observatory

*15. **Heterogeneous Combustion** (1964)
Hans G. Wolfhard
Institute for Defense Analyses
Irvin Glassman
Princeton University
Leon Green Jr.
Air Force Systems Command

*16. **Space Power Systems Engineering** (1966)
George C. Szego
Institute for Defense Analyses
J. Edward Taylor
TRW Inc.

*17. **Methods in Astrodynamics and Celestial Mechanics** (1966)
Raynor L. Duncombe
U.S. Naval Observatory
Victor G. Szebehely
Yale University Observatory

*18. **Thermophysics and Temperature Control of Spacecraft and Entry Vehicles** (1966)
Gerhard B. Heller
NASA George C. Marshall Space Flight Center

*19. **Communication Satellite Systems Technology** (1966)
Richard B. Marsten
Radio Corporation of America

*Out of print.

*20. **Thermophysics of Spacecraft and Planetary Bodies: Radiation Properties of Solids and the Electromagnetic Radiation Environment in Space** (1967)
Gerhard B. Heller
NASA George C. Marshall Space Flight Center

*21. **Thermal Design Principles of Spacecraft and Entry Bodies** (1969)
Jerry T. Bevans
TRW Systems

*22. **Stratospheric Circulation** (1969)
Willis L. Webb
Atmospheric Sciences Laboratory, White Sands, and University of Texas at El Paso

*23. **Thermophysics: Applications to Thermal Design of Spacecraft** (1970)
Jerry T. Bevans
TRW Systems

24. **Heat Transfer and Spacecraft Thermal Control** (1971)
John W. Lucas
Jet Propulsion Laboratory

25. **Communication Satellites for the 70's: Technology** (1971)
Nathaniel E. Feldman
The Rand Corporation
Charles M. Kelly
The Aerospace Corporation

26. **Communication Satellites for the 70's: Systems** (1971)
Nathaniel E. Feldman
The Rand Corporation
Charles M. Kelly
The Aerospace Corporation

27. **Thermospheric Circulation** (1972)
Willis L. Webb
Atmospheric Sciences Laboratory, White Sands, and University of Texas at El Paso

28. **Thermal Characteristics of the Moon** (1972)
John W. Lucas
Jet Propulsion Laboratory

29. **Fundamentals of Spacecraft Thermal Design** (1972)
John W. Lucas
Jet Propulsion Laboratory

30. **Solar Activity Observations and Predictions** (1972)
Patrick S. McIntosh and Murray Dryer
Environmental Research Laboratories, National Oceanic and Atmospheric Administration

31. **Thermal Control and Radiation** (1973)
Chang-Lin Tien
University of California at Berkeley

32. **Communications Satellite Systems** (1974)
P.L. Bargellini
COMSAT Laboratories

33. **Communications Satellite Technology** (1974)
P.L. Bargellini
COMSAT Laboratories

34. **Instrumentation for Airbreathing Propulsion** (1974)
Allen E. Fuhs
Naval Postgraduate School
Marshall Kingery
Arnold Engineering Development Center

35. **Thermophysics and Spacecraft Thermal Control** (1974)
Robert G. Hering
University of Iowa

36. **Thermal Pollution Analysis** (1975)
Joseph A. Schetz
Virginia Polytechnic Institute

37. **Aeroacoustics: Jet and Combustion Noise; Duct Acoustics** (1975)
Henry T. Nagamatsu, Editor
General Electric Research and Development Center
Jack V. O'Keefe, Associate Editor
The Boeing Company
Ira R. Schwartz, Associate Editor
NASA Ames Research Center

38. **Aeroacoustics: Fan, STOL, and Boundary Layer Noise; Sonic Boom; Aeroacoustic Instrumentation** (1975)
Henry T. Nagamatsu, Editor
General Electric Research and Development Center
Jack V. O'Keefe, Associate Editor
The Boeing Company
Ira R. Schwartz, Associate Editor
NASA Ames Research Center

39. **Heat Transfer with Thermal Control Applications** (1975)
M. Michael Yovanovich
University of Waterloo

SERIES LISTING

40. **Aerodynamics of Base Combustion** (1976)
S.N.B. Murthy, Editor
Purdue University
J.R. Osborn, Associate Editor
Purdue University
A.W. Barrows and J.R. Ward, Associate Editors
Ballistics Research Laboratories

41. **Communications Satellite Developments: Systems** (1976)
Gilbert E. LaVean
Defense Communications Agency
William G. Schmidt
CML Satellite Corporation

42. **Communications Satellite Developments: Technology** (1976)
William G. Schmidt
CML Satellite Corporation
Gilbert E. LaVean
Defense Communications Agency

43. **Aeroacoustics: Jet Noise, Combustion and Core Engine Noise** (1976)
Ira R. Schwartz, Editor
NASA Ames Research Center
Henry T. Nagamatsu, Associate Editor
General Electric Research and Development Center
Warren C. Strahle, Associate Editor
Georgia Institute of Technology

44. **Aeroacoustics: Fan Noise and Control; Duct Acoustics; Rotor Noise** (1976)
Ira R. Schwartz, Editor
NASA Ames Research Center
Henry T. Nagamatsu, Associate Editor
General Electric Research and Development Center
Warren C. Strahle, Associate Editor
Georgia Institute of Technology

45. **Aeroacoustics: STOL Noise; Airframe and Airfoil Noise** (1976)
Ira R. Schwartz, Editor
NASA Ames Research Center
Henry T. Nagamatsu, Associate Editor
General Electric Research and Development Center
Warren C. Strahle, Associate Editor
Georgia Institute of Technology

46. **Aeroacoustics: Acoustic Wave Propagation; Aircraft Noise Prediction; Aeroacoustic Instrumentation** (1976)
Ira R. Schwartz, Editor
NASA Ames Research Center
Henry T. Nagamatsu, Associate Editor
General Electric Research and Development Center
Warren C. Strahle, Associate Editor
Georgia Institute of Technology

47. **Spacecraft Charging by Magnetospheric Plasmas** (1976)
Alan Rosen
TRW Inc.

48. **Scientific Investigations on the Skylab Satellite** (1976)
Marion I. Kent and Ernst Stuhlinger
NASA George C. Marshall Space Flight Center
Shi-Tsan Wu
The University of Alabama

49. **Radiative Transfer and Thermal Control** (1976)
Allie M. Smith
ARO Inc.

50. **Exploration of the Outer Solar System** (1976)
Eugene W. Greenstadt
TRW Inc.
Murray Dryer
National Oceanic and Atmospheric Administration
Devrie S. Intriligator
University of Southern California

51. **Rarefied Gas Dynamics, Parts I and II** (two volumes) (1977)
J. Leith Potter
ARO Inc.

52. **Materials Sciences in Space with Application to Space Processing** (1977)
Leo Steg
General Electric Company

53. **Experimental Diagnostics in Gas Phase Combustion Systems** (1977)
Ben T. Zinn, Editor
Georgia Institute of Technology
Craig T. Bowman, Associate Editor
Stanford University
Daniel L. Hartley, Associate Editor
Sandia Laboratories
Edward W. Price, Associate Editor
Georgia Institute of Technology
James G. Skifstad, Associate Editor
Purdue University

54. **Satellite Communications: Future Systems** (1977)
David Jarett
TRW Inc.

55. **Satellite Communications: Advanced Technologies** (1977)
David Jarett
TRW Inc.

56. **Thermophysics of Spacecraft and Outer Planet Entry Probes** (1977)
Allie M. Smith
ARO Inc.

57. **Space-Based Manufacturing from Nonterrestrial Materials** (1977)
Gerard K. O'Neill, Editor
Princeton University
Brian O'Leary, Assistant Editor
Princeton University

58. **Turbulent Combustion** (1978)
Lawrence A. Kennedy
State University of New York at Buffalo

59. **Aerodynamic Heating and Thermal Protection Systems** (1978)
Leroy S. Fletcher
University of Virginia

60. **Heat Transfer and Thermal Control Systems** (1978)
Leroy S. Fletcher
University of Virginia

61. **Radiation Energy Conversion in Space** (1978)
Kenneth W. Billman
NASA Ames Research Center

62. **Alternative Hydrocarbon Fuels: Combustion and Chemical Kinetics** (1978)
Craig T. Bowman
Stanford University
Jorgen Birkeland
Department of Energy

63. **Experimental Diagnostics in Combustion of Solids** (1978)
Thomas L. Boggs
Naval Weapons Center
Ben T. Zinn
Georgia Institute of Technology

64. **Outer Planet Entry Heating and Thermal Protection** (1979)
Raymond Viskanta
Purdue University

65. **Thermophysics and Thermal Control** (1979)
Raymond Viskanta
Purdue University

66. **Interior Ballistics of Guns** (1979)
Herman Krier
University of Illinois at Urbana-Champaign
Martin Summerfield
New York University

67. **Remote Sensing of Earth from Space: Role of "Smart Sensors"** (1979)
Roger A. Breckenridge
NASA Langley Research Center

68. **Injection and Mixing in Turbulent Flow** (1980)
Joseph A. Schetz
Virginia Polytechnic Institute and State University

69. **Entry Heating and Thermal Protection** (1980)
Walter B. Olstad
NASA Headquarters

70. **Heat Transfer, Thermal Control, and Heat Pipes** (1980)
Walter B. Olstad
NASA Headquarters

71. **Space Systems and Their Interactions with Earth's Space Environment** (1980)
Henry B. Garrett and Charles P. Pike
Hanscom Air Force Base

72. **Viscous Flow Drag Reduction** (1980)
Gary R. Hough
Vought Advanced Technology Center

73. **Combustion Experiments in a Zero-Gravity Laboratory** (1981)
Thomas H. Cochran
NASA Lewis Research Center

74. **Rarefied Gas Dynamics, Parts I and II** (two volumes) (1981)
Sam S. Fisher
University of Virginia at Charlottesville

75. **Gasdynamics of Detonations and Explosions** (1981)
J.R. Bowen
University of Wisconsin at Madison
N. Manson
Université de Poitiers
A.K. Oppenheim
University of California at Berkeley
R.I. Soloukhin
Institute of Heat and Mass Transfer, BSSR Academy of Sciences

76. **Combustion in Reactive Systems** (1981)
J.R. Bowen
University of Wisconsin at Madison
N. Manson
Université de Poitiers
A.K. Oppenheim
University of California at Berkeley
R.I. Soloukhin
Institute of Heat and Mass Transfer, BSSR Academy of Sciences

77. **Aerothermodynamics and Planetary Entry** (1981)
A.L. Crosbie
University of Missouri-Rolla

78. **Heat Transfer and Thermal Control** (1981)
A.L. Crosbie
University of Missouri-Rolla

79. **Electric Propulsion and Its Applications to Space Missions** (1981)
Robert C. Finke
NASA Lewis Research Center

80. **Aero-Optical Phenomena** (1982)
Keith G. Gilbert and Leonard J. Otten
Air Force Weapons Laboratory

81. **Transonic Aerodynamics** (1982)
David Nixon
Nielsen Engineering & Research, Inc.

82. **Thermophysics of Atmospheric Entry** (1982)
T.E. Horton
The University of Mississippi

83. **Spacecraft Radiative Transfer and Temperature Control** (1982)
T.E. Horton
The University of Mississippi

84. **Liquid-Metal Flows and Magnetohydrodynamics** (1983)
H. Branover
Ben-Gurion University of the Negev
P.S. Lykoudis
Purdue University
A. Yakhot
Ben-Gurion University of the Negev

85. **Entry Vehicle Heating and Thermal Protection Systems: Space Shuttle, Solar Starprobe, Jupiter Galileo Probe** (1983)
Paul E. Bauer
McDonnell Douglas Astronautics Company
Howard E. Collicott
The Boeing Company

86. **Spacecraft Thermal Control, Design, and Operation** (1983)
Howard E. Collicott
The Boeing Company
Paul E. Bauer
McDonnell Douglas Astronautics Company

87. **Shock Waves, Explosions, and Detonations** (1983)
J.R. Bowen
University of Washington
N. Manson
Université de Poitiers
A.K. Oppenheim
University of California at Berkeley
R.I. Soloukhin
Institute of Heat and Mass Transfer, BSSR Academy of Sciences

88. **Flames, Lasers, and Reactive Systems** (1983)
J.R. Bowen
University of Washington
N. Manson
Université de Poitiers
A.K. Oppenheim
University of California at Berkeley
R.I. Soloukhin
Institute of Heat and Mass Transfer, BSSR Academy of Sciences

89. **Orbit-Raising and Maneuvering Propulsion: Research Status and Needs** (1984)
Leonard H. Caveny
Air Force Office of Scientific Research

90. **Fundamentals of Solid-Propellant Combustion** (1984)
Kenneth K. Kuo
The Pennsylvania State University
Martin Summerfield
Princeton Combustion Research Laboratories, Inc.

91. **Spacecraft Contamination: Sources and Prevention** (1984)
J.A. Roux
The University of Mississippi
T.D. McCay
NASA Marshall Space Flight Center

92. **Combustion Diagnostics by Nonintrusive Methods** (1984)
T.D. McCay
NASA Marshall Space Flight Center
J.A. Roux
The University of Mississippi

(Other Volumes are planned.)